Lecture Notes in Control and Information Sciences 293

Editors: M. Thoma · M. Morari

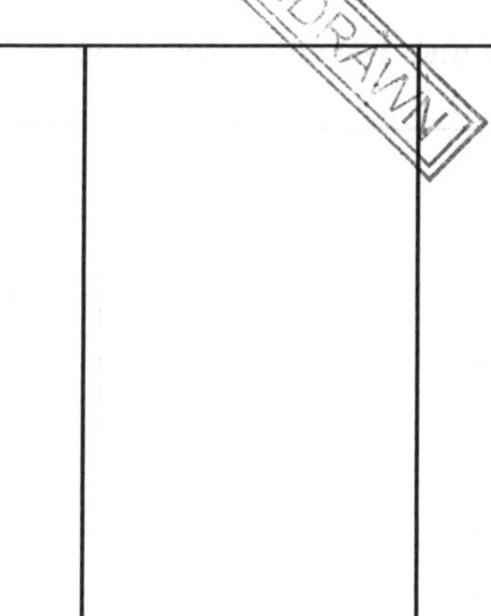

Springer

Berlin
Heidelberg
New York
Hong Kong
London
Milan
Paris
Tokyo

Guanrong Chen · David J. Hill · Xinghuo Yu (Eds.)

Bifurcation Control

Theory and Applications

With 125 Figures

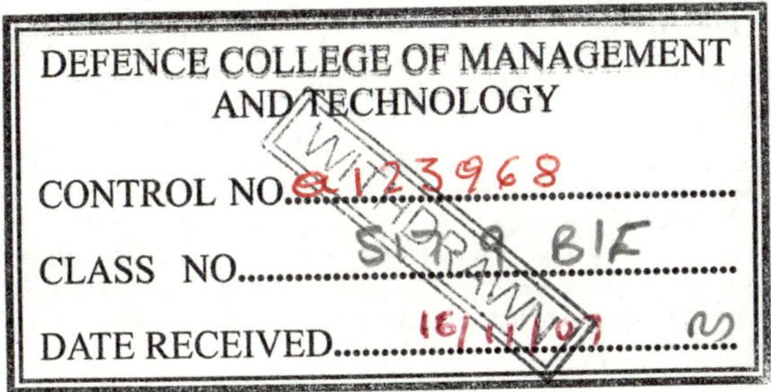

Springer

Editors

Prof. Guanrong Chen
Prof. David J. Hill
City University of Hong Kong
Department of Electronic Engineering
83 Tat Chee Avenue, Kowloon
Hong Kong SAR
P.R. China

Prof. Xinghuo Yu
Faculty of Engineering
Royal Melbourne Institute of Technology
GPO Box 2476V
Melbourne VIC 3001
Australia

ISSN 0170-8643

ISBN 3-540-40341-8 Springer-Verlag Berlin Heidelberg New York

Library of Congress Cataloging-in-Publication Data
Bifurcation control : theory and application / Guanrong Chen, David J. Hill, Xinghuo Yu (eds.).
p. cm. -- (Lecture notes in control and information sciences ; 293)
ISBN 3-540-40341-8 (alk. paper)
1. Control theory. 2. Bifurcation theory. I. Chen, G. (Guanrong) II. Hill, David J.
(David John), 1949- III. Yu, Xing Huo. IV. Series.

QA402.3.B52 2003
515'.64--dc21 2003054382

Springer-Verlag Berlin Heidelberg New York
a member of BertelsmannSpringer Science + Business Media GmbH

http://www.springer.de

Typesetting: Data conversion by the authors.
Final processing by PTP-Berlin Protago-TeX-Production GmbH, Berlin
Cover-Design: design & production GmbH, Heidelberg
Printed on acid-free paper 62/3020Yu - 5 4 3 2 1 0

Preface

Bifurcation control refers to the task of designing a controller that can modify the bifurcating properties of a given nonlinear system, so as to achieve some desirable dynamical behaviors. Typical bifurcation control objectives include delaying the onset of an inherent bifurcation, introducing a new bifurcation at a preferable parameter value, changing the parameter value of an existing bifurcation point, modifying the shape or type of a bifurcation chain, stabilizing a bifurcated solution or branch, monitoring the multiplicity, amplitude, and/or frequency of some limit cycles emerging from bifurcation, optimizing the system performance near a bifurcation point, or a combination of some of these objectives.

Bifurcation control not only is important in its own right, but also suggests an effective strategy for chaos control since bifurcation and chaos are usually "twins"; in particular, period-doubling bifurcation is a typical route to chaos in many nonlinear dynamical systems.

Both chaos control and bifurcation control suggest a new technology that promises to have a major impact on many novel, perhaps not-so-traditional, time- and energy-critical engineering applications. In addition to the vast area of chaos control applications, bifurcation control plays a crucial role in special dynamical analysis and crisis control of many complex nonlinear systems. The best known examples include high-performance circuits and devices (e.g., delta-sigma modulators and power converters), oscillation generation, vibration-based material mixing, chemical reactions, power systems collapse prediction and prevention, oscillators design and testing, biological systems modelling and analysis (e.g., the brain and the heart), and crisis management (e.g., jet-engine serge and stall), to name just a few. In fact, this new and challenging research and development area has become an attractive scientific inter-discipline involving control and systems engineers, theoretical and experimental physicists, applied mathematicians, and biomedical engineers alike.

There are many practical reasons for controlling various bifurcations. In a system where a bifurcating response is harmful and dangerous, it should be significantly reduced or completely suppressed. This task includes, for example, avoiding voltage collapse and oscillations in power networks, eliminating deadly cardiac arrhythmias, guiding disordered circuit arrays (e.g., multi-coupled oscillators and cellular neural networks) to reach a certain level of desirable pattern formation, regulating dynamical responses of some mechanical and electronic devices (e.g., diodes, laser machines, and machine tools), removing undesirable vibrations, and so on.

Bifurcation can also be useful and beneficial for some special applications, and it is interesting to see that there has been growing interest in utilizing the very nature of bifurcation, particularly in some engineering applications involving oscillations analysis and utilization. A prominent feature of bifurcation is its close relation with various vibrations (periodic oscillations or limit cycles), which sometimes are not only desirable but may actually be necessary. Mechanical vibrations and some material and liquid mixing processes are good examples in which bifurcations (and chaos) are very desirable. In biological systems, bifurcation control seems to be an essential mechanism employed by the human heart in carrying out some of its tasks particularly on atrial fibrillation. Some medical evidence lends support to the idea that control of certain bifurcating cardiac arrhythmias may contribute to the new design of a safer and more effective intelligent pacemaker. A further idea, suggested as useful in power systems, is to use the onset of a small oscillation as an indicator for proximity to collapse. In control systems engineering, the deliberate use of nonlinear oscillations has been applied effectively for system identification.

Motivated by many potential real-world applications, current research on bifurcation control has proliferated in recent years, along with the promising progress of chaos control. With respect to theoretical considerations, bifurcation control poses a substantial challenge to both system analysts and control engineers. This is due to the extreme complexity and sensitivity of bifurcating dynamics, which intrinsically is associated with the reduction in long-term predictability and short-term controllability of chaotic systems in general.

Notwithstanding many technical obstacles, both theoretical and practical developments in this area have experienced remarkable progress in the last decade. It is now known that bifurcations can be controlled via various methods. Some representative approaches employ linear or nonlinear state-feedback controls, perhaps with time-delayed feedback, apply a washout filter-aided dynamic feedback controller, use harmonic balance approximations, and utilize quadratic invariants in normal forms. Surprisingly, however, there exist no control-theory-oriented books written by control engineers for control engineers available in the market that are devoted to the subject of Bifurcation Control. In particular, there has been no exposure of these very active research topics in the Lecture Notes Series in Control and Information Science. This edited book, therefore, aims at filling in the gap and presenting current achievements in this challenging field at the forefront of research, with emphasis on the engineering perspectives, methodologies, and potential applications of bifurcation controls. It is intended as a combination of overview, tutorial and technical reports, reflecting state-of-the-art research of significant problems in this field. The anticipated readership includes university professors, graduate students, laboratory researchers and industrial practitioners, as well as applied mathematicians and physicists in the areas of electrical, mechanical, physical, chemical, and biomedical engineering and sciences.

We received enthusiastic assistance from several individuals in the preparation of this book. In particular, we are very grateful to Noel Patson, who helped a great deal in taking care of many painful editorial tasks. We would also like to thank Prof. T. Thoma and Dr. T. Ditzinger, Editors of Springer-Verlag, for their continued support and kind cooperation. Finally, we wish to express our sincere thanks to all the authors whose significant scientific contributions have directly led to the publication of this timely treatise.

Guanrong Chen, David J. Hill, Xinghuo Yu
Hong Kong and Melbourne, January, 2003

Contents

Bifurcation Control

Applications

Application of Bifurcation Analysis to Power Systems

Hsiao-Dong Chiang

School of Electrical and Computer Engineering
Cornell University, Ithaca, NY 14853, USA
chiang@ee.cornell.edu

Abstract. Electric power systems are physically some of the largest and most complex nonlinear systems in the world. Bifurcations are rather mundane phenomena in power systems. The pioneer work on the local bifurcation analysis of power systems can be dated back to the 1970's and earlier. Within the last 20 years or so nonlinear dynamical theory has become a subject of great interest to researchers and engineers in the power system community. Powerful computational tools for bifurcation analysis have been applied during this period to study important nonlinear problems arising in power systems, and in some cases, to relate this study to observed nonlinear phenomena in power systems. In this chapter, we will present an overview on the application of local bifurcation analysis and theory to (i) develop models explaining power system nonlinear behaviors and various power system instabilities such as voltage collapse and low-frequency oscillations, to (ii) develop a powerful global analysis tool based on continuation methods to trace power system quasi-steady-state behaviors due to load and generation variations in realistic power system models, and to (iii) develop performance indices for detecting and estimating local bifurcations of power systems. An overview on the extension of saddle-node bifurcation, Hopf bifurcation and limit-induced bifurcation to include the analysis of the system dynamics after the bifurcation is presented. In addition, the effects of un-modelled dynamics due to fast and slow variables on local bifurcations is presented.

1 Introduction

Electric power systems are comprised of a large number of components interacting with each other, exhibiting nonlinear dynamic behaviors with a wide range of time scales. Physically, an electric power system is an interconnected system composed of generating stations (which convert fuel energy into electricity), primary and secondary distribution substations (that distribute power to loads (consumers), and transmission lines, i.e., transmission network, that tie the generating stations and distribution substations together. The fundamental function of power systems is meeting customer load demands in a reliable and economical manner. To this end, various types of control devices, local and centralized, and protection systems are placed throughout the system. The local control devices attached to generating plants, such as excitation control system and turbine control system,

G. Chen, D.J. Hill, and X. Yu (Eds.): Bifurcation Control, LNCIS 293, pp. 1–28, 2003.

are automatic and relatively high speed. On the other hand, the local control devices, such as ULTC transformers, ULTC phase-shifters, synchronous var compensator (SVC), shunt capacitor (SC) installed in the transmission network are relatively low speed.

Electric power systems are physically some of the largest and most complex nonlinear systems in the world. Their nonlinear behaviors are difficult to analyze and predict due to several factors such as (i) the extraordinary size of the systems, (ii) the nonlinearity in the components and control devices in the systems, (iii) the dynamical interactions within the systems, (iv) the uncertainty in the load behaviors, (iv) the complexity and different time-scale of power system components (equipments and control devices). These complicating factors have forced power system engineers to analyze power systems through extensive computer simulations. Large-scale computer simulation programs are widely used in power utilities for studying power system steady-state behaviors and dynamic responses relative to disturbances. By nature, a power system continually experiences disturbances. These may be classified under two main categories: event disturbances and load disturbances. Event disturbances, i.e., contingencies, include loss of generating units or transmission components (lines, transformers, substations) due to short-circuits caused by lightning, high winds, failures such as incorrect relay operations or insulation breakdown, sudden large load changes, or a combination of such events. Event disturbances usually lead to a change in the network configuration of the power system du e to actions from protective relays and circuit breakers. Load disturbances, small or large, on the other hand, include the variations in load demands (e.g. the daily load cycle), termed load variations, the rescheduling of real power generations, the scheduled power transfers across the transmission network between two regions or two areas in the interconnected system, or a combination of the above three types of load disturbances. The network configuration usually remains unchanged after load disturbances. Power systems are planned and operated to withstand the occurrence of certain credible disturbances. A major activity in utility system planning and operations is to examine the impact of a set of credible disturbances on power system dynamical behaviors such as stability and to develop counter-measures.

A power system subject to load disturbances can be modelled as a set of parameter-dependent nonlinear differential and algebraic equations with parameter variation. Power systems are normally operated near a stable equilibrium point. When the system load parameters are away from their bifurcation values and their variations are occurring continuously but slowly, it is very likely that

- the stable equilibrium point of the underlying power system changes position but remains a stable equilibrium point, and
- the old stable equilibrium point lies inside the stability region of the new stable equilibrium point.

Consequently, the power system dynamics starting from the old stable equilibrium point will converge to the new stable equilibrium point and will make the system state track its new stable equilibrium point, whose position is changed continuously but slowly, and yet the system remains stable under this load disturbance. The typical ways in which a study power system may lose stability, under the influence of load variations are through the following:

- the stable equilibrium point and another equilibrium point coalesce and disappear in a saddle-node bifurcation as parameter varies, or
- the stable equilibrium point and another equilibrium point coalesce and exchange stability in a limit-induced bifurcation (a type of transcritical bifurcation) as parameter varies,
- the stable equilibrium point and an unstable limit cycle coalesce and disappear and an unstable equilibrium point emerges in a subcritical Hopf bifurcation as parameter varies,
- the stable equilibrium point bifurcates into an unstable equilibrium point surrounded by a stable limit cycle in a supercritical Hopf bifurcation as parameter varies.

It is now well recognized that bifurcations are rather mundane phenomena that can occur in many physical and man-made systems where nonlinearity is present. The pioneer work on the local bifurcation analysis of power systems can be dated back to the 1970's and earlier [62,63,41]. Within the last 20 years or so nonlinear dynamical theory has become a subject of great interest to researchers and engineers in the power system community. Powerful computational tools for bifurcation analysis have been applied during this period to study important nonlinear problems arising in power systems, and in some cases, to relate this study to observed nonlinear phenomena in power systems [45,46,59,53]. In addition, some counter-measures to avoid bifurcations have been developed to design control schemes for prevention of power system instabilities [23,24,52,73,34,35].

From the engineering viewpoint, one important task in performing bifurcation analysis to nonlinear systems, such as electric power systems, is the analysis of both the mechanism leading to disappearance of stable equilibrium points due to a bifurcation and the system dynamical behaviors after the bifurcation. After a bifurcation occurs, the system state will evolve according to the system dynamics. The dynamics after bifurcation determine whether the system remains stable or become unstable; and what is the type of system instability. Local bifurcation theory does not describe the dynamical behaviors after a bifurcation. We will review some work on the analysis of the system dynamics after typical local bifurcations in this chapter.

Electric power systems comprise a large number of components interacting with each other in nonlinear manners. The dynamical response of these components extends over a wide range of time scales. The different time-scale components of power systems all have their corresponding influences on power system dynamical responses. It has become convenient to divide the time span of dynamic response simulations into short-term (transient), mid-term and long-term, covering the post-disturbance times of up to a few seconds, 5 minutes and 20 minutes or so, respectively. Up to present, most power system models used for bifurcation analysis involve only short-term dynamical models (transient stability models). It raises the concern about the validity of short-term dynamical models, which have disregarded slow dynamics, for local bifurcation analysis. The effects of un-modelled dynamics on the local bifurcation analysis of a power system model is also discussed in this chapter.

P-V and Q-V curves have been widely used by power system analysis engineers to study voltage stability [60]. These curves represent one important aspect of the saddle-node bifurcation occurring in power systems due to variations of loads and generations. While global analysis tools based on continuation methods developed in the last decade can generate P-V and Q-V curves in a reliable manner, these tools may be too slow for certain power system on-line applications. To overcome this difficulty, a number of performance indices intended to measure the severity of the voltage stability problem have been proposed in the literature. We will examine several existing performance indices and discuss a performance index which has rendered practical applications. This performance index is based on the normal form theory of saddle-no de bifurcation point.

In this chapter, we will present an overview on the application of local bifurcation analysis and theory to (i) develop models explaining power system nonlinear behaviors and various power system instabilities such as voltage collapse and low-frequency oscillations, to (ii) develop a powerful global analysis tool based on continuation methods to trace power system quasi-steady-state behaviors due to load and generation variations in realistic power system models, and to (iii) develop powerful computational tools for detecting and estimating local bifurcations of power systems. An overview on the extension of saddle-node bifurcation, Hopf bifurcation and limit-induced bifurcation to include the analysis of the system dynamics after the bifurcation will be presented. In addition, the effects of un-modelled dynamics due to fast and slow variables on local bifurcations will be analyzed.

2 Local Bifurcations and Power System Behaviors

Recently, local bifurcation theory has being applied to interpret observed nonlinear dynamical behaviors in power systems. In some cases, local bifurcation theory has been extended to include the analysis of dynamics after a local bifurcation. A comprehensive bifurcation and chaos analysis of a 3-bus power system was carried out in [11]. Numerical bifurcation analysis of a simplified model of a 9-bus power system and a 39-bus power system were conducted in [12]. Other numerical bifurcation analysis of simple power systems can be found in, for example, [1,4]. It has been found that the bifurcation phenomena observed in the 3-bus and 9-bus power system are similar. These bifurcation phenomena have been observed in the 39-bus power system as well, including Hopf, period-doubling, and cyclic fold bifurcations. Furthermore, some bifurcation phenomena not appearing in the 3-bus and 9-bus systems have surfaced in the 39-bus system. These numerical studies favor the claim that various types of bifurcations can occur in real power systems.

Local bifurcation theory has been applied to provide an explanation for various observed power system nonlinear behaviors and power system instabilities such as voltage collapse and low-frequency electro-mechanical oscillations that occur in electric power networks. Abed and Varaiya [2] were probably the first to suggest a possible role for Hopf bifurcations in explaining the low-frequency electro-mechanical oscillation phenomena. Later Chen and Varaiya [10] numerically demonstrate that degenerate Hopf bifurcation can occur in a simple power system model. In [20], Dobson and Chiang investigated a generic mechanism leading to disappearance of stable equilibrium points due to a saddle-node bifurcation and the subsequent system dynamics after the bifurcation for one-parameter dynamical systems. A saddle-node voltage instability model to analyze the process of voltage instability was then proposed to explain voltage stability/instability due to slow load variations in three stages. Iravani and Semlyen investigate the transition of growing torsional oscillations into limit cycles occurring in torsional dynamics based on Hopf bifurcation [28]. Their studies indicate that the range of instability (growing oscillations) based on Hopf bifurcation is noticeably narrower than the one predicted by an eigen-analysis method.

Another dynamic phenomenon of concern in power systems that can be explained via Hopf bifurcation is subsynchronous resonance (SSR). This is a condition where the electric network exchanges energy with a turbine generator at one or more of the natural frequencies of the combined system below the synchronous frequency of the system. The IEEE SSR working group suggests that if the locus of a particular eigenvalue approaches or crosses the imaginary axis, then a critical condition is identified that will require the application of one or more SSR counter measures. The critical condition is closely related to the Hopf condition. To provide high-quality electricity, utilities must endeavor to minimize the effects of large fluctuating loads associated with large motors and furnaces on the transmission network. One of the effects, characterized by visible fluctuations in other customers' electric lighting, is called voltage flicker. This phenomenon is often categorized as cyclic and non-cyclic. From a dynamic system viewpoint, cyclic flickers may relate to limit cycles or quasi-periodic motion in power systems and non-cyclic flickers to quasi-periodic or chaotic trajectories.

Another advance of applying bifurcation analysis to power systems can be manifested in the development of several bifurcation-based models to explain several instances of recent power system voltage instability and/or collapse. This kind of blackout has occurred in several countries such as Belgium,

Canada, France, Japan, Sweden and the United States [49,60]. Voltage instability and/or collapse, a frequent concern on modern power systems, are generally caused by either of two types of system disturbances: load disturbances or contingencies, i.e., event disturbances. Among several examples of voltage collapse, the 1987 occurrence in Japan [43] was due to large load variations, while the collapse in Sweden in 1982 [71] was caused by a contingency. The dynamic process of voltage instability or collapse usually starts with a power system weakened by a contingency due to a transmission line or generator outage, or by an unusually high peak load (a high load variation), or by a combination of such events. The system may be further weakened due to an inappropriate transmission under-load tap-changer (ULTC) setting, or insufficient reactive power supports, or load restorations that have been temporary reduced because of low voltage. Three bifurcation-based voltage collapse models will be discussed in some details in this chapter.

3 Local Bifurcations in Power Systems

A power system model relative to a disturbance comprises a set of first-order differential equations and a set of algebraic equations

$$\begin{aligned} \dot{x} &= f(x, y, u, \lambda) \\ 0 &= g(x, y, \lambda) \end{aligned} \tag{1}$$

where $\lambda \in R^1$ is a parameter, x is a dynamic state variable and y a static "state" variable, such as the load variables of voltage magnitude and angle. The vector field depends on the value of parameter and will change its dimension accordingly. It describes the internal dynamics of devices such as generators, their associated control systems, certain loads, and other dynamically modelled components. The set of algebraic equations describe the electrical transmission system (the interconnections between the dynamic devices) and internal static behaviors of passive devices (such as static loads, shunt capacitors, fixed transformers and phase shifters). The differential equations (1) can describe as broad a range of behaviors as the dynamics of the speed and angle of generator rotors, flux behaviors in generators, the response of generator control systems such as excitation systems, voltage regulators, turbines, governors and boilers, the dynamics of equipments such as synchronous VAR compensators (SVCs), DC lines and their control systems, and the dynamics of dynamically modelled loads such as induction motors. The state variables x typically include generator rotor angles, generator velocity deviations (speeds), mechanical powers, field voltages, power system stabilizer signals, various control system internal variables, and voltages and angles at load buses (if dynamical load models are employed at these buses). The forcing functions u acting on the differential equations are terminal voltage magnitudes, generator electrical powers, and signals from boilers and automatic generation control systems. Some control system internal variables have upper bounds on their values due to their physical saturation effects. Let z be the vector of these constrained state variables; then the saturation effects can be expressed as

$$0 \leq z(t) \leq \hat{z} \tag{2}$$

We term the above model as a set of parameter-dependent differential and algebraic equations with hard constraints. A detailed description of equations (1) - (2) for each component can be found, for example, in [36,42,55]. For a 500-generator power system, the number of differential equations can easily reach as many as 10,000.

From a nonlinear dynamical system viewpoint, (1-2) is an one-parameter dynamical system while, in power system applications, it can represent a power system that operates with one of the following conditions:

1. the real (or reactive) power demand at one load bus varies while the others remain fixed,
2. both the real and reactive power demand at a load bus vary and the variations can be parameterized. Again the others remain fixed,
3. the real and/or reactive power demand at some collection of load buses varies and the variations can be parameterized while the others are fixed,
4. the real power transfer at one transmission corridor (e.g. interface transfer and import/export) varies while the others remain fixed,
5. the real power transfer at some collection of transmission corridors (e.g. interface transfer and import/export) varies while the others remain fixed.

The only typical ways in which a power system may lose stability (under the influence of one parameter variation) are through the saddle-node bifurcation, or limit-induced bifurcations or the Hopf bifurcation. In [21], it has been shown that for generic one-parameter dynamical systems, before a saddle-node bifurcation the equilibrium point $x_1(\lambda)$ is type-one. By type-one, we mean that the corresponding Jacobian matrix has exactly one eigenvalue with a positive real part and the rest of the eigenvalues have negative real parts. Furthermore, $x_1(\lambda)$ lies on the stability boundary of $x_s(\lambda)$. The Jacobian matrix, when evaluated at $x_s(\lambda)$, has all of its eigenvalues with only negative real parts and among them, one of the eigenvalues is close to zero. At the bifurcation occurring at say, the bifurcation value $\lambda = \lambda^*$, equilibrium points $x_s(\lambda)$ and $x_1(\lambda)$ coalesce to form an equilibrium point x^* $(= x_s(\lambda^*) = x_1(\lambda^*))$. The Jacobian matrix evaluated at x^* has one zero eigenvalue and the real parts of all t he other eigenvalues are negative. If the parameter λ increases beyond the bifurcation value λ^*, then x^* disappears and there are no other equilibrium points nearby.

Another local bifurcation peculiar to power systems is the so-called limit-induced bifurcation. Physically, the generation reactive power capability is limited. The reactive power capability of a generator can reach a limit due to the excitation current limit or the stator thermal limit. Power systems are vulnerable to voltage collapse when generation reactive power limits are reached. Given a load/generation variation pattern, the effect of reaching a generator reactive power limit is to immediately change the system equation. From a static analysis viewpoint the generator whose reactive power limit is reached may be simply modelled by replacing the equation describing a P-V bus by the equation describing a P-Q bus. In [21], numerical examples and general arguments were developed to show that a sufficiently heavily loaded but stable power systems can become immediately unstable via a transcritical bifurcation when a reactive power limit is encountered. We term this type of bifurcation as limit-induced bifurcation. We note that when a transcritical bifurcation occurs at say, the bifurcation value $\lambda = \lambda^*$, the stable equilibrium point $x_s(\lambda)$ and a type-one unstable equilibrium point $x_u(\lambda)$ coalesce to form an equilibrium point x^* $(= x_s(\lambda^*))$. The Jacobian matrix evaluated at x^* has a single, simple eigenvalue and the real parts of all the other eigenvalues are negative. After the bifurcation, the two equilibrium points change stability to become a type-one equilibrium point and a stable equilibrium point.

Hopf bifurcation can occur on generic one-parameter dynamical systems. Before a subcritical Hopf bifurcation, the unstable limit cycle $x_1^l(\lambda, t)$ lies on the stability boundary of $x_s(\lambda)$. The Jacobian matrix, when evaluated at $x_s(\lambda)$, has all of its eigenvalues with only negative real parts and among them, a pair of complex eigenvalues are close to zero. At the bifurcation occurring at say, the bifurcation value $\lambda = \lambda^*$, the equilibrium point $x_s(\lambda)$ and the unstable limit cycle $x_1^l(\lambda, t)$ coalesce to form an equilibrium point x^* $(= x_s(\lambda^*) = x_1^l(\lambda^*))$. The Jacobian matrix evaluated at x^* has two zero eigenvalues and the real parts of all the other eigenvalues are negative. If the parameter λ increases beyond the bifurcation value λ^*, then x^* becomes a type-two equilibrium point and there are no other equilibrium points or limit cycles nearby. As for the supercritical Hopf bifurcation, we make the following remarks. Before the bifurcation. The Jacobian matrix, when evaluated at $x_s(\lambda)$, has all of its eigenvalues with only negative real parts and among them, a pair of complex eigenvalues are close to zero. A t the bifurcation occurring at say, the bifurcation value $\lambda = \lambda^*$, the equilibrium point $x_s(\lambda)$ is an equilibrium point x^* $(= x_s(\lambda^*)$. The Jacobian matrix evaluated at x^* has two zero eigenvalues and the real parts of all the other eigenvalues are negative. If the parameter λ increases beyond the bifurcation value λ^*, then x^* becomes a unstable equilibrium point $x^*(\lambda)$ surrounded by a stable limit cycle.

4 Dynamics after Local Bifurcations

Local bifurcation theory has been developed to describe mechanisms leading to disappearance of stable equilibrium points due to a local bifurcation. Recently, local bifurcation theory has being extended to include the analysis of system dynamics after a local bifurcation. The knowledge of subsequent system dynamics is essential to determine whether the system remains stable or becomes unstable, and the type of system instability if the system is deemed unstable. This section briefly summarizes the model for analyzing the system dynamics after a saddle-node bifurcation. Models for analyzing the system dynamics after the limit-induced bifurcation and Hopf bifurcations are also described.

4.1 Saddle-node bifurcation

In [20], Dobson and Chiang investigated a generic mechanism leading to disappearance of stable equilibrium points due to a saddle-node bifurcation and the subsequent system dynamics for one-parameter dynamical systems. When a saddle-node bifurcation occurs at say, the bifurcation value $\lambda = \lambda^*$, equilibrium points $x_s(\lambda)$ and $x_1(\lambda)$ coalesce to form an equilibrium point x^* $(= x_s(\lambda^*) = x_1(\lambda^*))$. The Jacobian matrix evaluated at x^* has one zero eigenvalue and the real parts of all the other eigenvalues are negative. The eigenvector p that corresponds to the zero eigenvalue points in the direction along which the two vectors $x_s(\lambda)$ and $x_1(\lambda)$ approached each other. There is a curve made up of system trajectories which is tangent to eigenvector p at x^*. This curve is called the *center manifold* of x^* and is the union of a system trajectory W_-^c converging to x^*, the equilibrium point x^* and a system trajectory W_+^c diverging from x^*. Next, we consider the case that λ remains fixed at bifurcation value λ^*. When the system trajectory is near W_+^c at the moment that the bifurcation occurs and if λ remains fixed at its bifurcation value λ^*, then the system trajectory after the bifurcation moves near W_+^c. The system dynamics due to the bifurcation are then determined by the position of W_+^c in state space. If W_+^c is positioned so that some of the voltage magnitudes decrease along W_+^c, then we associate the movement along W_+^c with voltage collapse. This is the center manifold voltage collapse model due to saddle-node bifurcation. This model has two advantages from a computational point of view.

1. Since p is tangent to W_+^c at x^*, the initial direction of W_+^c near x^* is determined by p which can be computed f rom the Jacobian matrix at x^*.
2. Since W_+^c is a system trajectory, the dynamics of voltage collapse can be predicted by integrating system equations (1) starting on W_+^c near x^*.

4.2 Limit-induced bifurcation

In [21], Dobson and Lu studied a mechanism leading to disappearance of stable equilibrium points due to a limit-induced bifurcation and the subsequent system dynamics. Before the bifurcation, the system is operated around the stable equilibrium point $x_s(\lambda)$. At the limit-induced bifurcation, the stable equilibrium point $x_s(\lambda)$ and a type-one unstable equilibrium point $x_u(\lambda)$ coalesce to form an equilibrium point. The Jacobian matrix evaluated at the equilibrium point has a single, simple eigenvalue and the real parts of all the other eigenvalues are negative. The operating point becomes immediately unstable when the limit is reached and the system state will move away from the type-one unstable equilibrium point which has a one-dimensional unstable manifold, say W^u. Geometrically speaking, the unstable manifold W^u is tangent at the type-one equilibrium point to the system eigenvector associated with the positive eigenvalue. After the bifurcation, the system state will move along the unstable manifold W^u. It may converge to the near-by stable equilibrium point or diverge along the unstable manifold W^u.

4.3 Hopf bifurcation

A mechanism leading to disappearance of stable equilibrium points due to Hopf bifurcation and the subsequent system dynamics is presented below. When a Subcritical Hopf bifurcation occurs at say, the bifurcation value $\lambda = \lambda^*$, the stable equilibrium point $x_s(\lambda)$ and a unstable limit cycle $x_1^l(\lambda, t)$ coalesce to form an equilibrium point x^* $(= x_s(\lambda^*))$. The Jacobian matrix evaluated at x^* has two zero eigenvalues and the real parts of all the other eigenvalues are negative. The subspace spanned by the two eigenvectors p_1 and p_2 that correspond to the two zero eigenvalues points in the direction along which $x_s(\lambda)$ and $x_1^l(\lambda, t)$ approached each other. The subspace is the center manifold of x^*, say W^c. Next, we consider the case that λ remains fixed at bifurcation value λ^*. Recall that before the bifurcation occurs, the system state is tracking its stable equilibrium point. Therefore, at the moment the bifurcation occurs, the system state is in a neighborhood of x^*. Hence, if the system trajectory is near W^c at the moment that the bifurcation occurs and if λ remains fixed at its bifurcation value λ^*, then the system trajectory after the bifurcation moves along W^c. The system dynamics due to the bifurcation are then determined by the position of W^c in state space. If W^c is positioned so that some of the voltage magnitudes decrease along W^c, then we associate the movement along W^c with voltage collapse. This is the center manifold voltage collapse model due to subcritical Hopf bifurcation.

When a Supercritical Hopf bifurcation occurs at say, the bifurcation value $\lambda = \lambda^*$, the equilibrium point $x_s(\lambda)$ is an equilibrium point x^* $(= x_s(\lambda^*))$. The Jacobian matrix evaluated at x^* has two zero eigenvalues and the real parts of all the other eigenvalues are negative. If the parameter λ increases beyond the bifurcation value λ^*, then x^* becomes a unstable equilibrium point $x^*(\lambda)$ surrounded by a stable limit cycle, say $x_2^l(\lambda, t)$. Next, we consider the case that λ changes slowly after the bifurcation value λ^* and that at the moment the bifurcation occurs, the system state is in a neighborhood of x^*. Note that i t is very likely that the system state will lie inside the stability region of the stable limit cycle $x_2^l(\lambda, t)$, making the system trajectory attracted to $x_2^l(\lambda, t)$. Thus, the system dynamics after a supercritical Hopf bifurcation moves along the stable limit cycle $x_2^l(\lambda, t)$ until another bifurcation occurs on $x_2^l(\lambda, t)$ at another bifurcation value. So, the system dynamics due to the bifurcation are then determined by the position of the stable limit cycle $x_2^l(\lambda, t)$ in state space. If $x_2^l(\lambda, t)$ is positioned so that some of the voltage magnitudes oscillate along the stable limit cycle $x_2^l(\lambda, t)$ and become unacceptable, then we associate the movement with voltage instability. This is the *limit cycle voltage instability model* due to the supercritical Hopf bifurcation.

In summary, after a local bifurcation occurs, the system state will evolve according to the system dynamics as described above. The dynamics after bifurcation determine whether the system remains stable or becomes unstable, and the type of system instability. The analysis of a typical local bifurcation of a stable *equilibrium point* in a power system with slowly varying parameters has two parts:

1. Before the bifurcation when the quasi-static model applies
2. After the bifurcation when the corresponding dynamic model applies

We point out that the quasi-static model is not applicable after the bifurcation and cannot be used to explain the dynamical behaviors after the bifurcation.

4.4 Models for voltage collapse

Voltage collapse is characterized by a slow variation in the system operating point in such a way that voltage magnitudes at load buses gradually decrease until a sharp, accelerated change occurs. In this section, we present two bifurcation-based models to explain voltage collapse in power systems due to slow load variations. It will be then shown that one of the two models encompasses several existing models for explaining voltage collapse. Recall that "slow load variations" means the dynamics of load variations are relatively slower than the dynamics occurring in the state vector.

4.5 SAD voltage collapse model

Stage 1: the system is in quasi-steady state and is tracking a stable equilibrium point.
Stage 2: the system reaches its "steady-state" stability limit when t he stable equilibrium point undergoes a saddle-node bifurcation or a limit-induced bifurcation.
Stage 3: depending on the type of bifurcation encountered in Stage 2, the system dynamics after bifurcation are captured either by the center manifold trajectory of the saddle-node bifurcation point or by the unstable manifold of the limit-induced bifurcation point.

Stage 1 is related to the feasibility of "power flow" solutions (i.e., the existence of a system operating point in a feasible region). Stage 2 determines the steady-state stability limit based on the saddle-node bifurcation point or the limit-induced bifurcation point. Stage 3 describes the system dynamical behavior after bifurcation to assess whether the system, after bifurcation, remains stable or becomes unstable; and determines the types of system instability (voltage collapse and/or angle instability). Hence, the voltage collapse model describes both the static aspect (stages 1 and 2) and the dynamic aspect (stage 3) of the problem.

4.6 Hopf voltage collapse model

> Stage 1: the system is in quasi-steady state and is tracking a stable equilibrium point.
> Stage 2: the system reaches its steady-state stability limit when the stable equilibrium point undergoes a subcritical Hopf bifurcation.
> Stage 3: the system dynamics after bifurcation are captured by a two-dimensional center manifold.

The Hopf voltage collapse model also describes both the static aspect (stages 1 and 2) and the dynamic aspect (stage 3) of the problem. This model dictates that mathematically speaking, the steady-state stability limit may be determined by the subcritical Hopf bifurcation, instead of the Saddle-node bifurcation. One implication is that the load margins will be less that one might expect if the nose point was taken as the point of voltage collapse.

Since detecting Hopf bifurcation requires the knowledge of the eigenvalues of the system Jacobian, the traditional repetitive power flow approach cannot detect Hopf bifurcation.

The saddle-node bifurcation has been a widely accepted model for voltage collapse analysis. Most computational tools developed so far have been concentrated on the identification of saddle-node bifurcation point, also termed point of collapse [6]. The SAD voltage collapse model includes the saddle-node bifurcation point as a point of collapse. In fact, it can be shown that the SAD voltage collapse model encompasses many existing models used to explain voltage collapse such as the multiple power flow model, the power flow feasibility model, the static bifurcation model, the singular Jacobian model and the system sensitivity model. Indeed, Stage 1 is related to the feasibility of the power flow solution [37,40]. It has been shown that stage 2 itself is a generalization of many existing models used to explain voltage collapse [13].

From an algebraic point of view, the point $(x(\lambda_0), \lambda_0)$ abbreviated by (x_0, λ_0) is a saddle node of (1) if the following conditions hold:

1. $f(x_0, \lambda_0) = 0$.
2. $f_x(x_0, \lambda_0)$ has a simple eigenvalue 0.
3. $f_\lambda(x_0, \lambda_0) \notin$ Range space of $(f_x(x_0, \lambda_0))$.
4. there is a parameterization $(x(t), \lambda(t))$ with $x(t_0) = x_0$, $\lambda(t_0) = \lambda_0$ and $\frac{d^2\lambda(t_0)}{dt^2} \neq 0$.

We now use the above algebraic characterizations to examine several models proposed by several researchers for predicting voltage collapse. Note that the voltage collapse models based on the determinant of system Jacobian in [67,56] can be viewed as providing necessary conditions for the first two conditions for saddle nodes. We show in the following that the model based on the sensitivity of system Jacobian in [19] also provides a necessary condition for the first two conditions of saddle nodes. More specifically, we show that the sensitivity of state vector with respect to its parameter at a saddle node is infinity. Suppose that (x^*, λ^*) is a solution of (1) and $f_x(x^*, \lambda^*)$ is non-singular, the implicit function theorem guarantees the existence of a unique solution curve $(x(\lambda), \lambda)$ passing through (x^*, λ^*), i.e . $x^* = x(\lambda^*)$. And we have the following identity:

$$\frac{dx(\lambda)}{d\lambda} = - f_x(x(\lambda), \lambda)^{-1} f_\lambda(x(\lambda), \lambda) \tag{3}$$

Although the matrix $f_x(x(\lambda), \lambda)^{-1}$ does not exist at the saddle node $(x(\lambda^0),$ $\lambda^0)$, the property that $f_\lambda(x_0, \lambda_0) \notin$ Range space of $(f_x(x_0, \lambda_0))$ ensures that there is a unique solution curve passing through $(x(\lambda^0), \lambda^0)$ and $\frac{dx(\lambda)}{d\lambda}$ becomes infinite there and a small change in λ yields a large change in $\|x\|$. This result also explains the "knee" phenomenon in the P-V curve and Q-V curve.

The voltage collapse model based on multiple power flow solutions [61,64, 57] uses the presumption that the existence of a pair of very close power flow solutions indicates that the system is about to undergo a voltage collapse. We note that before the bifurcation, there are two equilibrium points (power flow solutions) close to each other. One is stable and the other is type-one. As these two points approach each other an annihilation occurs at the saddle node bifurcation, while at the same time the system Jacobian becomes singular.

The above two voltage collapse models can be extended to general slow time-varying one-parameter dynamical systems. The assumption that the parameter λ freezes at a bifurcation point may be inadequate to reflect real power system behavior. A more realistic model is to allow a slowly time-varying parameter both before and after the bifurcation. In particular, the assumption that the system parameter "freezes" at the bifurcation point of interest can be removed. The system dynamics after the bifurcation are then captured by the center manifold of the bifurcation point x^* with respect to (2).

$$\begin{aligned} \dot{x} &= f(x, \lambda) \\ \dot{\lambda} &= \epsilon g(x, \lambda) \end{aligned} \tag{4}$$

where ϵ is a small number and g is a locally Lipschitz function.

In order to extend the voltage collapse model to system (4), it will be necessary to examine the adequacy of the center manifold of x^* with respect to system (1) in capturing the dynamics after a bifurcation. In this regard, we examine the relationship between the trajectories of (1) and (4) which start at the same point near the center manifold of x^*. It can be shown that system trajectories of (4) follow the system trajectories of (1) during each choice of time interval provided the rate of parameter variation is sufficiently small as shown in the following:

Proposition 1
Let $x_1(t)$, $x_2(t)$ be the system trajectories of (1) and of (4) starting from the initial condition (x_1^0, t_0) and (x_2^0, t_0) respectively. Let U be a compact set of the state space containing the center manifold as far as it is of interest, and let K be the Lipschitz constant of (4) on the set U. If $x_1^0 = x_2^0 \in U$, then $|x_1(t) - x_2(t)| \leq \frac{\epsilon M}{K} [e^{K(t-t_0)} - 1]$, where M is a constant.

The above result shows that we may approximate the system dynamics after a local bifurcation point by applying (1) instead of (4). Hence, if the parameter changes slowly enough, then the solutions of (4) which lie near the center manifold of (1) at the time of bifurcation will subsequently track the center manifold trajectory of (1) in the state space. This analysis validates the voltage collapse model for power system with slowly variation loads.

5 Computational Tools

Power systems are subject to parameter variations. It is important to study the impacts of parameter variations on power system behaviors by tracing the quasi-steady-state of realistic power system models subject to parameter variations. A powerful global analysis tool based on continuation methods can meet this requirement. Continuation methods, sometimes called curve tracing or path following, are useful tools to generate solution curves for general nonlinear algebraic equations with a varying parameter. The theory of continuation methods has been studied extensively and has its roots in algebraic topology and differential topology. Continuation methods have four basic elements: parameterization, predictor, corrector and step-size control.

The application of continuation methods to power system analysis has been very actively investigated in recent decades, see for example [38,14,48,3, 7,15,32]. The most attractive feature of a continuation method is that it allows users to globally analyze a given power system relative to parameter variations in a reliable and efficient manner. In [12], a survey of existing and pioneering continuation methods applied to power system analysis, which may contain thousands of nonlinear algebraic equations with some limits on some of the state variables, is presented. This survey also includes a comparison among different implementations of continuation methods for power system applications according to predictor type, corrector type, step-size control strategy, parameterization schemes and modelling capability.

A widely used approach in the power industry to investigate potential voltage stability problems, with respect to a given parameter increase pattern, is the use of repetitive power flow calculations. The main advantages of a power system analysis tool based on continuation methods over the repetitive power flow calculations are the following:

- **Computation**
  1. it is more *reliable* than the repetitive power flow calculations in obtaining the solution curve and the nose point via the parameterization scheme;
  2. it is *faster* than the repetitive power flow calculations via an effective predictor-corrector, adaptive step-size selection algorithm and efficient I/O operations.
- **Function**
  1. it is more versatile than the repeated power flow approach via parameterizations such that general bus real and/or reactive loads, area real and/or reactive loads, or system-wide real and/or reactive loads, and real generation at P-V buses, e.g., determined by economic dispatch or participation factor, can vary.

Consider a comprehensive (static) power system mode expressed in the following form:

$$0 = f(x, \lambda) \tag{5}$$

where $\lambda \in R^1$ is a (controlling) parameter subject to variation. Using terminology from the field of nonlinear dynamical systems, system (5) is a one-parameter nonlinear system.

We next discuss an indirect method to simulate the approximate behavior of the power system (5) due to load and/or generation variation. Before reaching the "nose" point, the power system with a slowly varying parameter traces its operating point which is a solution of the following equation whose corresponding Jacobian has all eigenvalues with negative real parts:

$$\mathbf{f}(\mathbf{x}, \lambda) = 0, \ \ \mathbf{x} \in \mathbf{R}^n, \mathbf{f} \in \mathbf{R}^n, \lambda \in \mathbf{R} \tag{6}$$

These n equations of $n+1$ variables define in the $n+1$-dimensional space a one-dimensional curve $x(\lambda)$ passing through the operating point of the power system (x^0, λ^0). The indirect method is to start from (x^0, λ^0), and produce a series of solution points (x^i, λ^i) in a prescribed direction until the "nose" point is reached. However, it is known that the set of power flow equations near its "nose" point is ill-conditioned, making the Newton method diverge in the neighborhood of "nose" points. There are several possible means to resolve the numerical difficulty arising from the ill-conditioning. One effective way is as follows: First, treat the parameter λ as another state variable

$$\mathbf{x_{n+1}} = \lambda.$$

Second, introduce the arc-length s on the solution curve as a new parameter in the continuation process. This parameterization process gives

$$x = x(s), \quad \lambda = \lambda(s) = x_{n+1}.$$

The step-size along the arc-length s yields the following constraint:

$$\sum_{i=1}^{n}\{(x_i - x_i(s))^2\} + (\lambda - \lambda(s))^2 - (\Delta s)^2 = 0$$

Third, solve the following $n+1$ equations for the $n+1$ unknowns x and λ

$$f(x, \lambda) = 0 \tag{7}$$

$$\sum_{i=1}^{n}\{(x_i - x_i(s))^2\} + (\lambda - \lambda(s))^2 - (\Delta s)^2 = 0 \tag{8}$$

It can be shown that the above set of augmented power flow equations is well-conditioned, even at the "nose" point. These augmented power flow equations can be solved to obtain the solution curve passing through the "nose" point without encountering the numerical difficulty of ill-conditioning.

The task of computing maximum loading points (it saddle-node points or limit-induced bifurcation points) relative to a given load/generation variation pattern has important applications in power system operations and planning. The maximum loading points have a strong relationship with the operating points where voltage collapse may occur. A widely used approach in power industry to determine the maximum loading point, with respect to a given load/generation increase pattern, is the repetitive power flow calculations to generate the so-called P-V or Q-V curve relative to the variation pattern. An operating point on the curve is said to be the maximum loading point of the system if the point is the first point on the curve where power flow calculation does not converge. Note that, due to the its shape in the bifurcation diagram, the saddle-node bifurcation point is termed nose point in power engineering community. Depending on the physical meaning of the underlying parameter and the power network conditions, nose points have been physically related to maximum loading points, or to maximum transfer capability points, or to voltage collapse points.

Several issues arise regarding this approach [44]. These issues however can be resolved by applying the local bifurcation theory. First, the point where the power flow diverges (which is a numerical failure caused by a numerical method) does not necessarily represent the maximum loading points (which is a physical limitation). Second, the point where power flow calculations fail to converge may vary, depending on which numerical method was used in the power flow calculation. In other words, based on the criterion of power flow divergence, different numerical methods may come up with different calculated maximum loading points of the system while the maximum loading point physically is unique. We note that the set of power flow equations is ill-conditioned near nose points making Jacobian-based numerical methods such as the Newton method diverge in the neighborhood of nose points. It is well recognized that the Jacobian at a nose point has one zero eigenvalue, causing the set of power flow equations ill-conditioned near nose points. Recently, considerable progress has been made in calculating nose points in a reliable and exact way by using continuation methods and the characteristic equations of nose points. Continuation methods are reliable to overcome the singularity of the Jacobian near nose points and can provide partial initial conditions for solving the characteristic equations.

The standard formulation for the characteristic equations of nose points for a set of n-dimensional power flow equations is a set of (2n+1)-dimensional nonlinear equations. Solutions to the characteristic equations give the nose point (n-dimensional), the bifurcation value and the left or right eigenvector (n-dimensional corresponding to the zero eigenvalue. To solve the characteristic equations, continuation methods can only provide good initial conditions for an estimated nose point and an estimated bifurcation value. What is missing is a good initial guess for the eigenvector which is an additional factor affecting convergence to the solution. Another method which solves an extended (2n+1)-dimensional system of equations characterizing the saddle-node bifurcation point was proposed in [6] and more recently in [7]. The methods attempt to compute the saddle-node bifurcation point directly. The success of the above two methods depends greatly on a good initial guess of the desired saddle-node bifurcation point. Otherwise, the methods may diverge or converge to another saddle-node bifurcation point. This is because these methods are static in nature, they do not make use of any information on the particular branch of solutions and they do not confine their iterative process to the desired branch of solutions.

A simpler set of characteristic equations for nose points of power flow equations can be developed by exploring a decoupled parameter-dependent property of power flow equations. In [44], a test function was developed to characterize nose points of power flow equations. The test function possesses a monotonic property in the neighborhood of nose points that it is positive on one side of the bifurcation value while it is negative one the other side. Hence, it offers an effective way to bracket the parameter value during a search procedure of bifurcation values to guarantee a solution exists inside the bracket. This test function in conjunction with the set of power flow equations constitute a set of (n+1)-dimensional characteristic equations for saddle-node bifurcation points of general nonlinear equations with decoupled parameter [39]. Distinguishing features of the new set of characteristic equations are that they are of dimension n+1, instead of 2n+1, for n-dimensional power flow equations and that the required initial conditions (bifurcations point and bifurcation value) can be completely provided by the continuation method.

The task of computing Hopf bifurcations points has physical importance in power system analysis and control. This task, though more involved in computation, receives less attention than the task of computing nose points. For n-dimensional power flow equations, the standard formulation for the characteristic equations of Hopf bifurcation points is a set of (2n+2)-dimensional nonlinear equations. A simpler set of characteristic equations for Hopf bifurcation points of power flow equations can be developed by exploring a decoupled parameter-dependent property of power flow equations. The new set of characteristic equations is of dimension n+2 (even can be of n+1), instead of 2n+2 , for n-dimensional power flow equations. More research work is required in this task for power system applications.

We next discuss a practical package, CPFLOW (Continuation Power Flow), a comprehensive tool for tracing power system steady-state behavior due to parameter variations such as load variations, generation variations and control variations [15]. CPFLOW simulates a realistic operating condition or expected future operating conditions relative to parameter variations with activation of control devices during the process of parameter variations. The control devices include : (i) switchable shunts and static VAR compensators, (ii) ULTC transformers, (iii) ULTC phase shifters, (iv)static tap changer and phase shifters, (v) DC network.

CPFLOW can efficiently generate P-V, Q-V, and P-Q-V curves with the capability that the controlling parameter λ can be one of the following

- general bus (P and/or Q) loads + real power generation at P-V buses
- area (P and/or Q) loads + real power generation at P-V buses
- system (P and/or Q) loads + real power generation at P-V buses

CPFLOW, computationally based on the continuation method, can trace the power flow solution curve, with respect to any of the above three varying parameter, through the "nose" point (i.e. the saddle-node bifurcation point) or the limit-induced bifurcation point, without any numerical difficulty. CPFLOW can be used in a variety of applications such as (1) to analyze voltage problems due to load and/or generation variations (e.g. voltage dip, voltage collapse), (2) to evaluate maximum interchange capability and maximum transmission capability [31], (3) to simulate power system static behavior due to load and/or generation variations with/without control devices, and (4) to conduct coordination studies of control devices for steady-state security assessment.

CPFLOW's modelling capability is quite comprehensive. The current version of CPFLOW can handle power systems up to 43,000 buses. CPFLOW has been applied to a 40,000-bus power system with a complete set of operational limits and controls. CPFLOW provides three options of parameterization schemes including arc-length parameterization. In order to achieve computational efficiency, CPFLOW employs the tangent method in the first phase of solution curve tracing and the secant method in the second phase. However, if the number of corrector iterations becomes too large, then the predictor switches back to the tangent since it is more accurate than the secant. The Newton method is chosen in CPFLOW as the corrector. CPFLOW computes the arc-length in the state space, which automatically forces the predictor to take large steps on the "flat" part of the solution curve and small steps on the "curly" part.

6 Performance Indices for Assessing Voltage Collapse

We show in this section how local bifurcation theory can be applied to develop performance indices for assessing voltage collapse. While continuation power flow methods can generate P-V and Q-V curves in a reliable manner, they may be too slow for certain applications such as contingency selection and contingency analysis, design of preventive control for voltage collapse and on-line voltage security assessments. To overcome these difficulties, a number of performance indices intended to measure the severity of the voltage collapse problem have been proposed in the literature. They can be divided into two classes: state-space-based approach and the parameter-space-based approach.

The majority of performance indices developed for assessing voltage collapse adopt the state-space-based approach. Among them, the minimum singular value in [65], the eigenvalue pursued in [37] and the condition number in [54] of the system Jacobian intend to provide some measure of how far the system is away from the point at which the system Jacobian becomes singular. The performance index proposed in [61] and [64] is based on the angular distance between the current stable equilibrium point and the closest unstable equilibrium point in a Euclidean sense. the performance index proposed in [25,26] measures the energy distance between the current stable equilibrium point and the closest unstable equilibrium point using an energy function. These performance indices can be viewed as providing some measure of the "distance" between the current operating point and the bifurcation point. Note that all these performance indices are defined in the state space of power system models and they cannot directly answer questions such as: "Can the system withstand a 100 MVar increase on bus 20 without encountering voltage collapse?"

One basic requirement for useful performance indices is their ability to reflect the degree of direct mechanism leading the underlying system toward an undesired state. In the context of voltage collapse in power systems, a useful performance index must have the ability to measure the amount of load increase that the system can tolerate before collapse. The state-space-based performance indices, however, generally do not exhibit any obvious relation between their value and the amount of the underlying mechanism that the system can tolerate before collapse.

In order to provide a direct relationship between its value and the amount of load increases that the system can withstand before collapse, the performance index must be developed in the parameter space (i.e., the load/generation space). Development of performance indices in the parameter space is a relatively new concept which may have been spurred by the local bifurcation theory. In [16], a new performance index that provides a direct relationship between its value and the amount of load demand that the system can withstand before a saddle-node voltage collapse was developed. From an analytical viewpoint, this performance index is based on the normal form of saddle-node bifurcation points. It can be shown that, in the context of power flow equations, the power flow solution curve passing through the nose point is, at least locally, a quadratic curve. From a computational viewpoint, this performance index makes use of the information contained in the power flow solutions of the particular branch of interest. It only requires two power flow solutions. The first power flow solution is the current operating point which can be obtained from a state estimator. Only one additional power flow solution and its derivative are needed to compute this performance index. One of the features that distinguishes the proposed performance index is its development in the load-generation space (i.e. the parameter space) instead of the state space where the then existing performance indices were developed.

From an application viewpoint, the parameter-space-based performance indices can be readily interpreted by power system operators to answer questions such as: "Can the system withstand a simultaneous increase of 70 MW on bus 2 and 50 MVar on bus 6?". Moreover, the computation involved in the performance index is relatively inexpensive in comparison with those required in the state-space-based ones. A look-ahead performance index intended for on-line applications was developed in [17]. Given the following information; (1) the current operating condition, say obtained from the state estimator and the topological analyzer, (2) the near-term load demand at each bus, say obtained from short-term load forecaster and predictive data, and (3) the real power dispatch, say based on economic dispatch, the look-ahead performance index provides a look-ahead load margin measure (in MW and/or Mvar) which can be used to assess the system's ability to withstand both the forecasted load demands and real power variations. In addition, the index provides useful information as to how to derive effective load-shedding schemes to avoid voltage instability.

We note that the parameter-space-based performance indices can not take into account the physical limitations of typical control devices such as generator VAR limits and ULTC tap ratio limits; such that their computed load/generation margins may bear some 'distance' from the exact margins. Hence, the function of these performance indices is mainly for ranking the severity of a list of credible contingencies or for ranking the effectiveness of different control devices. Exact load/generation margins that accounts for all control devices and their physical limitations can be accurately calculated by using the continuation power flow approach. Recent work on the parameter-space-based performance indices can be found, for example in [33,8,27,29].

7 Persistence of Local Bifurcations under Unmodelled Dynamics

Many physical systems contain slow and fast dynamics. These slow and fast dynamics are not easy to model in practice. Even if these dynamics can be modelled properly, the resulting system model (the original model)is often ill-conditioned. These difficulties have motivated development of several model reduction or simplification approaches to derive reduced models from the original model. One popular model reduction approach (to derive a reduced model) is to neglect both the fast and slow dynamics in an appropriate way. On the other hand, traditional practice in system modelling has been to use the simplest acceptable model that captures the essence of the phenomenon under study. A common logic used in this practice is that the effect of a system component or control device can be neglected when the time scale of its response is very small or very large compared to the time period of interest [69,72].

Electric power systems comprise a large number of components interacting with each other in nonlinear manners. The dynamical response of these components extends over a wide range of time scales. For example, the difference between the time constants of excitation systems (fast control devices) and that of governors (slow control devices) is a couple orders of magnitudes. The dynamic behavior after a disturbance occurring on a power system involves all the system components to varying degrees. The degree of involvement from each component determines the appropriate system model necessary for simulating the dynamic behaviors after the disturbance. For instance, an extended power system dynamical mode l contain both fast variables, such as the damping flux in the direct and quadrature axis of generators, and slow variables, such as the field flux and the mechanical torque of generators. For simulating the dynamic behaviors of a power system after an event disturbance, the effect of these fast and slow variables can be neglected in the system modelling because the time scale of these variables is very small or very large compared to the time period of the disturbance of interest. A reduced system model is thus obtained from the original system model.

Several questions naturally arise regarding the validity of using the analysis based on a reduced system model to determine the behavior of the original system. These questions include the relation between the stability properties of the reduced system and those of the original system, between the trajectories of the reduced system and that of the original system, and so on. We consider a nonlinear dynamical system with slow and fast un-modelled dynamics of the form

$$\begin{aligned} \dot{x} &= f(x,y,z,\lambda) + \epsilon_1 f_0(x,y,z,\lambda,\epsilon_1,\epsilon_2,\tilde{\epsilon}_1,\tilde{\epsilon}_2) \\ \dot{y} &= \tilde{\epsilon}_1 g(x,y,z,\lambda,\epsilon_1,\epsilon_2,\tilde{\epsilon}_1,\tilde{\epsilon}_2) \quad \text{slow} \\ \epsilon_2 \dot{z} &= h(x,y,z,\lambda) + \tilde{\epsilon}_2 h_0(x,y,z,\lambda,\epsilon_1,\epsilon_2,\tilde{\epsilon}_1,\tilde{\epsilon}_2) \quad \text{fast} \end{aligned} \tag{9}$$

where $x \in R^n$, $y \in R^m$, $z \in R^p$, $\epsilon_1, \epsilon_2, \tilde{\epsilon}_1, \tilde{\epsilon}_2 \in R^+$, and f, f_0, g, h, h_0 are C^r with $r \geq 2$.

Associated with system (9), we define a reduced system which treats the fast variables z as instantaneous variables and the slow variables y as constants. This is done by setting $\epsilon_1, \epsilon_2, \tilde{\epsilon}_1, \tilde{\epsilon}_2 = 0$

$$\begin{aligned} \dot{x} &= f(x, y, z, \lambda) \\ \dot{y} &= 0 \\ 0 &= h(x, y, z, \lambda) \end{aligned} \tag{10}$$

We pose and study the following problems:
(p1) If the reduced system (10) has a saddle-node bifurcation point at $(x^*, y^*, z^*, \lambda^*) = (x^*, y^*, z^*(x^*, y^*, \lambda^*), \lambda^*)$ relative to the varying parameter λ, then does this imply that the original system (9) with the varying parameter λ also has a saddle-node bifurcation point in a neighborhood of $(x^*, y^*, z^*, \lambda^*)$? If the answer is yes, then
(p2) what is the relationship between these two saddle-node bifurcation points? Furthermore,
(p3) what is the relationship between the system behaviors after the saddle-node bifurcation of the reduced system (10) and that of the original system (9)?

We propose to solve the above three problems via the following three steps. In the first step, we consider a nonlinear dynamical system with slow un-modelled dynamics of t he form :

$$\begin{aligned} \dot{x} &= f(x, y^*, \lambda) + \tilde{\epsilon} f_0(x, y, \lambda, \epsilon, \tilde{\epsilon}) \\ \dot{y} &= \epsilon g(x, y, \lambda, \epsilon, \tilde{\epsilon}) \end{aligned} \tag{11}$$

where $x \in R^n$, and $y \in R^m$ is a slowly varying vector, $\epsilon, \tilde{\epsilon}$ are small numbers and $\lambda \in R$ is a parameter which is subject to variation. A reduced system associated with (11) can be derived by treating $y \in R^m$ as a constant vector:

$$\begin{aligned} \dot{x} &= f(x, y, \lambda) \\ \dot{y} &= 0 \end{aligned} \tag{12}$$

In this step several analytical results to address the above three issues can be developed. We consider in the second step a nonlinear dynamical system with fast un-modelled dynamics of the form:

$$\begin{aligned} \dot{x} &= f(x, y, \lambda, \epsilon) \\ \epsilon \dot{y} &= g(x, y, \lambda, \epsilon) \end{aligned} \tag{13}$$

where $x \in R^n$, $y \in R^m$, $\lambda, \epsilon > 0 \in R$, and f, g are C^r with $r \geq 2$. A reduced system by "neglecting" the fast dynamics y can be defined by setting $\epsilon = 0$ in (13)

$$\begin{aligned} \dot{x} &= f(x, y, \lambda, 0) \\ 0 &= g(x, y, \lambda, 0) \end{aligned} \tag{14}$$

In the third step we connect the analytical results derived in the first two steps to show that, under fairly general conditions, the general nonlinear system (with both fast and slow dynamics) (9) will encounter a saddle-node bifurcation relative to a varying parameter if the associated reduced system (10) (which is derived by neglecting both fast and slow dynamics) encounters a saddle-node bifurcation relative to the varying parameter. A error bound can be derived between the bifurcation point of the reduced system (10) and that of the original system (9). Furthermore, it can be shown that the system behaviors after the saddle-node bifurcation of the reduced system and that of the original system are close to each other in state space [18]. The general analytical results can be applied, among others, to justify the usage of simple power system models for analyzing voltage collapse in electric power systems. For instance, it provides a justification of the current practice that voltage collapse can be analyzed based on a simple model of synchronous machines (the so-called swing equation) rather than on a detailed model which includes the dynamics of several control devices.

8 Concluding Remarks

We have presented in this chapter an overview on the application of local bifurcation analysis and theory to (i) develop models explaining power system nonlinear behaviors and various power system instabilities such as voltage collapse and low-frequency oscillations, to (ii) develop a powerful global analysis tool based on continuation methods to trace power system quasi-steady-state behaviors due to load and generation variations in realistic power system models, and to (iii) develop performance indices for detecting and estimating local bifurcations of power systems. Furthermore, an overview on the extension of saddle-node bifurcation, Hopf bifurcation and limit-induced bifurcation to include the analysis of the system dynamics after the bifurcation has been presented.

Electric power systems comprise a large number of components whose dynamical response extends over a wide range of time scales. Up to present, most power system models used for bifurcation analysis involve only short-term dynamical models (transient stability models). It raises the concern about the validity of short-term dynamical models, which have disregarded slow dynamics, for local bifurcation analysis. The effects of un-modelled dynamics due to fast and slow state variables on the local bifurcation analysis of a power system model has been also discussed in this chapter.

During the last two decades, numerical bifurcation analysis of power system models has been a subject of great interest to researchers and engineers in the power system community. These numerical studies seem to favor the claim that various types of bifurcations can occur in real power systems. Several bifurcation-based models have been developed to provide an explanation for various observed power system nonlinear behaviors and power system instabilities. Furthermore, these numerical studies support the observation that the complexity of power system dynamic behaviors is related more to the nonlinearity of individual power system models than to the dimensionality of the system. However, these numerical studies only establish a presumption.

The next logical step is to investigate the nature, extent and significance of these (local) bifurcations in realistic power system models; if not in real power systems. To this regard, several issues must be addressed. The first issue, related to its nature, is whether the model used reflects a realistic power system. The second issue is under what conditions can realistic power system models encounter bifurcations (such as saddle-node bifurcation, Hopf bifurcation). The third issue, related to its extent, is whether the regions in the parameter space as well as in the state space where bifurcation can occur lie near normal operating points of power systems. The forth issue, related to its significance, is whether the magnitudes of dynamical behaviors after bifurcations are observable in power system behaviors. Other issues remained to be addressed include the following

- Under what conditions can realistic power system models encounter global bifurcations?
- Under what conditions can realistic power system models encounter limit-induced bifurcations?
- How can bifurcation affect power system nonlinear behaviors?
- How to evaluate the merits of each explanation of power system instabilities when there are several competing explanations?
- What kind of actions can be taken to prevent bifurcations?

The above issues related to the nature, extent and significance of bifurcations in realistic power system models can only be addressed using both powerful computational tools and analytical tools. This presents a great challenge for researchers to develop a highly effective computational environment for analyzing bifurcations in large-scale power systems, which are described by a large set of nonlinear equations with parameter-dependent differential and algebraic equations with hard limits.

References

1. Abed, E. H., Wang, H. O., Alexander, J. C., Hamdan, A. M. A., Lee, H. C. (1993) Dynamic bifurcations in a power system model exhibiting voltage collapse. Int. J. of Bifur. Chaos, **3**:1169–1176
2. Abed, E. H., Varaiya, P. P. (1984) Nonlinear oscillations in power system. Int. J. of Electr. Power Energy Syst., **6**:37–43
3. Ajjarapu, V. A., Lee, B. (1992) The continuation power flow: A tool for steady state voltage stability analysis. IEEE Trans. Power Syst., **7**:416–423
4. Ajjarapu, V. A., Lee, B. (1992) Bifurcation theory and its application to nonlinear dynamical phenomena in an electric power system. IEEE Trans. Power Syst., **7**:424–431
5. Budd, C. J., Wilson, J. P. (2002) Bogdanov-Takens bifurcation points and Sil'nikov homoclinicity in a simple power-system model of voltage collapse. IEEE Trans. Circ. Syst.-I, **49**:575–590

6. Cañizares, C. A., Alvarado, F. L. (1992) Point of collapse methods applied to AC/DC systems. IEEE Trans. Power Syst., **7**:673–683
7. Cañizares, C. A., Alvarado, F. L. (1993) Point of collapse and continuation methods for large AC/DC systems. IEEE Trans. Power Syst., **8**:1–8
8. Canizares, C. A., de Souze, A. C. Z., Quintana, V. H. (1995) Comparison of performance indices for detection of proximity to voltage collapse. In Porc. IEEE PES Summer Power Meet., Paper 95, SM 583–585 PERS
9. Baillieul, J., Byrnes, C. I. (1982) Geometric critical point analysis of lossless power system models. IEEE Trans. Circ. Syst. **29**:724–737
10. Chen, R. L., Varaiya, P. P. (1988) Degenerate Hopf bifurcation in power systems. IEEE Trans. Circ. Syst., **35**:818–824
11. Chiang, H. D., Liu, C. W., Varaiya, P. P., Wu, F. F., Lauby, M. G. (1993) Chaos in a simple power system. IEEE Trans. Power Syst., **8**:1407–1417
12. Chiang, H. D., Conneen, T. P., Flueck, A. J. (1994) Bifurcations and chaos in electric power systems: Numerical studies. J. Franklin Institute, **331B**:1001–1036
13. Chiang, H. D. (1999) Power system stability. Wiley Encyclopedia of Electrical and Electronics Engineering, ed. by Webster J. G., New York: Wiley, 104–137
14. Chiang, H. D., Ma, W., Thomas, R. J., Thorp, J. S. (1990) A tool for analyzing voltage collapse in electric power systems. In Proc. 10th Power System Computation Conference, Graz, Austria, August 1990
15. Chiang, H. D., Flueck, A. J., Shah, K. S., Balu, N. (1995) CPFLOW: A Practical tool for tracing power system steady-state stationary behavior due to load and generation variations. IEEE Trans. Power Syst., **10**:623–634
16. Chiang, H. D., Jean-Jumeau, R. (1995) Toward a practical performance index for predicting voltage collapse in electric power system. IEEE Trans. Power Syst., **10**:584–592
17. Chiang, H. D., Wang, C. S., Flueck, A. J. (1997) Look-ahead voltage and load margin contingency selection functions for large-scale power systems. IEEE Trans. Power Syst., **12**:173–180
18. Fekih-Ahmed, L., Chiang, H. D. (1993) Persistence of saddle-node bifurcations for general nonlinear systems with slow unmodelled dynamics. J. of Circ., Sys. Sign. Proc., **12**:533–555
19. Navarro-Perez, R., Cory B. J., Short, M. J. (1989) Voltage collapse proximity analysis using reactive area identification. In Proc. Bulk Power System Voltage Phenomena - Voltage Stability and Security, EPRI Report EL-6183, 7-41/7-58
20. Dobson, I., Chiang, H. D. (1989) Towards a theory of voltage collapse in electric power systems. Syst. Contr. Lett., **13**:253–262
21. Dobson, I., Lu, L. (1992) Voltage collapse precipitated by the immediate change in stability when generator reactive power limits are encountered. IEEE Trans. Circ. Syst.-I, **39**:762–766
22. Dobson, I. (1992) Observations on the geometry of saddle node bifurcation and voltage collapse in electrical power system. IEEE Trans. Circ. Syst., **39**:240–243
23. Dobson, I., Lu, L. (1992) Computing an optimum direction in control space to avoid saddle node bifurcation and voltage collapse in electric power systems. IEEE Trans. Auto. Contr., **37**:1616–1620
24. Dobson, I., Lu, L. (1993) New method for computing a closest saddle node bifurcation and worst case load power margin for voltage collapse. IEEE Trans. Power Syst., **8**:905–913

25. Overbye, T. J., DeMarco, C. L. (1991) Voltage security enhancement using an energy based sensitivity. IEEE Trans. Power Syst., **9**:1196–1202
26. Overbye, T. J., Dobson, I., DeMarco, C. L. (1994) Q-V curve interpretations of energy measures for voltage security. IEEE Trans. Power Syst., **9**:331–340
27. Ejebe, G. C., Irisarri, *et al.* (1996) Methods for contingency screening and ranking for voltage stability analysis of power systems. IEEE Trans. Power Syst., **11**:350–356
28. Iravani, M. R., Semlyen, A. (1992) Hopf bifurcations in torsional dynamics. IEEE Trans. Power Syst., **7**:28–36
29. Jia, Z., Jeyasurya, B. (2000) Contingency ranking for on-line voltage stability assessment. IEEE Trans. Power Syst., **15**:1093–1097
30. Mercede, F., Chow, J. C., Yan, H., Fischl, R. (1988) A framework to predict voltage collapse in power systems. IEEE Trans. Power Syst., **3**:1807–1813
31. Flueck, A. J., Chiang, H. D., Shah, K. S. (1996) Investigating the installed real power transfer capability of a large scale power system under a proposed multi-area interchange schedule using CPFLOW. IEEE Trans. Power Syst., **11**:883–889
32. Flueck, A. J., Dondeti, J. R. (2000) A new continuation power flow tool for investigating the nonlinear effects of transmission branch parameter variations. IEEE Trans. Power Syst., **15**:223–227
33. Flueck, A. J., Chiang, H. D., Wang, C. S. (1997) A novel method of look-ahead contingency ranking for saddle-node bifurcation. In Proc. 20th Int. Conf. Power Industry Computer Applications, Columbus, OH, 266–271
34. Feng, Z., Ajjarapu, V., Maratukulam, D. J. (1998) A practical minimum load shedding strategy to mitigate voltage collapse. IEEE Trans. Power Syst., **13**:1285–1291
35. Feng, Z., Ajjarapu, V., Long, B. (2000) Identification of voltage collapse through direct equilibrium tracing. IEEE Trans. Power Syst., **15**:342–349
36. Fouad, A. A., Vittal, V. (1991) Power System Transient Stability Analysis: Using the Transient Energy Function Method. Englewood Cliffs, NJ: Prentice-Hall
37. Galiana, F. D. (1984) Load flow feasibility and the voltage collapse problem. In Proc. 23th IEEE Conf. Decision Control, Las Vegas, 485–487
38. Huneault, M., Fahmideh-Vojdani, A, Juman, M., Calderon, R., Galiana, F. G. (1985) The continuation method in power system optimization: Applications to economy security functions. IEEE Trans. Power Syst., **104**:114–124
39. Kataoka, Y. (1992) An approach for the regularization of a power flow solution around the maximum loading point. IEEE Trans. Power Syst., **7**:1068–1077
40. Kessel, P., Glavistch, H. (1986) Estimating the voltage stability of a power system. IEEE Trans. Power Delivery, **1**:346–354
41. Korsak, A. J. (1972) On the question of uniqueness of stable load flow solutions. IEEE Trans. Power Syst., **91**:1093–1100
42. Kundur, P. (1994) Power System Stability and Control. New York: McGraw-Hill
43. Kurita, A., Sakurai, T. (1988) The power system failure on July 23, 1987 in Tokyo. In Proc. 27th IEEE Conf. Decision Control, Austin, TX, 2093–2097
44. Jean-Jumeau, R., Chiang, H. D. (1995) A more efficient formulation for computation of the maximum loading points in electric power systems. IEEE Trans. Power Syst., **10**:635–641

45. Kwatny, H. G., Pasrija, A. K., Bahar, L. Y. (1986) Static bifurcations in electric power networks: Loss of steady state stability and voltage collapse. IEEE Trans. Circ. Syst., **33**:989–991
46. Kwatny, H. G., Fischl, R. F., Nwankpa, C. O. (1995) Local bifurcation in power systems: Theory, computation, and application. Proceedings of the IEEE, **83**:1456–1483
47. Makarov, Y. V., Hill, D. J., Dong, Z. Y. (2000) Computation of bifurcation boundaries for power systems: A new D-plane method. IEEE Trans. Circ. Syst.-I, **47**:536–544
48. Iba, K., Suzuli, H., Egawa, M., Watanabe, T. (1991) Calculation of the critical loading condition with nose curve using homotopy continuation method. IEEE Trans. Power Syst., **6**:584–593
49. NERC Report (1991) Survey of the voltage collapse phenomena. Summary of the Interconnection Dynamics Task Force's Survey on the Voltage Collapse Phenomena. Princeton, NJ
50. Shahrestani, S., Hill, D. J. (2000) Global control with application to bifurcating power systems. Syst. Contr. Lett., **41**:145–155
51. Van Cutsem, T., Vournas, C. D. (1998) Voltage Stability of Electric Power Systems. Boston, MA: Kluwer
52. Van Cutsem, T. (2000) Voltage instability: Phenomena, countermeasures, and analysis methods. Proceedings of the IEEE, **88**:208–227
53. Varaiya, P. P., Wu, F. F., Chiang, H. D. (1992) Bifurcation and chaos in power systems: A survey. EPRI Report TR-100834
54. Pai, M. A., O'Grady, M. G. (1989) Voltage collapse analysis with reactive generation and voltage dependent load constraints. Electric Machines Power Syst., **17**:379–390
55. Sauer, P. W, Pai, M. A. (1998) Power System Dynamics and Stability. Upper Saddle River, NJ: Prentice-Hall
56. Sekine, Y., Yokoyama, A., Kumano, T. (1989) A method for detecting a critical state of voltage collapse. In Proc. of Bulk Power System Voltage Phenomena - Voltage Stability and Security, EPRI Report EL-6183, 5-65/5-72
57. Yokoyama, A., Sekine, Y. (1989) A static voltage stability index based on multiple load flow solutions. In Proc. of Bulk Power System Voltage Phenomena - Voltage Stability and Security, EPRI Report EL-6183, 7-111/7-123
58. Seydel, R. (1988) From Equilibrium to Chaos: Practical Bifurcation and Stability Analysis. New York: Elsevier
59. Tan, C. W., Varghese, M., Varaiya, P. P., Wu, F. F. (1995) Bifurcation, chaos, and voltage collapse in power systems. Proceedings of the IEEE, **83**:1484–1496
60. Taylor, C. W. (1994) Power System Voltage Stability. New York: McGraw-Hill
61. Tamura, Y., Mori, H., Iwamoto, S. (1983) Relationship between voltage stability and multiple load flow solutions in electric power systems. IEEE Trans. Power Syst., **102**:1115–1125
62. Tavora, C. J., Smith, O. J. M. (1972) Equilibrium analysis of power systems. IEEE Trans. Power Syst., **91**:1131–1137
63. Tavora, C. J., Smith, O. J. M. (1972) Stability analysis of power systems. IEEE Trans. Power Syst., **91**:1138–1144
64. Tamura, Y. (1988) Voltage instability proximity index based on multiple load-flow solutions in ill-conditioned power systems. In Proc. 27th IEEE Conf. Decision Control, Austin, TX, 2114–2119

65. Tiranuchit, A., Thomas, R. J. (1988) A posturing strategy against voltage instabilities in electric power systems. IEEE Trans. Power Syst., **3**:87–93
66. Thomas, R. J., Barnard, R. D., Meisel, J. (1971) The generation of quasi steady-state load flow trajectories and multiple singular point solutions. IEEE Trans. Power Syst., **90**:1967–1974
67. Venikov, V. A., Stroev, V. A., Idelchick, V. I., Tarasov, V. I. (1975) Estimation of electric power systems steady-state stability in load flow calculations. IEEE Trans. Power Syst., **94**:1934–1038
68. Venkatasubramanian, V., Schattler, V. H., Zaborszky, J. (1995) Local bifurcations and feasibilty regions in differential-algebraic systems. IEEE Trans. Auto. Contr., **40**:1992–2013
69. Venkatasubramanian, V., Schattler, H., Zaborszky, J. (1993) Homoclinc orbits and the persistence of the saddle connection bifurcation in the large power system. In Proc. IEEE ISCAS, Chicago, 2648–2651
70. Venkatasubramanian, V., Schattler, H., Zaborszky, J. (1995) Dynamics of large constrained nonlinear systems – A taxonomy theory. Proceedings of the IEEE, **83**:1530–1560
71. Walve, K. (1986) Modelling of power system components at severe disturbances. In Proc. Int. Conf. Large High Voltage Electric Systems, CIGRE paper 38–18
72. Yorino, N., Sasaki, H., Masuda, Y., Tamura, Y., Kitagawa, M., Oshimo, A. (1994) On voltage stability from the viewpoint of singular perturbation theory. Electr. Power Energy Syst., **16**:409–417
73. Yorino, N., Harada, S., Cheng, H. Z. (1997) A method to approximate a closest loadability limit using multiple load flow solutions. IEEE Trans. Power Syst., **12**:424–429

Bifurcation Analysis with Application to Power Electronics

Chi K. Tse and Octavian Dranga

Department of Electronic and Information Engineering
Hong Kong Polytechnic University
P. R. China
ncktse@polyu.edu.hk

Abstract. The problem of sudden loss of stability (more precisely, sudden change of operating behaviour) is frequently encountered in power electronics. A classic example is the current-mode controlled dc/dc converter which suffers from unwanted subharmonic operations when some parameters are not properly chosen. For this problem, power electronics engineers have derived an effective solution approach, known as *ramp compensation,* which has become the industry standard for current-mode control of dc/dc converters. In this chapter, the problem is reexamined in the light of bifurcation analysis. It is shown that such an analysis allows convenient prediction of stability boundaries and facilitates the selection of parameter values to guarantee stable operation. It also permits new phenomena to be discovered. An example is given at the end of the chapter to illustrate how some bizarre operation in a power-factor-correction (PFC) converter can be systematically explained.

1 Introduction

The term "stability" has a variety of meanings. In the strict mathematical sense, one may state its meaning in terms of rigorously defined conditions. In engineering, however, stability is often interpreted as a condition in which the system being examined is operating in the expected regime. For instance, in power electronics, we refer a stable operation to a specific periodic operation. When a power converter fails to maintain its operation in this expected manner, it is considered unstable, even though it may be operating in a perfectly predictable regime such as a subharmonic or quasi-periodic regime. In conventional power electronics, all those subharmonic, quasi-periodic and chaotic operations are regarded as being undesirable and should be avoided. Thus, the traditional design objective must include the prevention of any bifurcation within the intended operating range. In other words, any effective design automatically has to avoid the occurrence of bifurcation for the range of variation of the parameters [1].

Bifurcations and chaos have been observed and analysed for various kinds of power electronics circuits [2]–[4]. For systems that have been shown to bifurcate when a certain parameter is changed, the design problem is, in a sense, addressing the "control of bifurcation". Such a design problem can therefore be solved on the basis of bifurcation analysis. One of our objectives in this chapter is to examine the traditional stability problem from a bifurcation analysis perspective. We will study, to some depth, dc/dc converters under current-mode control, which has been

G. Chen, D.J. Hill, and X. Yu (Eds.): Bifurcation Control, LNCIS 293, pp. 29–48, 2003.

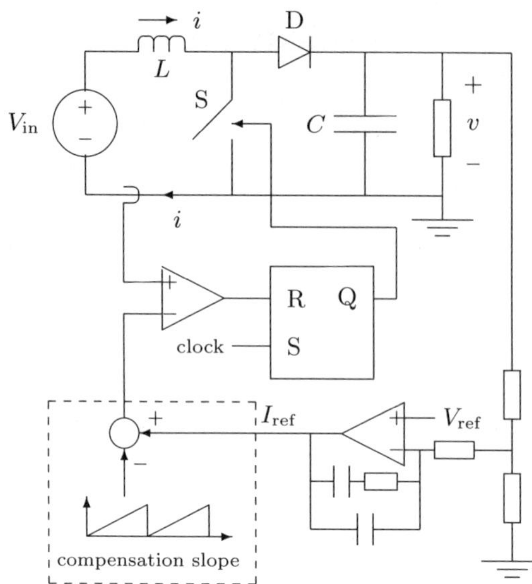

Fig. 1. Boost converter under current-mode control

a mature control technique applied in power electronics [5]–[8]. Specifically we will show that the widely known *ramp compensation* technique is effectively a means of controlling bifurcation, although it was never understood as such in the power electronics community. One useful extension of this work is in the observation of the subtle effect of controlling bifurcation (ramp compensation) on the overall system dynamics. As we will show in this chapter, dynamical response can be undesirably affected if excessive design margin is applied to avoid bifurcation. Hence, in deriving effective control methods, consideration should be given to maintain adequately fast response as well as sufficient clearance of bifurcation. In this chapter we will demonstrate how an "adaptive avoidance of bifurcation" can be achieved by a simple variable ramp compensation scheme. Furthermore, bifurcation analysis may lead to discovery of new phenomena, as we will illustrate using the same current-mode controlled converter but applying it to a power-factor-correction application.

The rest of the chapter is organized as follows. In the next section, we briefly review the circuit operation of the boost converter under a typical current-mode control and the instability condition of the inner current loop in terms of period-doubling bifurcation. In Section 3, the conventional ramp compensation for stabilization of the inner current loop is considered in the light of "avoiding bifurcation". Useful design curves will be provided for steady-state design. The effect of the use of compensating ramp on the dynamics of the overall closed-loop system is then considered formally, followed by experimental verifications. We also show how one can make fruitful use of the basic concept of controlling bifurcation to derive a simple effective control method that can ensure adequate margin from bifurcation as well as sufficient transient speed. Finally, in Section 4, we examine the same converter circuit when it is used for a power-factor-correction (PFC) application. Based on results of our bifurcation analysis, we demonstrate an interesting practical behaviour of this popular type of PFC converters, to which traditional theory offers no simple explanation.

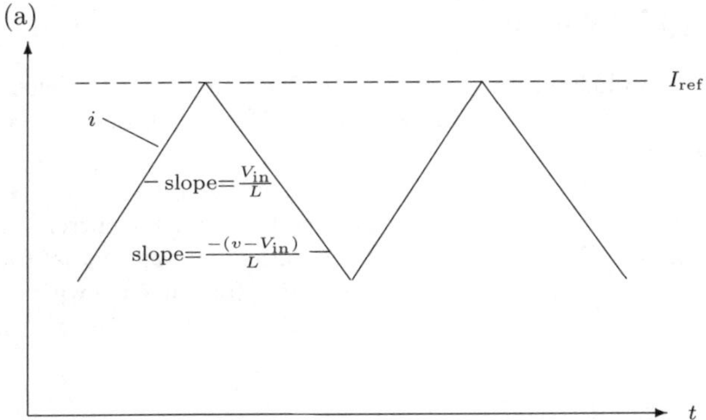

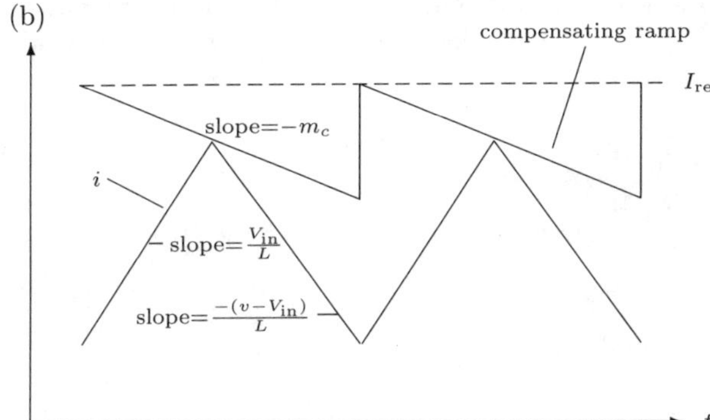

Fig.2. Illustration of current-mode control showing inductor current (a) without and (b) with ramp compensation

2 Review of Operation and Bifurcation Analysis

2.1 Basic operation

Consider the boost converter shown in Fig. 1. The switch is turned on periodically, and off according to the output of a comparator that compares the inductor current with a reference level I_{ref}. Specifically, while the switch is on, the inductor current i climbs up, and as it reaches I_{ref}, the switch is turned off, thereby causes the inductor current to ramp down until the next periodic turn-on instant. Thus, the average inductor current is programmed approximately by I_{ref}. In the closed-loop system, I_{ref} is controlled via a feedback loop which attempts to keep the output voltage fixed by adjusting I_{ref}.

An important feature of the current-mode control is the presence of an inner current loop. It is now widely known that this inner loop becomes unstable when the duty ratio (designed steady-state value) exceeds 0.5 [7,8]. The usual practical (conventional) remedy is to introduce a compensating ramp to the loop, as shown in Fig. 1. The essential operation is illustrated by the waveforms shown in Fig. 2.

2.2 Period-doubling bifurcation

The aforementioned inner-loop instability can in fact be examined from the viewpoint of nonlinear dynamics. A handy starting point is the iterative function that describes the inductor current dynamics.

We begin with the typical period-doubling bifurcation in the boost converter without ramp compensation [2,4,9]. We let i_n and i_{n+1} be the inductor current at $t = nT$ and $(n+1)T$ respectively. Denote also the output voltage (voltage across the output capacitor) by v. By inspecting the slopes of the inductor current in Fig. 2 (a), we get

$$\frac{I_{\text{ref}} - i_{n+1}}{(1-D)T} = \frac{v - V_{\text{in}}}{L} \quad \text{and} \quad \frac{I_{\text{ref}} - i_n}{DT} = \frac{V_{\text{in}}}{L} \tag{1}$$

where D is the duty ratio which is defined as the fractional duration of a switching period when switch S is closed. Combining the above equations, we have the following iterative function:

$$i_{n+1} = \left(1 - \frac{v}{V_{\text{in}}}\right) i_n + \frac{I_{\text{ref}} v}{V_{\text{in}}} - \frac{(v - V_{\text{in}})T}{L} \tag{2}$$

If we are interested in the inner current loop dynamics near the steady state, we may write

$$\delta i_{n+1} = \left(\frac{-D}{1-D}\right) \delta i_n + O(\delta i_n^2) \tag{3}$$

Clearly, the characteristic multiplier or eignenvalue, λ, is given by

$$\lambda = \frac{-D}{1-D} \tag{4}$$

which must fall between −1 and 1 for stable operation. In particular, the first period-doubling occurs when $\lambda = -1$ which corresponds to $D = 0.5$. Consistent with what is well known in power electronics, current-mode controlled converters must operate with the duty ratio set below 0.5 in order to maintain a stable period-1 operation [10].

In the application of current-mode control, the error signal derived from the output voltage is often used to modify I_{ref} directly (not the duty ratio as in the case of voltage mode PWM control). It is thus helpful to look at the period-doubling bifurcation in terms of the current reference I_{ref}. Specifically we can express the "criterion of a bifurcation-free operation", $D < 0.5$, in terms of I_{ref} by using the steady-state equation relating R, D and I_{ref}. For the boost converter, the equivalent criterion of a bifurcation-free operation is

$$I_{\text{ref}} < \frac{V_{\text{in}}}{R} \left[\frac{DRT}{2L} + \frac{1}{(1-D)^2} \right]_{D=0.5} = I_{\text{ref,c}} \tag{5}$$

which can be derived from the power-balance equation

$$\left(I_{\text{ref}} - \frac{\Delta i}{2}\right) V_{\text{in}} = \frac{V_{\text{in}}^2}{(1-D)^2 R} \tag{6}$$

where $\Delta i = DTV_{\rm in}/L$ and all symbols have their usual meanings. The critical value (upper bound) of $I_{\rm ref}$ for the uncompensated case is thus given by

$$I_{\rm ref,c} = \frac{V_{\rm in}}{R}\left(\frac{RT}{4L} + 4\right) \tag{7}$$

Hence, period-doubling occurs when $I_{\rm ref}$ exceeds the above-stated limit. To prevent period-doubling, we must therefore control $I_{\rm ref}$. Indeed, the use of a compensating ramp, as we will see, is to raise the upper bound of $I_{\rm ref}$, thereby widening the operating range.

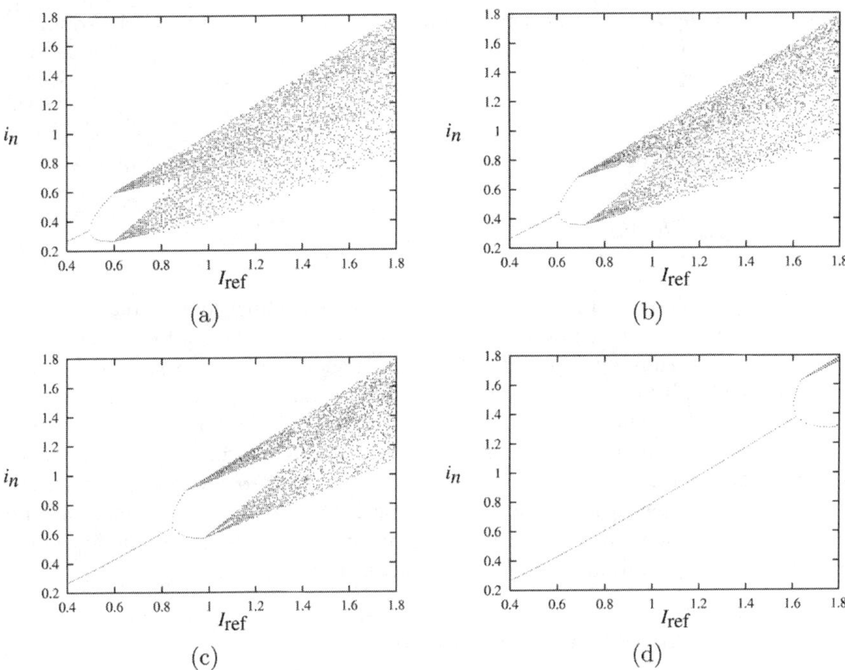

Fig.3. Bifurcation diagrams obtained numerically for the boost converter under current-mode control, showing the "delaying" of the onset of bifurcation by ramp compensation. (a) No ramp compensation; (b) with compensating ramp $m_c = 0.1V_{\rm in}/L$; (c) $m_c = 0.3V_{\rm in}/L$; (d) $m_c = 0.8V_{\rm in}/L$. For all cases, $C = 20\ \mu$F, $L = 1.5$ mH, $R = 40\ \Omega$, $V_{\rm in} = 5$ V and $T = 100\ \mu$s.

3 Control of Bifurcation by Ramp Compensation

3.1 Design to avoid bifurcation

With compensation, the reference current is first subtracted from an artificial ramp before it is used to compare with the inductor current, as shown in Fig. 2. By inspecting the inductor current waveform, we obtain the modified iterative function for the inner loop dynamics as

$$\delta i_{n+1} = \left(\frac{M_c}{1+M_c} - \frac{D}{(1-D)(1+M_c)} \right) \delta i_n + O(\delta i_n^2) \tag{8}$$

where $M_c = m_c L / V_{\text{in}}$ is the normalized compensating slope, and m_c is defined in Fig. 2. Now, using (8), we get the eigenvalue or characteristic multiplier, λ, for the compensated inner loop dynamics as

$$\lambda = \frac{M_c}{1+M_c} - \frac{D}{(1-D)(1+M_c)} \tag{9}$$

Hence, by putting $\lambda = -1$, the critical duty ratio, at which the first period-doubling occurs, is obtained, i.e.,

$$D_c = \frac{M_c + 0.5}{M_c + 1} \tag{10}$$

Using (5) and the above expression for D_c, we get the critical value of I_{ref} for the compensated system as

$$I_{\text{ref},c} = \frac{V_{\text{in}}}{R} \left[\frac{RT}{2L} \frac{M_c + 0.5}{M_c + 1} + 4(M_c + 1)^2 \right] > I_{\text{ref}} \tag{11}$$

Note that $I_{\text{ref},c}$ increases monotonically as the compensating slope increases. Hence, it is obvious that compensation effectively provides more margin for the system to operate without running into the bifurcation region. Figure 4 shows some plots of the critical value of I_{ref} against R, for a few values of M_c. The choice of the magnitude of the compensating ramp constitutes a design problem which aims at avoiding bifurcation. In a likewise manner, we may consider the input voltage variation and produce a similar set of design curves that provide information on the choice of the compensating slope for ensuring no bifurcation for a range of input voltage. This is shown in Fig. 5 Also, for a general reference, the boundary curves in terms of normalized parameters are shown in Fig. 6.

3.2 Effects on dynamical response

The transient response of a power converter can be compromised if bifurcation is kept too remote in order to give a large safe margin, especially when the operating range required is very wide, since guaranteeing "no bifurcation" for a wide range of parameter values would inevitably make the safe margin excessively large at one extreme end of the range.

It is therefore of interest to study the effect of the presence of compensating ramp on the closed-loop dynamics of the *overall system*. We will take a simple averaging approach to derive the eigenvalues of the stable closed-loop system, mainly to reveal the transient speed for different values of the compensating slope. Specifically we can write down the normalized state equations for the boost converter as

$$\frac{dx}{d\tau} = \frac{-x}{\gamma} + \frac{(1-d)y}{\gamma} \tag{12}$$

$$\frac{dy}{d\tau} = \frac{-(1-d)x}{\zeta} + \frac{E}{\zeta} \tag{13}$$

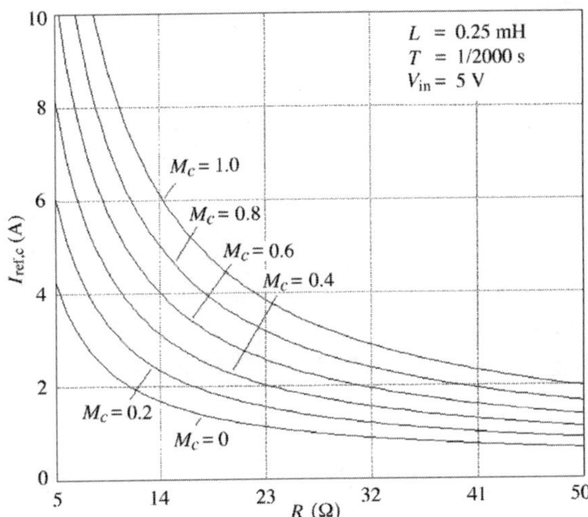

Fig.4. Specific boundary curves $I_{\mathrm{ref,c}}$ versus R for current-mode controlled boost converter without compensation and with normalized compensating slope $M_c = 0.2$, 0.4, 0.6, 0.8 and 1

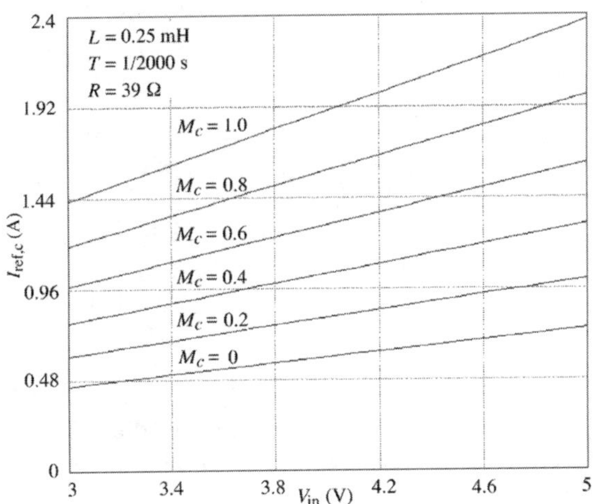

Fig.5. Specific boundary curves $I_{\mathrm{ref,c}}$ versus V_{in} for current-mode controlled boost converter without compensation and with compensation slope $M_c = 0.2$, 0.4, 0.6, 0.8 and 1

where the normalized variables and parameters are defined by $x = v/V_{\mathrm{ref}}$, $y = i/(V_{\mathrm{ref}}/R)$, $E = V_{\mathrm{in}}/V_{\mathrm{ref}}$, $\tau = t/T$, $\gamma = CR/T$, and $\zeta = L/RT$. Here, we choose the steady-state output voltage as V_{ref}. The closed-loop control can be modelled

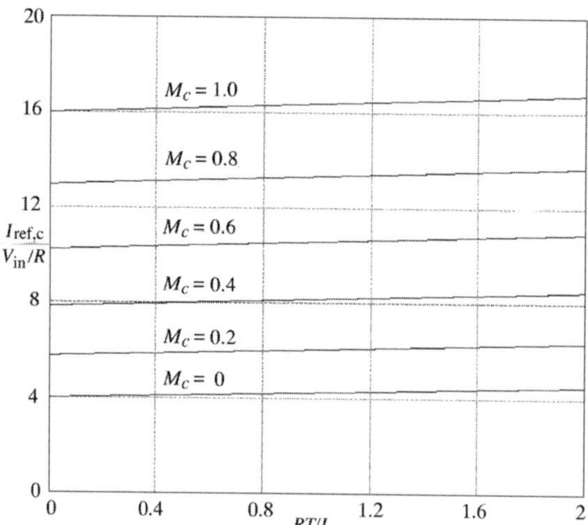

Fig.6. Specific boundary curves plotted with normalized parameters

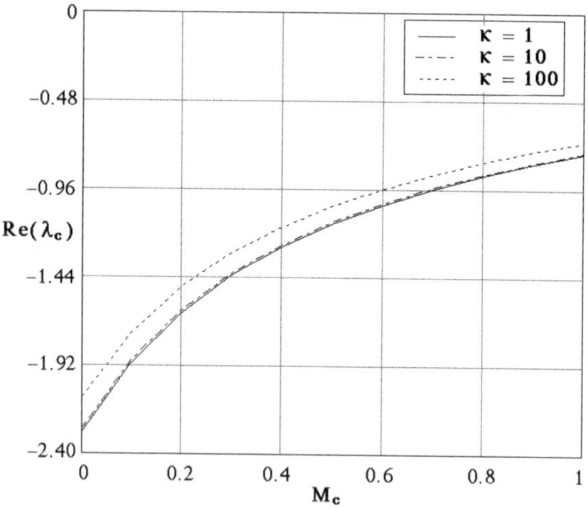

Fig.7. Plots of $\text{Re}(\lambda_c)$ versus M_c for $\zeta = 0.128$, $\gamma = 343$ (corresponding to $C = 440\ \mu\text{F}$, $L = 250\ \mu\text{H}$, $R = 39\ \Omega$ and $T = 1/20000$ s) and $E = 3.5/8$

approximately by (see Fig. 2)

$$i + \frac{\Delta i}{2} \approx \overbrace{\left[\frac{P_o}{V_{\text{in}}} + I_{\text{offset}} - k(v - V_{\text{ref}})\right]}^{I_{\text{ref}}} - m_c dT \tag{14}$$

where P_o is the output power (i.e., $P_o = V_{\text{ref}}^2/R$), k is the voltage feedback gain, and I_{offset} accounts for the steady-state shift due to the difference between the average and the peak value of the inductor current. This control equation can readily be translated, in terms of the normalized parameters, into

$$d = 1 - \frac{2\zeta\left(\frac{1}{E} - y - \kappa(x-1)\right) - 2M_cE}{x - E(1-2M_c)} \tag{15}$$

where $\kappa = kR$. Hence, putting (15) into the state equations, we get the closed-loop state equations which can then be used to study the closed-loop dynamics. Specifically, we can obtain the Jacobian matrix J_F as

$$J_F = \begin{bmatrix} \dfrac{2\zeta\left(2Y - \frac{1}{E}\right) + 2M_cE}{E(1+M_c)\gamma} & \dfrac{\left(\frac{2\zeta}{E} + 2M_cE\right)}{E(1+2M_c)\gamma} \\ \dfrac{-2\kappa}{E(1+2M_c)} & \dfrac{-2}{E(1+2M_c)} \end{bmatrix}_{x=X,y=Y} \tag{16}$$

Note that in the steady state, $x = X = 1$ and $y = Y = 1/E$. Suppose the eigenvalues of the closed-loop system, λ_c, are complex. The real part of λ_c can be easily found as

$$\mathrm{Re}(\lambda_c) = -\frac{E^2(1+2M_c) + 2E(\gamma + M_c) - 2\kappa\zeta}{2E^2(1+2M_c)\gamma} \tag{17}$$

In practice, $E < 1$ and $\gamma \gg 1$. Also, for stable operation, κ has to be kept small enough so that $\mathrm{Re}(\lambda_c) < 0$. Under such condition, we can readily show that $\frac{d}{dM_c}\left|\mathrm{Re}(\lambda_c)\right| < 0$. In other words, the transient becomes slower as M_c increases. Some plots of $\mathrm{Re}(\lambda_c)$ versus M_c are shown in Fig. 7.

Consider the current-mode controlled boost converter. We first observe that for a higher input voltage, the system is more remote to bifurcation. In fact, we can use (11) to determine if a current-mode controlled boost converter, which is designed for a certain I_{ref} (corresponding to a given power level), may bifurcate and be chaotic for a given input voltage.

Remarks — It should be stressed that the overall dynamics is modelled by (12) and (13), while the inner current loop dynamics is described by (8). Inconsistent conclusion may be drawn from studying the two dynamical equations. Specifically, from (8), we observe that increasing M_c will make the inner loop dynamics "faster". However, the foregoing analysis of the overall system dynamics reveals that *for some range of parameter values,* the system actually becomes slower as M_c increases. Obviously, (8) is inadequate for the purpose of examining the overall system dynamics.

3.3 Experimental measurements

An experimental prototype of a boost converter under current-mode control has been constructed, as schematically shown in Fig. 1. The circuit parameters are: $L = 250$ μH, $C = 440$ μF, $R = 39\ \Omega$, $v = 8$ V (steady-state), and $T = 50$ μs (i.e., 20 kHz). The feedback circuit has an appropriate integral control to adjust the steady-state level of I_{ref} in the event of a change in V_{in}. Such an arrangement is common in current-mode control of dc/dc converters. Now, from $(1-D)v = V_{\mathrm{in}}$, we know that the uncompensated system will walk out of the stable region if V_{in} is reduced to below about 4 V, since the output voltage is kept at 8 V by the integral control. We may apply compensation to restore stability.

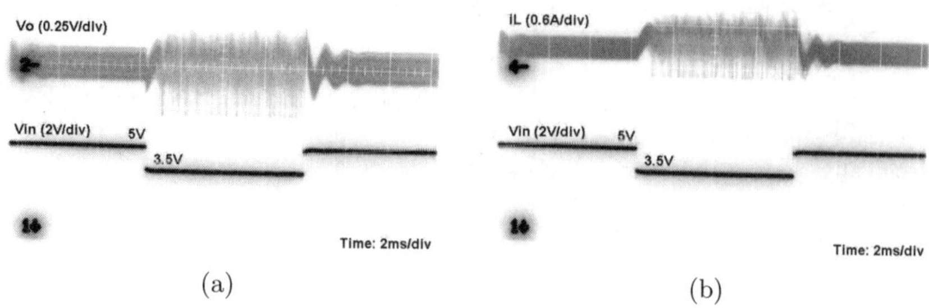

Fig.8. Measured output voltage v (a) and inductor current i (b) of boost converter under current-mode control without compensation showing chaos as input voltage drops to 3.5 V

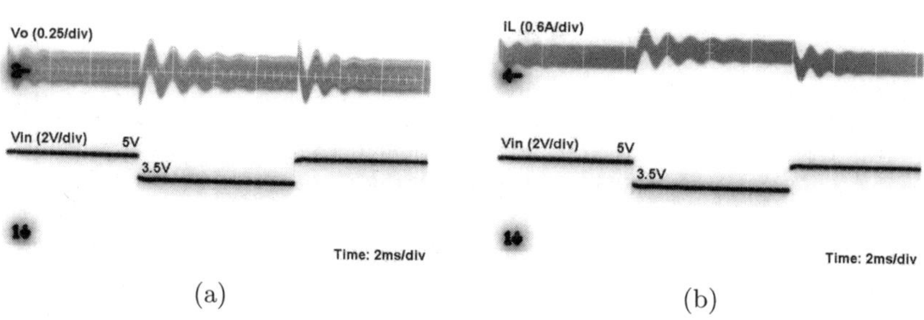

Fig.9. Measured output voltage v (a) and inductor current i (b) of boost converter under current-mode control with fixed ramp compensation $M_c = 0.2$ showing stabilized operation

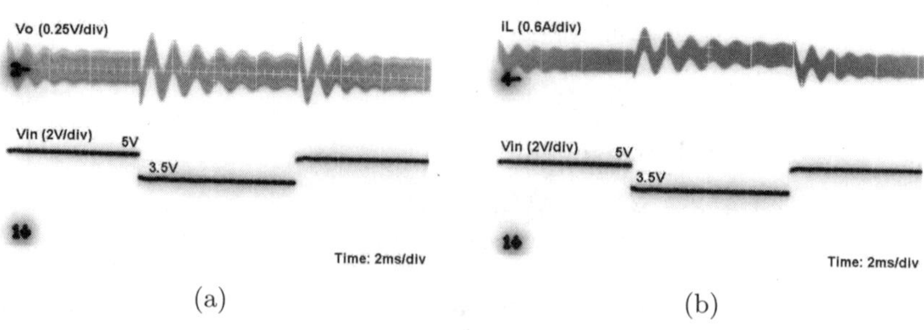

Fig.10. Measured output voltage v (upper) and inductor current i (lower) of boost converter under current-mode control with fixed ramp compensation $M_c = 0.8$ showing stabilized operation but slow transient

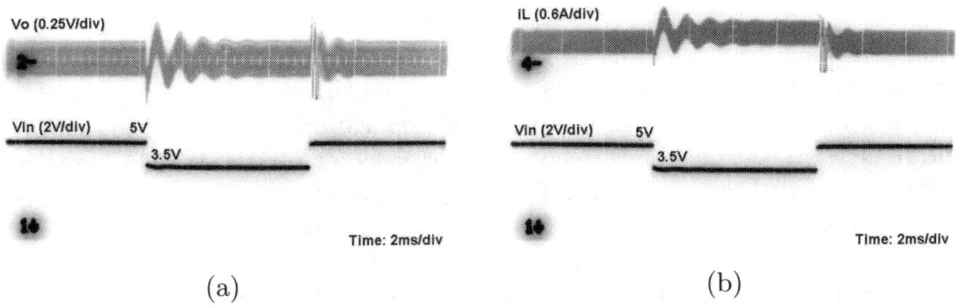

(a) (b)

Fig.11. Measured output voltage v (upper) and inductor current i (lower) of boost converter under current-mode control with variable ramp compensation showing improved transient for 5 V input

The results show that, without compensation (Fig. 8), the system becomes chaotic when the input voltage falls to 3.5 V. Moreover, with compensation, the system remains stable. We further verify, from Figs. 9 and 10, that excessive compensation lengthens the response time. Further elaboration will be given in the next subsection.

3.4 Variable ramp compensation

In order to keep bifurcation away while maintaining fast response, the control should incorporate a special function that dynamically adjusts the compensating ramp. The aim is to give just enough compensation under all input voltage conditions. Thus, the controller may contain, in addition to a conventional proportional-integral gain, a variable ramp generator providing necessary, but not excessive, compensation. For this simplified scenario (i.e., fixed load and output voltage), the compensating ramp needs only be controlled according to

$$M_c(V_{\rm in}) \geq \frac{v}{2V_{\rm in}} - 1 \quad \text{or} \quad m_c(V_{\rm in}) \geq \frac{v - 2V_{\rm in}}{2L} \tag{18}$$

which is derivable from (10). In our experiment, a variable-ramp control circuit has been constructed in discrete form. Figure 11 shows the measured waveforms for the boost converter under such control.

The series of waveforms shown in Figs. 9 through 11 serve to illustrate effect of applying ramp compensation to the system dynamics. Specifically, from (18), the value of M_c needed for a 3.5 V input is about 0.14, and no compensation is at all needed for a 5 V input. Thus, with $M_c = 0.8$ (Fig. 10), the system is over-compensated and hence suffers a slower transient compared to the case with $M_c = 0.2$ (Fig. 9). Furthermore, even for $M_c = 0.2$, the system is still over-compensated when the input is 5 V. Thus, we can see a much faster response with the variable ramp compensation (Fig. 11) since it applies just enough compensation for the 3.5 V input and none for the 5 V input.

4 Application Example: Power-Factor-Correction Boost Converter

In the foregoing, we have described the application of ramp compensation in controlling bifurcation in dc/dc converters under current-mode control. In practice, the current-mode controlled converter also finds application in shaping the input current. In fact, the so-called boost rectifier or power-factor-correction (PFC) converter is effectively a current-mode controlled boost converter [11]. The circuit schematic is shown in Fig. 12. In this case, instead of setting $I_{\rm ref}$ constant for a fixed load, we let $I_{\rm ref}$ vary according to the input voltage waveform. Thus, the input current is being directly programmed to track the waveform of the input voltage. The result is a nearly unity power factor.

4.1 Bifurcation analysis

The bifurcation analysis described earlier is directly applicable to the case of the PFC boost converter. Effectively, since $I_{\rm ref}$ follows the input voltage, its waveform is a rectified sine wave whose frequency is much lower than the switching frequency. Typically the frequency of this sine wave is 50 or 60 Hz. Thus, the situation is analogous to the case of applying a *time-varying ramp compensation* to a current-mode controlled boost converter. Suppose the input voltage is given by

$$v_{\rm in}(t) = \hat{V}_{\rm in}\,|\sin\omega_m t| \tag{19}$$

where ω_m is the line angular frequency. For algebraic brevity we express the input voltage in terms of the phase angle θ, i.e., $v_{\rm in}(\theta) = \hat{V}_{\rm in}\,|\sin\theta|$.

As shown in Fig. 13, when the input voltage is in its first quarter cycle (i.e., $0 \le \theta < \pi/2$), the value of $I_{\rm ref}$ increases, which is equivalent to applying a negative compensating ramp to $I_{\rm ref}$ (i.e., $M_c < 0$). Moreover, when the input voltage is in its second quarter cycle (i.e., $\pi/2 < \theta \le \pi$), the value of $I_{\rm ref}$ decreases, which is equivalent to applying a positive compensating ramp to $I_{\rm ref}$ (i.e., $M_c > 0$). At $\theta = \pi/2$, there is no ramp compensation. Therefore, based on the earlier analysis, we can conclude that the system has asymmetric regions of stability for the two quarter cycles.

Specifically, the second quarter cycle (i.e., $\pi/2 \le \theta < \pi$) should be more remote from period-doubling[1] because of the presence of ramp compensation. To be precise, we need to find the critical phase angle, θ_c, at which period-doubling occurs. Since the duty ratio is equal to $1 - v_{\rm in}/v$ and M_c is $-(dI_{\rm ref}/dt)L/\hat{V}_{\rm in}|\sin\theta|$, we have, from (10),

$$|\sin\theta_c| = \frac{v + 2L\dfrac{dI_{\rm ref}}{dt}}{2\hat{V}_{\rm in}}. \tag{20}$$

Moreover, if the power factor approaches one, we have

$$I_{\rm ref} \approx \hat{I}_{\rm in}\,|\sin\theta| \qquad \text{for } 0 \le \theta \le \pi \tag{21}$$

[1] The term period-doubling here refers to the switching period being doubled.

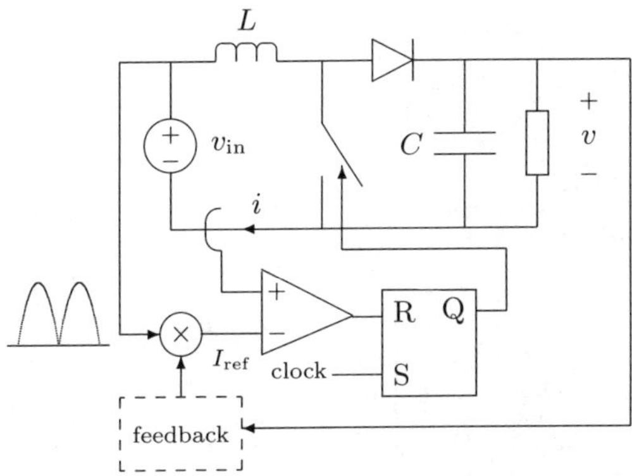

Fig.12. Schematic of the PFC boost converter showing direct programming of the input current. I_{ref} is a rectified sine wave whose amplitude is adjusted by the "feedback" network to match the power level.

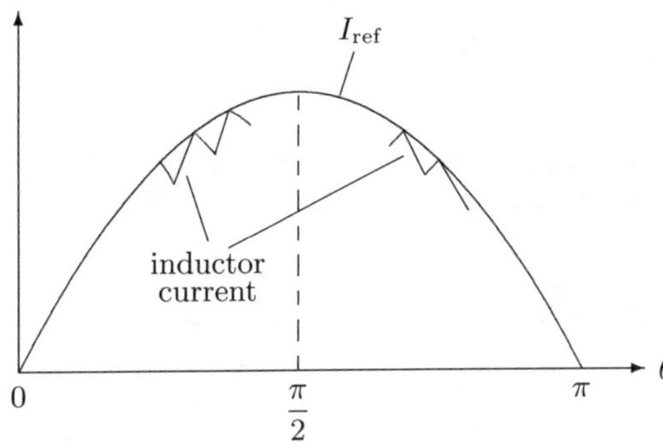

Fig.13. Programming of input current waveform in PFC boost converter. For $0 \leq \theta < \pi/2$, an effective negative ramp compensation is applied (i.e., $M_c < 0$), whereas for $\pi/2 < \theta \leq \pi$, an effective positive ramp compensation is applied (i.e., $M_c > 0$).

where $\hat{I}_{\mathrm{in}}$ is the peak input current. For brevity we restrict the analysis to the range $[0, \pi]$, understanding that the waveform repeats for every interval $[k\pi, (k+1)\pi]$, for all integers k. Thus, we have

$$\frac{dI_{\mathrm{ref}}}{dt} \approx \omega_m \hat{I}_{\mathrm{in}} \cos\theta \qquad \text{for } 0 \leq \theta \leq \pi. \tag{22}$$

Hence, from (20), we have

$$\theta_c = 2 \arctan\left(\frac{2\hat{V}_{\mathrm{in}} \pm \sqrt{4\hat{V}_{\mathrm{in}}^2 - v^2 + 4\omega_m^2 \hat{I}_{\mathrm{in}}^2 L^2}}{v - 2\omega_m \hat{I}_{\mathrm{in}} L}\right). \tag{23}$$

Furthermore, incorporating the power equality, i.e., $\hat{V}_{\mathrm{in}}\hat{I}_{\mathrm{in}}/2 = v^2/R$ (assuming 100% efficiency), and defining two parameters r_v and τ_L as

$$r_v = \frac{v}{\hat{V}_{\mathrm{in}}} \tag{24}$$

$$\tau_L = \frac{L}{R}, \tag{25}$$

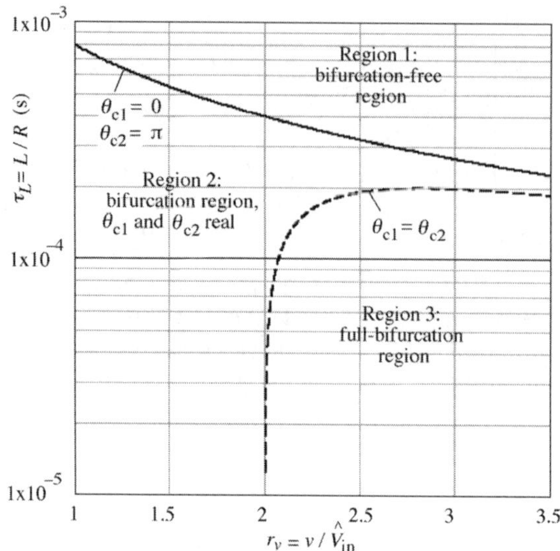

Fig.14. Bifurcation regions in parameter space (mains frequency is 50 Hz). Upper boundary curve is $\tau_L = 1/4\omega_m r_v$ and lower curve is $\tau_L = \sqrt{(r_v^2 - 4)/16\omega_m^2 r_v^4}$.

the critical phase angle given in (23) can be written in the following compact form:

$$\theta_c = 2 \arctan \left(\frac{2 \pm \sqrt{4 - r_v^2 + 16\omega_m^2 \tau_L^2 r_v^4}}{r_v - 4\omega_m \tau_L r_v^2} \right). \tag{26}$$

By inspecting (26), we clearly see that the voltage ratio $r_v = v/\hat{V}_{in}$ and the parameter $\tau_L = L/R$ control the bifurcation behaviour. For clarity, we denote the two real solutions (if exist) by θ_{c1} and θ_{c2}. Specifically, we can identify three regions in the parameter space (see Fig. 14):

- **Region 1 (bifurcation-free region):** We can readily show that if

$$\tau_L > \frac{1}{4\omega_m r_v}, \tag{27}$$

the solutions given by (26) are essentially outside of the range of interest. In fact, at the boundary $\tau_L = 1/4\omega_m r_v$, we simply have $\theta_{c1} = 0$ and $\theta_{c2} = \pi$, which correspond to a bifurcation-free operation for all time.
- **Region 2 (bifurcation region):** We also observe from (26) that if $4 - r_v^2 + 16\omega_m^2 \tau_L^2 r_v^4$ is non-negative in addition to satisfying (27), i.e.,

$$\sqrt{\frac{r_v^2 - 4}{16\omega_m^2 r_v^4}} \le \tau_L < \frac{1}{4\omega_m r_v}, \tag{28}$$

then there are two real solutions for θ_c. Under this condition, period-doubling occurs for intervals $[0, \theta_{c1}]$ and $[\theta_{c2}, \pi]$. Moreover, as θ_{c1} and θ_{c2} get closer

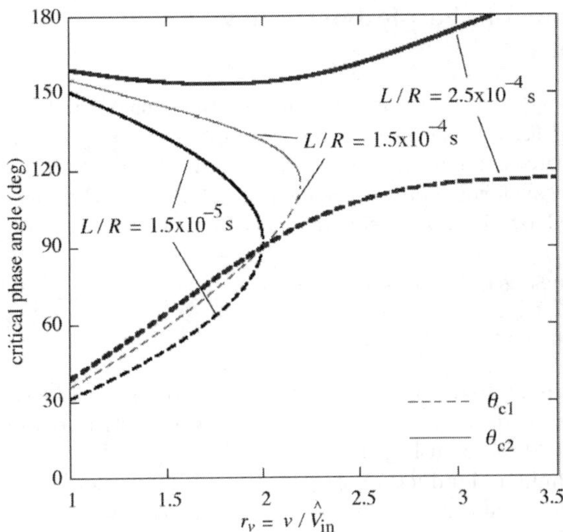

Fig.15. Critical phase angles versus $v/\hat{V}_{\text{in}}$ for the PFC boost converter

to each other, the stable interval diminishes. At the lower boundary $\tau_L = \sqrt{r_v^2 - 4/16\omega_m^2 r_v^4}$, the two real solutions merge together, i.e., $\theta_{c1} = \theta_{c2}$, and period-doubling bifurcation cannot be avoided.

- **Region 3 (full-bifurcation region):** If τ_L is below the boundary of Region 2, i.e.,

$$\tau_L \leq \sqrt{\frac{r_v^2 - 4}{16\omega_m^2 r_v^4}}, \tag{29}$$

the stable interval has disappeared altogether.

In Fig. 15, we plot the critical phase angle as a function of r_v (i.e., $v/\hat{V}_{\text{in}}$). We summarize as follows some important observations regarding the occurrence of period-doubling in the input current waveform during the half mains cycle, i.e., $0 \leq \theta \leq \pi$.

1. To guarantee operation in the bifurcation-free regime (Region 1 in Fig. 14), we need a sufficiently large τ_L, i.e., either a sufficiently large L or small R. This is actually the preferred operation in practice.
2. For values of $v/\hat{V}_{\text{in}}$ where real solutions of θ_c exist, the converter fails to maintain the expected bifurcation-free operation for intervals of time corresponding to $\theta < \theta_{c1}$ and $\theta > \theta_{c2}$.
3. If θ_{c1} is greater than 90°, the converter would have gone into period-doubling for the whole first quarter cycle. Likewise, if θ_{c2} is less than 90°, the converter would have gone into period-doubling for the whole second quarter cycle.
4. Referring to Fig. 14, if τ_L is smaller than a certain value (about 0.0001 for a mains frequency of 50 Hz), period-doubling is unavoidable and will occur for all values of $v/\hat{V}_{\text{in}} > 2$.

4.2 Bifurcation behaviour of the pfc boost converter by computer simulations

In this subsection we verify the above findings by computer simulations of the PFC boost converter depicted in Fig.12. Consistent with the assumption of unity power factor used in the foregoing theoretical analysis, the reference current waveform $I_{\rm ref}$ is generated according to the waveform template defined in (21), where the peak input current $\hat{I}_{\rm in}$ is determined by the power equality condition, i.e., $\hat{V}_{\rm in}\hat{I}_{\rm in}/2 = v^2/R$.

The circuit component values used in the simulations are:

$$L = 2\text{ mH}, C = 470\ \mu\text{F and } R = 135\ \Omega.$$

The switching frequency and mains frequency are 50 kHz and 50Hz, respectively. This choice of component values leads to the theoretical curves of the critical phase angles corresponding to $\tau_L = 0.000015$ s in Fig. 15.

In Fig. 16, we present the simulated inductor current waveform for an operation in Region 2 (bifurcation region), where real solutions of θ_c exist. The peak input voltage is $110\sqrt{2}$ V, and the reference output voltage is 220 V, which correspond to $v/\hat{V}_{\rm in} = \sqrt{2}$ in Fig. 15. Indeed, period-doubling bifurcation can be observed during a half mains cycle in the inductor current waveform, as shown in Fig. 16 (upper). In order to see the period-doubling more clearly, we sample the waveform at a rate equal to the switching frequency, as shown in Fig. 16 (lower), where the two critical phase angles and the corresponding bifurcations can be clearly identified. Between these two points the sampled values of the current follow accurately the sinusoidal shape of the reference current. A close-up view of the waveform around the critical points is shown in Fig. 17. Furthermore, Fig. 18 compares the values of the critical phase angles found by simulations and those obtained analytically. They are in very good agreement.

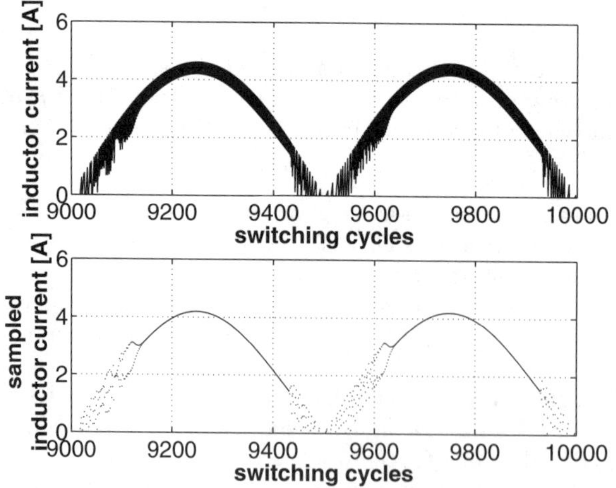

Fig.16. Upper: simulated inductor current time-domain waveform; Lower: same waveform sampled at the switching frequency (lower) for Region 2 at $v/\hat{V}_{\rm in} = \sqrt{2}$.

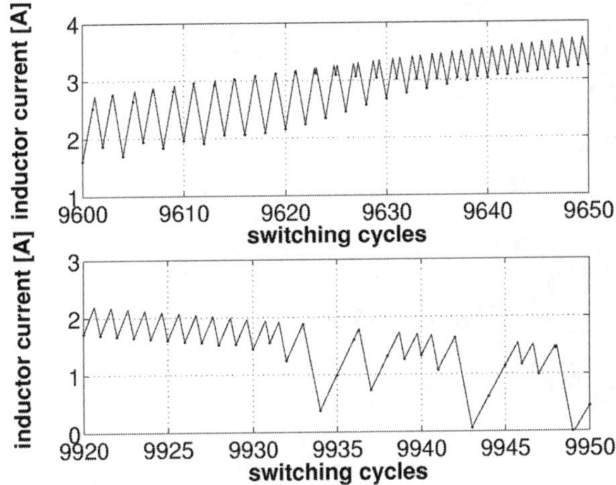

Fig.17. Close-up view of simulated inductor current waveform near the critical points for Region 2 at $v/\hat{V}_{\rm in} = \sqrt{2}$.

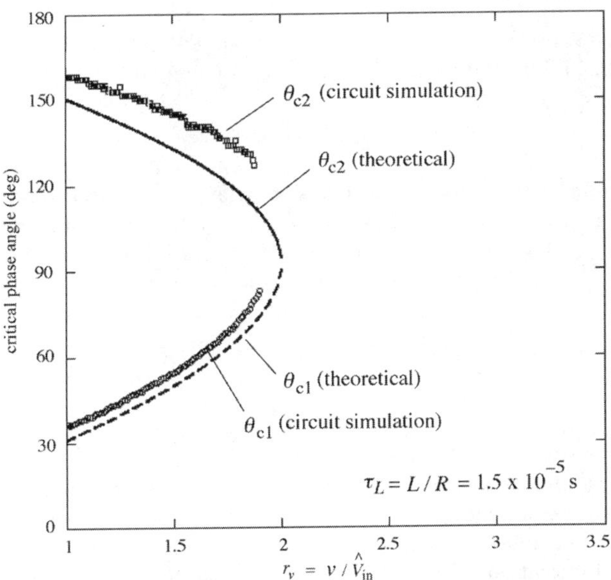

Fig.18. Critical phase angles obtained by simulations and analysis for τ_L = 0.000015 s.

The bifurcation region characterized by the presence of a critical point in each quarter mains cycle persists until the left-hand side critical phase angle θ_{c1} reaches its maximum, i.e., 90°, corresponding to the peak current value. As mentioned in the preceding section, when θ_{c1} becomes greater than 90°, the whole first quarter cycle should have turned into period-doubling and possible chaos for some intervals. This result is indeed confirmed by the simulated inductor current waveform shown in Fig. 19, obtained for $v/\hat{V}_{\rm in} = 2$.

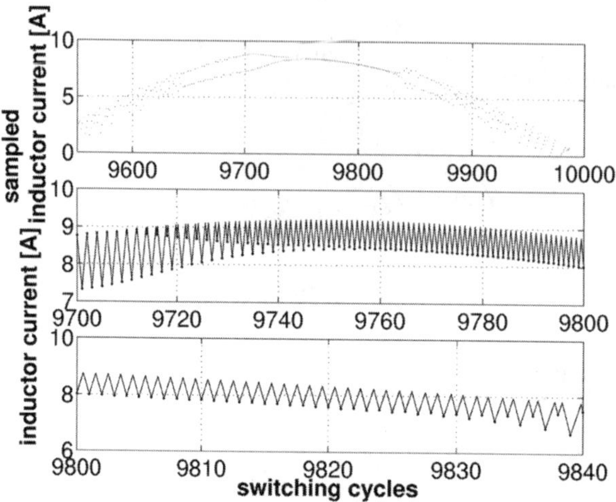

Fig.19. Upper: simulated sampled inductor current when one critical phase angle reaches 90° for $v/\hat{V}_{\rm in} = 2$; Middle and lower: close-up views of simulated waveform near critical points.

Finally, in order to confirm the theoretical conclusion regarding the occurrence of full-bifurcation in Region 3, we present the simulation results for parameters satisfying (29). As shown in Fig. 20, the system operates in full-bifurcation with the stable interval replaced completely by period-doublings and chaos.

5 Conclusions

In this chapter we present bifurcation analysis of power converters and illustrate how such an analysis can systematically explain some previously observed phenomena. The conventional current-mode control strategy, in particular, is reexamined in the light of "avoiding bifurcation". Under this viewpoint, the function of ramp compensation can be regarded as a means to keep the system sufficiently remote from the first bifurcation point, thereby maintaining stability in the traditional sense. It has also been shown that excessive bifurcation clearance is accompanied by undesirably slow dynamical response. Finally, as an application example, we consider a power-factor-correction boost converter and perform an bifurcation analysis to study the effects of various parameters on the stability of the converter. It has been shown that bifurcation analysis uncovers an interesting bifurcation behaviour which has not been previously detected.

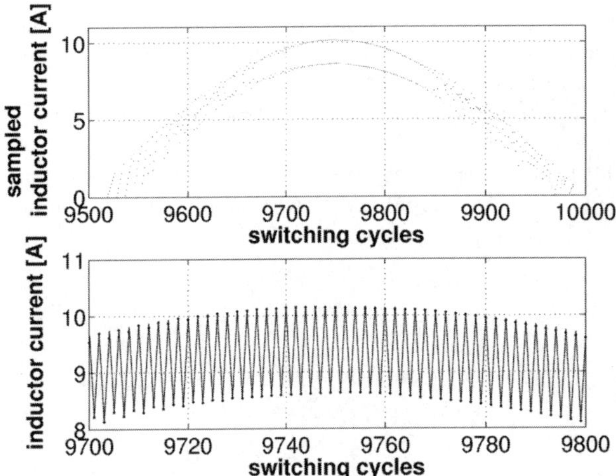

Fig.20. Upper: simulated sampled inductor current for Region 3 at $v/\hat{V}_{\rm in} = 2.5$; Lower: close-up view of the inductor current waveform.

Acknowledgment.
We wish to thank Herbert Iu for helpful discussions on the bifurcation of power-factor-correction converters. The assistance of Velibor Pjevalica in performing the experimental work is also gratefully acknowledged.

References

1. Kapitaniak, T. (1996) Controlling Chaos. London: Academic Press
2. Deane, J. H. B. (1992) Chaos in a current-mode controlled boost dc/dc converter. IEEE Trans. Circ. Syst.-I, **39**(8):680–683
3. Tse, C. K., di Bernardo, M. (2002) Complex behavior in switching power converters. Proceedings of the IEEE **90**(5):768–781
4. Banerjee, S., Verghese, G. (2001) Nonlinear Phenomena in Power Electronics: Attractors, Bifurcations, Chaos, and Nonlinear Control. New York: IEEE Press
5. Capel, A., Ferrante, G., O'Sullivan, D., Weinberg, A. (1978) Application of the injected-current control model for the dynamic analysis of switching regulators with a new concept of LC^3 modulator. In Proc. IEEE Power Electr. Spec. Conf. Rec. 135–147
6. Holland, B. (1984) Modelling, analysis and compensation of the current-mode converter. In Proc. Powercon **11**:I-2-1–I-2-6
7. Redl, R., Sokal, N. O. (1985) Current-mode control, five different types, used with the three basic classes of power converters. In Proc. IEEE Power Electr. Spec. Conf. Rec. 771–775
8. Kislovski, A. S. (1985) Introduction to Dynamical Analysis of Switching DC/DC Converters. Berne: EWV Engineering

9. Chan, W. C. Y., Tse, C. K. (1997) Study of bifurcations in current-programmed dc/dc boost converters: from quasiperiodicty to period-doubling. IEEE Trans. Circ. Syst.-I, **43**(12):1129–1142
10. Krein, P. T. (1998) Elements of Power Electronics. New York: Oxford Univ. Press
11. Dixon, L. H. Jr. (1990) High power factor preregulator for off-line power supplies. In Proc. Unitrode Switching Regulated Power Supply Design Manual, Paper 12 SEM-700

Distance to Bifurcation in Multidimensional Parameter Space: Margin Sensitivity and Closest Bifurcations

Ian Dobson

Electrical and Computer Engineering Dept
University of Wisconsin-Madison
Madison WI 53706 USA
dobson@engr.wisc.edu

Abstract. The problem of operating or designing a system with robust stability with respect to many parameters can be viewed as a geometric problem in multidimensional parameter space of finding the position of nominal parameters λ_0 relative to hypersurfaces at which stability is lost in a bifurcation. The position of λ_0 relative to these hypersurfaces may be quantified by numerically computing the bifurcations in various directions in parameter space and the bifurcations closest to λ_0. The sensitivity of the distances to these bifurcations yield hyperplane approximations to the hypersurfaces and optimal changes in parameters to improve the stability robustness. Methods for saddle-node, Hopf, transcritical, pitchfork, cusp, and isola bifurcation instabilities and constraints are outlined. These methods take full account of system nonlinearity and are practical in high dimensional parameter spaces. Applications to the design and operation of electric power systems, satellites, hydraulics and chemical process control are summarized.

1 Introduction

Consider a system modelled by smooth parameterized differential equations

$$\dot{z} = f(z, \lambda), \qquad z \in \mathbb{R}^n, \quad \lambda \in \mathbb{R}^m \tag{1}$$

where z are states and λ are parameters. We write z_0 for a particular equilibrium of (1) and assume that z_0 is asymptotically stable at the fixed parameter vector $\lambda_0 \in \mathbb{R}^m$. λ_0 arises in applications as the parameters at which the system is currently operated or nominal parameters for engineering design purposes. As λ varies in the parameter space $\mathbb{R}^m$, the equilibrium z_0 varies in the state space $\mathbb{R}^n$ and may disappear or become unstable in a bifurcation. The set $\Sigma \subset \mathbb{R}^m$ of parameters at which z_0 disappears or becomes unstable in a bifurcation determines the limits in parameter space at which the system may be stably operated at z_0. Σ is part of the bifurcation set and typically includes hypersurfaces in $\mathbb{R}^m$ and their intersections.

The robust stability of the equilibrium z_0 requires that the parameters are sufficiently far from Σ. For a one dimensional parameter space ($m = 1$), Σ is a point and is easily specified. For a two dimensional parameter space, Σ includes one dimensional curves and it is clear, at least graphically, how to specify Σ and stay sufficiently far from Σ. However, most practical engineering problems and some

G. Chen, D.J. Hill, and X. Yu (Eds.): Bifurcation Control, LNCIS 293, pp. 49–66, 2003.

scientific problems contain many parameters so that m is large and the parameter space $\mathbb{R}^m$ is high dimensional. Methods that are practical for handling a few parameters ($m = 1, 2, 3$) are often impractical when m is larger. This is sometimes called the "curse of dimensionality". Specifying or approximating even one hypersurface of Σ is generally very difficult for large m. This chapter outlines and informally describes numerical methods and associated calculations suitable for ensuring robust stability of the equilibrium z_0 in these high dimensional parameter spaces.

2 Margin to Bifurcation and Its Sensitivity

2.1 Margin to bifurcation in a given direction

Given the fixed parameter value λ_0, an obvious question concerns the spatial relation of λ_0 to Σ. It is usually impractical to describe portions of Σ analytically, but progress can be made by numerically finding a point on Σ in a given direction from λ_0. Let k be a unit vector in $\mathbb{R}^m$ defining a direction of parameter change from λ_0 and let c be the amount of change so that along the ray defined by k,

$$\lambda = \lambda_0 + kc \tag{2}$$

One can then move along the ray by increasing c from zero and numerically compute the first occurrence of a bifurcation along the ray. Suppose that this first bifurcation happens when $c = M$ and $\lambda = \lambda_*$, where

$$\lambda_* = \lambda_0 + kM \tag{3}$$

The numerical computation of M can be done by a continuation method [38]. Since k is a unit vector, the margin (or distance) M from λ_0 to Σ in the direction k is

$$M(\lambda_0) = |\lambda_* - \lambda_0| \tag{4}$$

Note that the margin M is a function of λ_0. (The ray direction k is held constant, and, as λ_0 varies, the first bifurcation along the ray λ_* and the margin M also vary.)

By restricting the calculation to the ray (2) we are solving a one dimensional problem that gives a limited amount of information about Σ; namely the point λ_* on Σ nearest to λ_0 in the direction k as shown in Fig. 1. However, we will see that more information about Σ near λ_* can be efficiently obtained by a modest amount of further computation.

2.2 Hypersurfaces of Σ and their normal vector N

We make the generic assumption that the bifurcation encountered at $\lambda_* \in \Sigma$ is a codimension one bifurcation satisfying certain transversality conditions (see Sect. 4.1). These transversality conditions guarantee that Σ is locally a smooth hypersurface near λ_* of qualitatively similar bifurcations and that ray (2) passes through the hypersurface transversally at λ_*. Let the equation of the hypersurface near λ_* be $\psi(\lambda) = 0$. (In practice, the hypersurface is specified implicitly by larger systems of equations in λ and other variables. The other variables can in principle be eliminated to yield $\psi(\lambda) = 0$.)

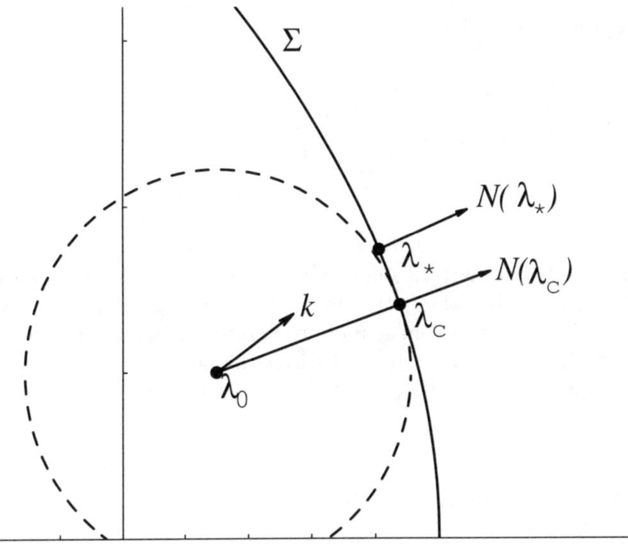

Fig.1. Parameter space geometry

The linear approximation to Σ at λ_* is a hyperplane H and the orientation of H can determined by its normal vector

$$N(\lambda_*) = \psi_\lambda|_{\lambda_*} \tag{5}$$

where ψ_λ is notation for the derivative, or gradient of ψ with respect to λ. ($N(\lambda_*)$ can be computed using formulas given in Sect. 4.1 from derivatives of (1).) The transversality condition requiring that the ray in direction k passes through the hypersurface transversally at λ_* is

$$N(\lambda_*)k \neq 0 \tag{6}$$

Here $N(\lambda_*)$ is a row vector and k is a column vector so that $N(\lambda_*)k$ is a scalar obtained by matrix multiplication.

2.3 Margin sensitivity

We will now show that the normal vector $N(\lambda_*)$ also determines the sensitivity of the margin M with respect to the parameters λ_0. (Here "sensitivity" is engineering terminology for derivative: sensitivity of M with respect to λ_0 is simply the partial derivative M_{λ_0}.) The condition on the first bifurcation along the direction k as a function of λ_0 is

$$\psi(\lambda_0 + kM(\lambda_0)) = 0 \tag{7}$$

Differentiating with respect to λ_0 yields

$$\psi_\lambda + \psi_\lambda k M_{\lambda_0} = 0 \tag{8}$$

Evaluating at λ_* and using (5) and (6) gives

$$M_{\lambda_0} = \frac{-N(\lambda_*)}{N(\lambda_*)k} \tag{9}$$

The sensitivity (9) can also be derived informally to show its straightforward geometric content [11]. Suppose that we approximate Σ to first order near λ_* by its tangent hyperplane H. Now the problem reduces to finding the sensitivity to λ_0 of the distance $M = |\lambda_* - \lambda_0|$ of the point λ_0 to the hyperplane H in the direction k. Because the direction k is fixed, M is proportional to the perpendicular distance D from λ_0 to H:

$$M = |\lambda_* - \lambda_0| = \frac{D}{\cos(\text{angle between } N(\lambda_*) \text{ and } k)} = \frac{D}{N(\lambda_*)k/|N(\lambda_*)|} \tag{10}$$

Since the optimum direction to move away from a hyperplane is normal to the hyperplane, the sensitivity $D_{\lambda_0} = -N(\lambda_*)/|N(\lambda_*)|$ and (9) follows.

An example of the content of (9) is that the optimum direction to maximize the distance of a point in a room to the floor of the room along a fixed direction is always straight up.

Sensitivity and optimization with respect to a subset of parameters

The parameters in λ can often be divided into two types:

$$\lambda = (\lambda^{\mathrm{r}}, \lambda^{\mathrm{p}}) \tag{11}$$

where the parameters in λ^{r} vary along the rays of interest and the parameters in λ^{p} are held constant along the rays of interest. Typically the parameters in λ^{r} vary to cause bifurcation but are not controlled by the operator or designer whereas λ^{p} contains design or operational parameters that are controlled by the operator or designer. We are interested in changing the controllable parameters λ^{p} to ensure sufficient robustness with respect to the uncontrollable parameters λ^{r}.

The normal vector may be similarly partitioned:

$$N = (N^{\mathrm{r}}, N^{\mathrm{p}}) \tag{12}$$

Then k has no λ^{p} components so that $N(\lambda_*)k = N^{\mathrm{r}}(\lambda_*)k$, and the sensitivity of the margin with respect to the parameters in λ^{p} is [20]

$$M_{\lambda_0^{\mathrm{p}}} = \frac{-N^{\mathrm{p}}(\lambda_*)}{N^{\mathrm{r}}(\lambda_*)k} \tag{13}$$

The sensitivities (13) can be used to select parameters λ^{p} to optimally increase the margin M along the parameters λ^{r} [10,21]. More generally, one can apply an optimization method to maximize M with respect to the parameters λ^{p} [2]; at the optimum the sensitivity (13) vanishes.

Margin to annihilation of a pair of bifurcations Some bifurcations create structures that are destroyed by further bifurcations. For example, as parameters are varied from λ_0 along a ray, a Hopf bifurcation at margin M^{create} can create a periodic orbit that is later destroyed by a further Hopf bifurcation at margin M^{destroy}. A margin M^{cd} in parameter space between the creation at λ_{create} and destruction at λ_{destroy} is

$$M^{\text{cd}} = |\lambda_{\text{destroy}} - \lambda_{\text{create}}| = M^{\text{destroy}} - M^{\text{create}} \tag{14}$$

and its sensitivity is easily computed by two applications of (9) as

$$M^{\text{cd}}_{\lambda_0} = M^{\text{destroy}}_{\lambda_0} - M^{\text{create}}_{\lambda_0} \tag{15}$$

M^{cd} measures the size of the "window" over which the structures created by the bifurcations exist. The margin sensitivity $M^{\text{cd}}_{\lambda_0}$ can be used to optimally change parameters to drive M^{cd} to zero and eliminate both bifurcations by making them annihilate each other. The use of this method to suppress a periodic orbit by annihilation of a pair of Hopf bifurcations is suggested in [12].

3 Closest Bifurcations

One way to assess the proximity of λ_0 to Σ in the parameter space $\mathbb{R}^m$ is to obtain the points of Σ which are closest to λ_0. Indeed, the distance from λ_0 to the closest points of Σ measures the robustness of the system to parameter variations when it is operated at λ_0 and the directions of the closest points of Σ from λ_0 are "worst case" directions for parameter variations leading to disappearance or instability of x. There is an inherent difficulty in describing the geometry of the multidimensional hypersurfaces in Σ and computing the points of Σ locally closest to a fixed parameter value λ_0 is a useful way to summarize the spatial relation of λ_0 to Σ while avoiding the difficult task of describing all of Σ. We call the bifurcations at the points of Σ locally closest to λ_0 *closest bifurcations*.

Closest bifurcations are local minima of the distance to λ_0 and are not necessarily globally closest bifurcations. This is an important observation in practice, because there are often multiple hypersurfaces of Σ corresponding to bifurcations of different eigenvalues (modes) or different types of bifurcations. For example, there could be hypersurfaces of Σ corresponding to several saddle-node and Hopf bifurcations of several different eigenvalues. One can try to obtain globally closest bifurcations by trying to find all the locally closest bifurcations [1,25,16]. It is also possible that a hypersurface of Σ corresponding to the bifurcation of a particular eigenvalue is corrugated so that there are several closest bifurcations on that hypersurface.

If one does in practice find a globally closest bifurcation by computing all the locally closest bifurcations, then this solves a robust stability problem in the parameter space $\mathbb{R}^m$ for a nonlinear system. For then all parameters within the sphere with center λ_0 and radius equal to the distance to the globally closest bifurcation yield stability at the equilibrium x. (This robust stability problem is hard even for linear systems.)

In defining the closest bifurcation problem, one needs to choose a distance function and a normalization of the parameter space, and this is discussed in Sect.

4.2. This Section assumes Euclidean distance. Define the distance to bifurcation $\Delta : \Sigma \to \mathbb{R}$ by $\Delta(\lambda_*) = |\lambda_* - \lambda_0|$. Then a closest bifurcation to λ_0 is a local minimum of Δ. If Δ has a local minimum at λ_c, then λ_c is a critical point of Δ and $\lambda_c - \lambda_0$ is parallel to $N(\lambda_c)$ as shown in Fig. 1. This observation explains the key role of the normal vector $N(\lambda_c)$ in computing a closest bifurcation.

3.1 Iterative method

We describe how computation of the margin M in a given direction k and computation of the normal vector N of Σ may be iterated to compute the direction k_c and parameter value λ_c of a closest bifurcation and hence a locally worst case margin $M^c = |\lambda_c - \lambda_0|$. The procedure is as follows:

0. Let k_0 be an initial guess for the direction k_c.
1. Given k_{i-1}, compute the first bifurcation along the ray given by k_{i-1}; that is, compute M_i and λ_i so that $\lambda_i = \lambda_0 + k_{i-1}M_i \in \Sigma$.
2. Compute the normal vector $N(\lambda_i)$ to Σ at λ_i.
3. Set $k_i = N(\lambda_i)$.
4. Iterate steps 1,2,3 until convergence of k_i to a value k_c. Then $\lambda_c = \lambda_0 + M^c k_c$.

The direction k_c of a locally closest bifurcation is parallel to the normal vector $N(\lambda_c)$ of Σ at λ_c and it follows that k_c is a fixed point of the iteration. The quickest way to grasp how the iteration works is to try it with pencil and paper in the case of Σ an ellipse and λ_0 an interior point of the ellipse.

In the special case of Σ being a hyperplane, the iteration converges in one step. Indeed, the iteration can be understood as minimizing $|\lambda_c - \lambda_0|$ on a series of hyperplane approximations to Σ. At each iteration, $k_i = N(\lambda_i)$ indicates the direction of the point closest to λ_0 on the hyperplane H_i tangent to Σ at λ_i.

If the iteration converges exponentially to a fixed point k_c, then the corresponding λ_c necessarily gives a closest bifurcation [7]. However, the iteration does not converge to all closest bifurcations; it converges locally only to closest bifurcations at which Σ is 'not too concave' at λ_c (that is, the minimum principal curvature of Σ at λ_c must exceed $-|\lambda_c - \lambda_0|^{-1}$) [7]. For example, the iteration converges locally to closest bifurcations if Σ is convex or if Σ is only slightly concave.

3.2 Direct methods

There are well known direct methods for computing bifurcations when one parameter is varied [38]. These direct methods work by numerically solving systems of equations that simultaneously require equilibrium and bifurcation conditions to be satisfied. Newton's method is often used to solve the equations.

Direct methods can be extended to compute closest bifurcations by allowing the parameter to be the vector $\lambda \in \mathbb{R}^m$ and augmenting the equations with an equation requiring $\lambda_c - \lambda_0$ to be parallel to $N(\lambda_c)$ [7]. However, solving these extended equations for λ_c only ensures that λ_c is a turning point of $\Delta(\lambda_*) = |\lambda_* - \lambda_0|$. To confirm that λ_c is a local minimum of Δ, it is necessary to verify a condition involving the curvature of Σ at λ_c. The condition is

$$|\lambda_c - \lambda_0| < (k^{\max})^{-1} \qquad \text{if} \quad k^{\max} > 0 \tag{16}$$

where $k^{\max}$ is the maximum principal curvature of Σ at λ_c [41]. That is, the radius $|\lambda_c - \lambda_0|$ of the sphere centered on λ_0 must be exceeded by the minimum radius of curvature $(k^{\max})^{-1}$ of Σ at λ_c. The curvature condition (16) is always satisfied if λ_0 is close enough to Σ or if $k^{\max}$ is negative.

When Newton methods are used to solve the direct method equations, the direct method has the advantage of a quadratic convergence rate and the disadvantage of requiring a good initial guess. The direct method has the disadvantage of requiring a check on the curvature of Σ, which is complicated to compute [5,7]. The iterative method has only linear convergence, but when it does converge exponentially, the solution is guaranteed to be a closest bifurcation [7].

3.3 Optimization and continuation formulations

Finding a closest bifurcation can be stated as minimizing the distance $|\lambda_c - \lambda_0|$ in parameter space subject to equilibrium and bifurcation constraints. Solution of this optimization problem with Lagrange multipliers yields equations equivalent to the direct method. Indeed, both the direct and indirect methods described above are particular approaches to solving the optimization problem. Another approach is given by Makarov et al. [29], who formulate the closest Hopf or saddle-node bifurcation problem as a constrained optimization problem minimizing the size of the real part of the eigenvalue of interest.

In optimization formulations, it is important to note that the normal vector N appears as a Lagrange multiplier [4,21]. Inequality constraints can be handled in a natural way in optimization formulations [4].

Optimization formulations also suggest other solution methods and we mention three examples: For closest saddle-node bifurcations, a generalized reduced gradient method of solving the optimization problem can be seen as generalizing one parameter continuation methods to find closest bifurcations [28,2]. The closest saddle-node bifurcation can be found by locating one point on Σ and then following a path on Σ by homotopic deformation of the equations to closest saddle-node equations [28]. Genetic algorithm sharing function optimization methods for closest Hopf and saddle-node bifurcations are initially explored in [16].

Suppose that the parameters $\lambda = (\lambda^r, \lambda^p)$ are divided into uncontrolled parameters λ^r that vary to cause the bifurcation and parameters λ^p controlled by the designer or operator as described in Sect. 2.3. Then one can formulate the problem of finding bifurcations that minimize the distance in parameters λ^r and maximize the distance in parameters λ^p. That is, one finds bifurcations that maximize with respect to λ^p the distance to closest bifurcations with respect to λ^r [2].

One can also regard sufficient robustness with respect to parameter variation as an inequality constraint when maximizing a cost function ϕ [32]. In the simple case of only one closest bifurcation being of concern, suppose that the required parameter robustness is that the distance to this closest bifurcation should exceed $M^c_{\min}$. Then the optimization has the form [32]

$$\max_{z,\lambda_0,M} \phi(z,\lambda_0) \quad \text{such that} \quad \begin{cases} 0 = f(z,\lambda_0) \\ 0 = F(z,\lambda_c,y,N) \\ \lambda_c = \lambda_0 + MN \\ M \geq M^c_{\min} \end{cases} \tag{17}$$

where F are equations to compute the normal vector N at the closest bifurcation given by λ_c and y contains auxiliary variables.

3.4 Sensitivity of distance to closest bifurcation

A closest bifurcation λ_c to nominal parameters λ_0 is a function of λ_0 that may be written, with some abuse of notation, as $\lambda_c(\lambda_0)$. The distance M^c to the closest bifurcation is also a function of λ_0:

$$M^c(\lambda_0) = |\lambda_c(\lambda_0) - \lambda_0| \tag{18}$$

The sensitivity of M^c with respect to λ_0 is [7,17]

$$M^c_{\lambda_0} = \frac{-N(\lambda_c)}{|N(\lambda_c)|} \tag{19}$$

That is, the optimum direction to move away from a closest bifurcation is antiparallel to the normal vector N. The minus sign of (19) follows from an assumption that N points in a direction towards the opposite side of Σ than λ_0.

If the nominal system is stable, but the distance to a closest bifurcation is too small to give sufficient robustness to parameter variation, then sensitivity (19) gives an optimal direction to change parameters to give sufficient distance to the closest bifurcation [11,26,31]. If the nominal system is unstable, (19) can give an optimum direction to stabilize the system [26]. Sensitivities (19) and (9) can also be used to eliminate insensitive parameters from the parameter space by fixing them at nominal values [26].

If there are several locally closest portions of Σ near λ_0, then one can maneuver with respect to several locally closest bifurcations using the corresponding sensitivities (19) [29,32].

4 Various Bifurcations and Detailed Considerations

4.1 Codimension one bifurcations

For a general description of the bifurcations discussed here, see [22,40,44]. The generic codimension one bifurcations are, roughly speaking, the bifurcations that are typically and robustly encountered when moving along smooth one dimensional curves in the parameter space [3,22,44]. The bifurcations that are codimension one depend on the class of systems being considered.

If the differential equations (1) are assumed to be smooth, then the generic codimension one bifurcations are saddle-node and Hopf bifurcations satisfying the transversality and critical eigenvalue conditions shown in Table 1 [39] (discussion of state space compactness is omitted; see [9,39]). It is intuitive that saddle-node bifurcations are codimension one, since they are defined by one additional condition on the equilibrium; namely that an eigenvalue of the linearization evaluated at the equilibrium be zero. Therefore saddle-node bifurcation of an equilibrium is defined by one equation and forms hypersurfaces in parameter space that can be typically and robustly encountered by curves in parameter space. The same argument applies to Hopf bifurcation because Hopf bifurcation is also defined by one additional condition on the equilibrium; namely that the real part of a complex pair of eigenvalues be zero.

The generic bifurcations change if the class of systems considered is restricted. (Choosing the class of systems is an important modelling decision.) If the system always has an equilibrium at the origin, then generic codimension one bifurcations of the origin are transcritical or Hopf bifurcations. If, in addition, the system has odd symmetry, then generic codimension one bifurcations of the origin are pitchfork or Hopf bifurcations.

Table 1. Formulas for vectors N normal to Σ

	N	transversality conditions
saddle-node	wf_λ	$N \neq 0,\ wf_{zz}(v,v) \neq 0$
Hopf	$\mathbf{Re}\{w(f_{z\lambda} - f_{zz}f_z^{-1}f_\lambda)v\}$	$N \neq 0,$ see [22,19]
transcritical	$wf_{z\lambda}v$	$N \neq 0,\ wf_{zz}(v,v) \neq 0$
pitchfork	$wf_{z\lambda}v$	$N \neq 0,\ wf_{zzz}(v,v,v) \neq 0$

- All quantities are evaluated at the bifurcation.
- All bifurcations are assumed to have unique, simple eigenvalues on the imaginary axis with an associated left eigenvector w and right eigenvector v.

Normal vector formulas and transversality conditions for generic codimension one bifurcations are shown in Table 1. The formulas only depend on higher derivatives of f and the eigenstructure of f_z, all evaluated at the bifurcation. Notice that $N \neq 0$ is always one of the transversality conditions. One of the themes of this chapter is the geometric interpretation and exploitation of this transversality condition as a vector normal to Σ.

The normal vector formulas for Hopf, transcritical and pitchfork bifurcations are sensitivities of the bifurcating eigenvalue. For example, for a Hopf bifurcation, the hypersurfaces of Σ are defined by the bifurcating eigenvalue μ having zero real part:

$$0 = \mathbf{Re}\{\mu\} = \mathbf{Re}\{wf_z v\} \tag{20}$$

The Hopf normal vector formula in Table 1 is obtained by differentiating (20) and evaluating at the bifurcation [7].

Σ can include several hypersurfaces corresponding to different bifurcations. For example, there could be several hypersurfaces corresponding to saddle-node bifurcations of particular eigenvalues and several hypersurfaces corresponding to Hopf bifurcations of particular complex pairs of eigenvalues. These hypersurfaces can intersect in surfaces of lower dimension (higher codimension), but the intersections would not be generically encountered by one dimensional curves in the parameter space.

4.2 Parameter space normalization and distance function

In many applications, the parameter space is inhomogeneous, consisting of different types of parameters measured in different units. Since the distance function and the closest bifurcation depend on the normalization (scaling) of the parameter space, it is important to normalize the parameter space in some sensible way to make the different parameters more comparable [26].

If the parameters are normalized by their nominal values, then distances in parameter space are nondimensional and correspond to percentage changes in parameters (e.g. [11,26,31]). This normalization (sometimes called per unit) is often routinely applied to the differential equation model (1) before any stability analysis.

A useful normalization introduced in [26] and also used in [32] establishes a range for each parameter by considering the parameter uncertainty or tolerances and then normalizes each parameter by the length of its range. If the parameters are assumed to each have uniformly distributed uncertainty over their ranges, then distances in parameter space correspond to percentage changes in uncertainty. Moreover, if the parameters have independent uncertainties normally distributed about their nominal values, then a closest bifurcation can be interpreted as a most likely bifurcation.

Sometimes parameter space distance measures other than Euclidean (L_2 norm) are appropriate. For example, suppose the size of parameter changes is measured by the sum of absolute values of the changes (L_1 norm). This can arise when each parameter change has a cost proportional to the change. Then the least cost optimum direction of parameter change to increase the margin to bifurcation is to change the control corresponding to the largest element of M_{λ_0} [36].

4.3 Constraints

Inequality constraints on the state variables and parameters can be addressed in the same framework as the stability limits due to bifurcations [21,32]. Suppose that the inequality constraint is

$$0 \leq g(z, \lambda) \tag{21}$$

where g is a scalar equation. (Multiple constraints are usually considered but it is convenient for explanation to first assume a single constraint.) We augment the parameter space limitations Σ due to bifurcations with the hypersurface corresponding to the violation of the constraint. This hypersurface can be locally specified by the $n+1$ equations

$$0 = h(z, \lambda) \qquad \text{where} \qquad h(z, \lambda) = \begin{cases} f(z, \lambda) \\ g(z, \lambda) \end{cases} \tag{22}$$

assuming that h_z has rank n so that the z variables can in principle be eliminated by the implicit function theorem.

Suppose that the inequality constraint (21) is satisfied at the nominal equilibrium z_0 with parameter λ_0. The continuation process described in Sect. 2 that finds the first bifurcation along a ray in parameter space can be modified to also detect and locate violations of the constraint (21). (Alternatively, if an optimization formulation is used, then the inequality constraints can be included in a standard way.) If λ_* on Σ corresponds to violation of the constraint, then $|\lambda_* - \lambda_0|$ is the distance to the constraint in parameter space.

The margin sensitivity and closest bifurcation or constraint computations require the normal vector N to Σ at λ_*. Since the $n+1 \times n$ matrix $h_z|_*$ has rank n, there is a row vector w' unique up to a scaling that satisfies

$$0 = w' h_z|_* \tag{23}$$

The normal vector to Σ at λ_* is [21,32]

$$N = w' h_\lambda|_* \tag{24}$$

In the case of multiple constraints of the form (21), they will generically become binding one at a time and, given the detection of which constraint is binding by the continuation software, the computations reduce to the single constraint case.

4.4 Simplifications for saddle-node bifurcation

Computing the parameter space geometry of most bifurcations requires the differential equation model (1). However, static (algebraic) equations suffice for the saddle-node bifurcation [8]. That is, $0 = f(z, \lambda)$ can be used instead of $\dot{z} = f(z, \lambda)$. Even more useful is that simpler static equations $0 = h(z, \lambda)$ equivalent to $0 = f(z, \lambda)$ can be used. In applications, it is easier to obtain reliable data for static equations.

This simplification arises because the saddle-node bifurcation involves two equilibria of $\dot{z} = f(z, \lambda)$ coalescing and disappearing and this also happens to the corresponding solutions of the static equations $0 = h(z, \lambda)$. The bifurcation is detected by the singularity (zero eigenvalue) of either f_z or h_z and the direction of the normal vector may be computed using either $N = w f_\lambda$ or $N = w' h_\lambda$, where w' is the left eigenvector of h_z corresponding to the zero eigenvalue of h_z.

Thus static equations may be used to compute margins to saddle-node bifurcations, the sensitivities of these margins, and closest saddle-node bifurcations. In particular, differential equation models may be reduced to static equations and then simplified without affecting the results of these calculations. Alternatively, if the differential equation models are not well known, the computations may be done with static models and conclusions may be drawn while assuming a general class of sensible underlying differential equation models [8].

However, care may be needed to ensure that the static equation solutions correspond to stable equilibria of the underlying differential equation models. This can sometimes be done by prior knowledge or experience that the nominal operating point is stable, and then excluding the possibility that Hopf bifurcation occurs as the parameters change. In this case, the solution obtained by continuation corresponds to a stable equilibrium until a saddle-node bifurcation occurs.

When working with static equations, it is probably better to call saddle-node bifurcation a fold bifurcation, because of the lack of dynamics (and hence saddles and nodes) in static equations. Moreover there is a distinction: A combined saddle-node Hopf bifurcation (Bogdanov–Takens point) of a differential equation is not a generic saddle-node bifurcation but it does correspond to a generic fold bifurcation of the corresponding static equations. Thus finding a fold bifurcation of static equations usually implies a generic saddle-node bifurcation of the underlying differential equations, but exceptionally this saddle-node bifurcation could also coincide with a Hopf bifurcation.

4.5 Transcritical or pitchfork bifurcations of hamiltonian systems

For Hamiltonian systems with the origin always an equilibrium, the generic codimension bifurcation is either a transcritical bifurcation or, if the system has an odd symmetry, a pitchfork bifurcation. The margin sensitivity and closest bifurcation calculations can be carried out by writing the system equations in the form (1), but it is much easier and more practical to exploit the Hamiltonian structure [31]. For the transcritical or pitchfork bifurcations, Σ is defined by the Hessian of the Hamiltonian having a zero eigenvalue. It is also convenient to derive the Hamiltonian h as a function of Lagrange coordinates $z = (q, \dot{q})$. Then the normal vector formula is

$$N = v^T h_{zz\lambda}|_* v \tag{25}$$

where v is the right eigenvector corresponding to the zero eigenvalue of h_{zz}. An application to the robust design of satellites is summarized in Sect. 5.2.

4.6 Codimension two bifurcations

Mönnigmann and Marquardt [32] have introduced a way to extend the margin sensitivity and closest bifurcation methods to codimension 2 bifurcations. Suppose that there are m parameters $\lambda = (\lambda^1, \lambda^2, \cdots, \lambda^m) \in \mathbb{R}^m$. We choose λ^1 as a distinguished parameter [19]. We are interested in the behavior of the sets of systems parameterized by λ^1 as the other parameters vary in $\mathbb{R}^{m-1}$. This may be thought of as studying the behavior of bifurcation diagrams with respect to λ^1 as the other parameters vary. In this context, the bifurcations that were codimension 2 in the parameter space $\mathbb{R}^m$ are codimension one in the parameter space $\mathbb{R}^{m-1}$. Now the margin sensitivity and closest bifurcation methods apply to hypersurfaces of Σ in $\mathbb{R}^{m-1}$.

For example, cusp bifurcations separate bifurcation diagrams in λ_1 that have hysteresis due to a pair of saddle-node bifurcations from bifurcation diagrams in λ_1 that do not have hysteresis. The cusp bifurcations satisfying transversality conditions correspond to hypersurfaces in $\mathbb{R}^{m-1}$. Another example is that isola bifurcations separate bifurcation diagrams in λ_1 that have a "bubble" of two additional solutions from bifurcation diagrams in λ_1 lacking the bubble. The isola bifurcations satisfying transversality conditions also correspond to hypersurfaces in $\mathbb{R}^{m-1}$. It can be useful to operate or design the system to avoid hysteresis or bubbles appearing in the bifurcation diagram and the margin sensitivity and closest bifurcation methods can be applied to these hypersurfaces in $\mathbb{R}^{m-1}$. Equations to compute the normal vector to these hypersurfaces are given in [32].

5 Applications

This Section briefly reviews applications of margin sensitivity and closest bifurcation methods in electric power systems, satellites, hydraulics and chemical engineering.

5.1 Large scale electric power systems

Large scale electric power transmission systems are a challenging application for bifurcation computations because of the large number (hundreds or thousands) of state variables and parameters and their nonlinear hybrid system structure.

Voltage collapse blackouts Voltage collapse is an instability of electric power systems that leads to declining voltages and blackout. It occurs when power systems are heavily loaded and can cause blackout across large regions of countries. Voltage collapse blackouts are primarily associated with loss of an operating equilibrium in a saddle-node bifurcation [4,9].

The parameter space includes the powers injected by generators or powers consumed by loads at each node in the network as well as a multitude of other controls available to the operators of the power system. The operators can adjust the powers injected by generators but are very reluctant to reduce the powers consumed by loads unless there is an emergency.

Since there is some uncertainty in the parameters, particularly the powers consumed at the loads, it is necessary to operate the power system with some margin to voltage collapse, or sufficiently far away from the saddle-node bifurcation in parameter space. This can be done by using a load power forecast to estimate the future load changes and then measuring the margin in parameter space assuming that direction of load increase with continuation software. If the margin is too small, then the sensitivity of the margin with respect to controls can help to select controls that are effective in increasing the margin [10,20,27,29,36,46].For optimization approaches to the margin sensitivity computation, see [4,21,45].

If a reliable load forecast is not available, then the closest bifurcation calculation in load power parameter space gives a "worst case" direction of load increase in which to measure the margin [11]. Several closest saddle-node bifurcations can be found by randomly sampling initial search directions [1]. One can also maximize the distance with respect to controllable parameters to a closest bifurcation that is closest with respect to the uncontrollable load parameters [2]. Closest bifurcation methods to restore an equilibrium solution removed by a saddle-node bifurcation are described in [4,34].

We briefly comment on the history of the closest saddle-node bifurcation problem in static equations for electric power systems. The idea of computing a closest instability in a real power injection parameter space for quadratic power systems equations is due to Galiana and Jarjis [17]. Using a hypothesis that Σ is convex, Galiana and Jarjis parameterize Σ with the normal vector N to Σ and define a measure D which is the perpendicular distance from the operating real power injections λ_0 to the tangent hyperplane of Σ with normal N. Minimizing D with conjugate gradient methods yields a closest instability and this computation is illustrated in a 6 node power system. Jarjis and Galiana [23] minimize a non-Euclidean distance to instability in a load power and voltage magnitude parameter space using Fletcher-Powell methods. Constrained minimization in the load power parameter space is also considered. Makarov [28,30] reviews or references similar ideas by Kontorovich et al., Vasin, and Venikov et al. published in the Russian literature starting in the early 1980s. Jung et al. [24] suggest a gradient projection optimization method to compute a closest saddle-node bifurcation and Sekine et al. [37] attempt to compute a closest saddle-node bifurcation by gradient descent on the determinant of the Jacobian.

Hopf bifurcation Hopf bifurcation of a power system operating equilibrium leads to oscillatory instabilities and it is desirable to design system parameters to ensure a sufficiently large margin to Hopf bifurcation. The margin is often measured in a parameter space that includes load powers and controllable or tunable parameters. The margin is measured assuming a direction of load increase with continuation software. These margin sensitivities with respect to the controllable parameters can be used to choose parameter changes to increase the loading margin.

Dobson et al. [12] compute Hopf margin sensitivities of a simple power system with a voltage regulator and a dynamic load model. There are 7 states and 18 parameters. The normal vector computation for Hopf margin sensitivity essentially computes the sensitivity of the critical eigenvalue. For eigenvalue sensitivity computations on larger power system models see [13,35,42].

Makarov et al. [29] compute closest Hopf and saddle-node bifurcations in a power system model with 4 states and 8 parameters. The sensitivity of the distance to the closest bifurcations are used to select effective controls to maneuver with respect to two locally closest bifurcations.

Transfer capability Transfer capability is the additional amount of bulk electrical power that can be transferred over the transmission network before an operational limit is encountered [14,15]. The transfer capability depends on where the power is generated and consumed in the network. The operational limits include state constraints such as thermal power flow limits and voltage magnitude constraints as well as stability constraints such as voltage collapse (saddle-node bifurcation), oscillations (Hopf bifurcation) and transient stability.

The transfer capability problem may be posed in a parameter space of generated (or consumed) powers as finding the distance in a given direction (determined by the transfer) to the operational limits Σ. To make the computations practical, one simplification so that static equations can be used is to address only the state constraint and saddle-node bifurcation limits (saddle-node bifurcations can be addressed with static equations as explained in Sect. 4.4). These computations are practical for power system static equations with thousands of states and parameters [14,21]. To try these computations on the web see [15]. The sensitivity of the transfer capability with respect to parameters is important in maximizing the transfer capability and handling uncertainty in the parameters [14,15,21].

5.2 Design of Hamiltonian satellites

Mazzoleni and Dobson [31] consider the robust stability design of an artificial satellite modelled as Hamiltonian system. The closest transcritical or pitchfork bifurcation to a nominal design point is computed as indicated in Sect. 4.5. The sensitivity of the margin to this closest bifurcation is used to modify the design to achieve sufficient stability robustness with respect to parameter variations. The computation is illustrated in a 6 dimensional parameter space with nondimensionalized parameters. The satellite is a flexible dual-spin satellite with a 10 dimensional state space.

5.3 Hydraulic control systems

Kremer and Thompson [26] and Kremer [25] apply closest bifurcation methods to design hydraulic control systems for robustness with respect to Hopf bifurcations. The distinctions between design and operating parameters and the parameter normalization needed to obtain sensible results are emphasized [26]. The methods are illustrated in hydraulic control systems with 3 and 9 states and parameter spaces of dimensions 7 and 24.

Nominal design parameters are obtained in [26] by checking stability for a series of values of other parameters that represent different operating conditions. If instability is found for any of these operating conditions, a closest Hopf bifurcation to the parameters giving instability is found and the sensitivity of the distance to this closest Hopf bifurcation is used to find a parameter change that optimally corrects the instability. For each operating mode, the closest Hopf bifurcation to the nominal design is found and, if the distance to this closest Hopf bifurcation is not sufficient, the sensitivity of this distance is used to modify the nominal design parameters. If limits on design parameters are encountered during the process, then these limits are enforced so that the closest Hopf bifurcations computed are the closest bifurcations subject to these limits. Thus closest bifurcation and sensitivity calculations are used in an iterative design procedure to maneuver in design parameter space while accounting for multiple and uncertain operating parameters.

Different modes (eigenvalue pairs) becoming unstable correspond to different hypersurfaces of Σ and Kremer [25] explores the parameter space with closest bifurcation computations with different initial directions of search to try to locate all the closest bifurcations and corresponding modes of practical interest. Given suitable convexity properties in each hypersurfaces of Σ, this procedure seems in practice to be successful in finding all the locally closest Hopf bifurcations and hence the globally closest Hopf bifurcation.

5.4 Chemical engineering processes

Mönnigmann and Marquardt [32] optimize the cost function with respect to design parameters of several chemical engineering processes subject to output temperature constraints and parametric robust stability constraints. The robust stability constraints require the closest Hopf or saddle-node bifurcations to be sufficiently distant as summarized in Sect. 3.3. The methods are illustrated on a fermenter and a continuously stirred tank reactor, with 2 states and 2-10 parameters. The continuously stirred tank reactor is also used to illustrate the robust avoidance of hysteresis in the bifurcation diagram with respect to Damköhler number as briefly explained in Sect. 4.6. Optimization of a continuous polymerization process with a differential-algebraic model with 4 dynamic states and 15 algebraic states is addressed in [33].

6 Conclusions

This chapter summarizes general methods of assuring stability robustness of an equilibrium with respect to many parameters. The methods provide numerical computations to explore and quantify the proximity of nominal parameters λ_0 to the hypersurfaces of the bifurcation set and other constraints. Because of the "curse of dimensionality", it is usually impractical to analytically or numerically specify all the points of these hypersurfaces in the high dimensional parameter spaces that are routine in realistic applications. However, the approach of finding points on the hypersurfaces in a given direction, together with the hyperplane approximations determined by their normal vectors does allow progress to be made. In particular, margin sensitivities and locally closest bifurcations can be computed and these allow maneuvering in parameter space so that sufficient parameter space distance to instability or constraint can be maintained. These methods have already been tested in a variety of applications and there is ample scope for further developments.

References

1. Alvarado, F. L., Dobson, I., Hu, Y. (1994) Computation of closest bifurcations in power systems. IEEE Trans. Power Syst., **9**(2):918–928
2. Canizares, C. A. (1998) Calculating optimal system parameters to maximize the distance to saddle node bifurcations. IEEE Trans. Circ. Syst.-I, **45**(3):225–237
3. Chow, S. N., Hale, J. (1982) Methods of bifurcation theory. New York: Springer-Verlag
4. van Cutsem, T., Vournas, C. (1998) Voltage Stability of Electric Power Systems. Boston: Kluwer
5. Dai, R. -X., Rheinboldt, W. C. (1990) On the computation of manifolds of foldpoints for parameter-dependent problems. SIAM J. Numerical Analysis, **27**(2):437–446
6. Dobson, I. (1992) Observations on the geometry of saddle node bifurcation and voltage collapse in electric power systems. IEEE Trans. Circ. Syst.-I, **39**(3):240–243
7. Dobson, I. (1993) Computing a closest bifurcation instability in multidimensional parameter space. J. Nonl. Sci., **3**(3):307–327
8. Dobson, I. (1994) The irrelevance of load dynamics for the loading margin to voltage collapse and its sensitivities. In Proc. Bulk Power System Voltage Phenomena III, Voltage stability, security & control, ECC/NSF workshop, Davos, Switzerland, 509–518
9. Dobson, I., Chiang, H.-D. (1989) Towards a theory of voltage collapse in electric power systems. Syst. Contr. Lett., **9**:253–262
10. Dobson, I., Lu, L. (1992) Computing an optimum direction in control space to avoid saddle-node bifurcation and voltage collapse in electric power systems. IEEE Trans. Auto. Contr., **37**(10):1616–1620
11. Dobson, I., Lu, L. (1993) New methods for computing a closest saddle node bifurcation and worst case load power margin for voltage collapse. IEEE Trans. Power Syst., **8**(3):905–913
12. Dobson, I., Alvarado, F. L., DeMarco, C. L. (1992) Sensitivity of Hopf bifurcations to power system parameters. In Proc. 31st IEEE Conf. Decision Control, Tucson, AZ, 2928–2933
13. Dobson, I. *et al.* (1999) Avoiding and suppressing oscillations. PSerc Publication 00-01, available from www.pserc.wisc.edu

14. Dobson, I. *et al.* (2001) Electric power transfer capability: concepts, applications, sensitivity, uncertainty. PSerc Publication 01-34, available from www.pserc.wisc.edu
15. Dobson, I. *et al.* (2001) Transfer capability calculator and tutorial web site: www.pserc.cornell.edu/tcc
16. Dong, Y. D., Makarov, Y. V., Hill, D. J. (1998) Analysis of small signal stability margins using genetic optimization. Electric Power Systems Research, **46**:195–204
17. Galiana, F. D., Jarjis, J. (1978) Feasibility constraints in power systems. In Proc. IEEE PES Summer meeting, Los Angeles, **A 78**:560-565
18. Galiana, F. D., Zeng, Z. (1991) Analysis of the load flow behaviour near a Jacobian singularity. In Proc. IEEE Power Industry Computer Applications Conf., 149–155
19. Golubitsky, M., Schaeffer, D. G. (1985) Singularities and Groups in Bifurcation Theory, vol 1. New York: Springer Verlag
20. Greene, S., Dobson, I., Alvarado, F. L. (1997) Sensitivity of the loading margin to voltage collapse with respect to arbitrary parameters. IEEE Trans. Power Syst., **12**(1):262–272
21. Greene, S., Dobson, I., Alvarado, F. L. (2002) Sensitivity of transfer capability margins with a fast formula. IEEE Trans. Power Syst., **17**(1):34–40
22. Guckenheimer, J., Holmes, P. (1983) Nonlinear Oscillations, Dynamical Systems and Bifurcations of Vector Fields. New York: Springer Verlag
23. Jarjis, J., Galiana, F. D. (1981) Quantitative analysis of steady state stability in power networks. IEEE Trans. on Power Apparatus Syst., **100**:318–326
24. Jung, T. H., Kim, K. J., Alvarado, F. L. (1990) A marginal analysis of the voltage stability with load variations. In Proc. Power Systems Computation Conference, Graz, Austria
25. Kremer, G. G. (2001) Enhanced robust stability analysis of large hydraulic control systems via a bifurcation-based procedure. J. Franklin Institute, **338**(7):781–809
26. Kremer, G. G., Thompson, D. F. (1998) A bifurcation-based procedure for designing and analysing robustly stable non-linear hydraulic servo systems. In Proc. Inst. Mechanical Engineers, Part I – J. Syst. Contr. Eng., **212**(15):383–394
27. Kumano, T., Yokoyama, A., Sekine, Y. (1990) Fast monitoring and optimal preventive control of voltage instability. In Proc. Power Systems Computation Conference, Graz, Austria
28. Makarov, Y. V., Hiskens, I. A.(1994) A continuation method approach to finding the closest saddle node bifurcation point. In Proc. Bulk Power System Voltage Phenomena III, Voltage stability, security & control, ECC/NSF workshop, Davos, Switzerland, 333–347
29. Makarov, Y. V., Wu, Q., Hill, D. J., Popović, D. H., Dong, Z. (1997) Coordinated steady-state voltage stability assessment and control. In Proc. 4th Int. Conf. Advances in Power System Control, Operation and Management, Conf. Publ. No. 450, **1**:248–253
30. Makarov, Y. V., Hill, D. J., Dong, Z. Y. (2000) Computation of bifurcation boundaries for power systems: A new Delta-plane method. IEEE Trans. Circ. Syst.-I, **47**(4):536–544
31. Mazzoleni, A., Dobson, I. (1995) Closest bifurcation analysis and robust stability design of flexible satellites. J. Guidance, Contr. Dynam., **18**(2):333–339

32. Mönnigmann, M., Marquardt, W. (2002) Normal vectors on manifolds of critical points for parametric robustness of equilibrium solutions of ODE systems. J. Nonl. Sci., **12**(2):85–112
33. Mönnigmann, M., Marquardt, W. (2002) Process optimization with guaranteed robustness and flexibility. Technical report LPT-2002-01, available from www.lfpt.rwth-aachen.de
34. Overbye, T. J. (1995) Computation of a practical method to restore power flow solvability. IEEE Trans. Power Syst., **10**(1):280–287
35. Ozcan, I. A., Schattler, H. (1999) On the calculation of the feasibility boundary for differential-algebraic systems. In Proc. IEEE Conf. Decision Control, Phoenix, AZ, 2580–2586
36. Popović, D. H., Hill, D. J., Wu, Q. (2002) Optimal voltage security control of power systems. Electr. Power Energy Syst., **24**(4):305–320
37. Sekine, Y., Yokoyama, A., Kumano, T. (1989) A method for detecting a critical state of voltage collapse. In Proc. Bulk Power System Voltage Phenomena Part III, Voltage stability and security, EPRI Report EL-6183, Potosi, MO, USA
38. Seydel, R. (1988) From Equilibrium to Chaos: Practical Bifurcation and Stability Analysis. New York: Elsevier
39. Sotomayor, J. (1973) Generic bifurcations of dynamical systems. In Dynamical Systems, ed. by Peixoto, M. M., New York: Academic Press
40. Strogatz, S. (1994) Nonlinear Dynamics and Chaos with applications in physics, biology, chemistry, and engineering. Reading, MA: Addison-Wesley
41. Thorpe, J. A. (1979) Chapter 16, Elementary Topics in Differential Geometry. New York: Springer-Verlag
42. Wang, K. W., Chung, C. Y., Tse, C. T., Tsang, K. M. (2000) Multimachine eigenvalue sensitivities of power system parameters. IEEE Trans. Power Syst., **15**(2):741–747
43. Wang, R., Lasseter, R. H. (2000) Re-dispatching generation to increase power system security margin and support low voltage bus. IEEE Trans. Power Syst., **15**(2):496–501
44. Wiggins, S. (1990) Introduction to Applied Nonlinear Dynamical Systems and Chaos. New York: Springer Verlag
45. Wu, T., Fischl, R. (1993) An algorithm for detecting the contingencies which limit the inter-area megawatt transfer. In Proc. North Amer. Power Symp., Washington DC, 222–227
46. Wu, Q., Popović, D. H., Hill, D. J., Parker, C. J. (2001) Voltage security enhancement via coordinated control. IEEE Trans. Power Syst., **16**(1):127–135

Static Bifurcation in Mechanical Control Systems

Harry G. Kwatny[1], Bor-Chin Chang[1], and Shiu-Ping Wang[2]

[1] Drexel University, Philadelphia, PA 19104, USA
hkwatny@coe.drexel.edu
[2] 202*nd* Arsenal, C.S.F., Nankang, Taipei 115, Taiwan

Abstract. Feedback regulation of nonlinear dynamical systems inevitably leads to issues concerning static bifurcation. Static bifurcation in feedback systems is linked to degeneracies in the system zero dynamics. Accordingly, the obvious remedy is to change the system input-output structure, but there are other possibilities as well. In this paper we summarize the main results connecting bifurcation behavior and zero dynamics and illustrate a variety of ways in which zero structure degeneracy can underly bifurcation behavior. We use several practical examples to illustrate our points and give detailed computational results for an automobile that undergoes loss of directional and cornering stability.

1 Introduction

Many important problems in the operation of technological systems can be interpreted as static bifurcations, i.e., bifurcations associated with a change in the equilibrium point structure of the underlying equations. Examples include stall in aircraft, voltage collapse in power networks, loss of cornering stability in ground vehicles, furnace implosion in power plants, and rotating stall in jet engine compressors. In each of these cases the bifurcation occurs while attempting to regulate certain plant outputs. Consequently, the bifurcation takes on unique characteristics associated with a control system (with an input-output structure) as opposed to a simple dynamical system.

The basic tools used to investigate static bifurcations in feedback systems are the usual ones: the Implicit Function Theorem and Lyapunov-Schmidt reduction [7], and the Newton-Raphsom-Seydel method [18] for locating bifurcation points. By applying these tools in the context of a control system we find new control theoretic interpretations of the bifurcation that suggest remedies.

We will consider control systems of the form

$$\begin{aligned} \dot{x} &= f(x,u,\mu) \\ y &= g(x,\mu) \\ z &= h(x,\mu) \end{aligned} \tag{1}$$

where $x \in R^n$ is the system state, $u \in R^m$ is the control, $y \in R^q$ is the measurement, and $z \in R^p$ is the regulated output (performance variables). $\mu \in R^k$ is a parameter vector that may be composed of plant parameters, exogenous constant disturbances, and/or set points. We assume that f, g, h are smooth (sufficiently differentiable).

G. Chen, D.J. Hill, and X. Yu (Eds.): Bifurcation Control, LNCIS 293, pp. 67–81, 2003.

The control problem is to design a feedback regulator that stabilizes a desired equilibrium point corresponding to $z = 0$.

As we will see in Section 2, static bifurcations in regulators are always associated with a degeneracy in the linearized system zero dynamics. Several specific examples are given. Such degeneracies include loss of linear observability or controllability. But this does not imply that the system fails to be observable or controllable in a nonlinear sense. We describe these connections in Section 3. Section 3 also explains the computations we use to locate and analyze bifurcation behavior. Section 4 gives a detailed analysis of automobile directional instability and cornering instability.

2 Characterizing Bifurcations in Control Systems

2.1 Necessary conditions for static bifurcation

A triple (x^*, u^*, μ^*) is an *equilibrium point* of the open loop dynamics (1) if

$$F(x^*, u^*, \mu^*) := \begin{bmatrix} f(x^*, u^*, \mu^*) \\ h(x^*, \mu^*) \end{bmatrix} = 0 \tag{2}$$

Ordinarily, we obtain equilibria by specifying, μ^* and solving (2) for x^*, u^*. Then $y^* = g(x^*, \mu^*)$. Typically, we expect that (2) will have solutions only if $m \geq p$. Since the number of controls can always be reduced, we henceforth assume $m = p$.

Definition 1. Consider the set, $\mathcal{E}$, of points that satisfy (2),

$$\mathcal{E} = \left\{ (x^*, u^*, \mu^*) \in R^{n+m+k} \,|F(x^*, u^*, \mu^*) = 0 \right\} \tag{3}$$

The set $\mathcal{E}$ is called the *open loop equilibrium manifold.*

Remark 1. If

$$\text{rank} \left[D_x F \ D_u F \ D_\mu F \right] = n + m$$

on $\mathcal{E}$, then $\mathcal{E}$ is a regular manifold of dimension k in R^{n+m+k}.

Definition 2. An equilibrium point $(x^*, u^*, \mu^*) \in \mathcal{E}$ is *regular* if there is a neighborhood of μ^* on which there exist unique, continuously differentiable functions $\bar{x}(\mu)$, $\bar{u}(\mu)$ with $x^* = \bar{x}(\mu^*)$, $u^* = \bar{u}(\mu^*)$ satisfying

$$F\left(\bar{x}(\mu), \bar{u}(\mu), \mu\right) = 0$$

Otherwise, it is a *(static) bifurcation point.*

Remark 2. Notice that the implicit function theorem implies that an equilibrium point is regular if

$$\det \left[D_x F(x^*, u^*, \mu^*) \ D_u F(x^*, u^*, \mu^*) \right] \neq 0 \tag{4}$$

In view of Remark 2 we can obtain a useful interpretation of static bifurcations in control systems. Consider the linearization of Eq. (1) at the equilibrium point $(x^*, u^*, \mu^*) \in \mathcal{E}$ and define the matrices

$$A = \frac{\partial f}{\partial x}(x^*, u^*, \mu^*), \ B = \frac{\partial f}{\partial u}(x^*, u^*, \mu^*), \ C = \frac{\partial h}{\partial x}(x^*, \mu^*) \tag{5}$$

Eq. (4) is equivalent to

$$\det \begin{bmatrix} A & B \\ C & 0 \end{bmatrix} \neq 0 \Leftrightarrow \operatorname{Im} \begin{bmatrix} -A & B \\ -C & 0 \end{bmatrix} = R^{n+m}$$

where the minus sign is introduced for convenience. Recall that $m = p$. Then, in view of Remark 2, the following result is obvious.

Lemma 1. *An equilibrium point (x^*, u^*, μ^*) is a static bifurcation point only if*

$$\operatorname{Im} \begin{bmatrix} -A & B \\ -C & 0 \end{bmatrix} \neq R^{n+m}$$

Remark 3. It is important to emphasize that Lemma 1 is a *necessary* but not *sufficient* condition for static bifurcation.

The necessary condition for a static bifurcation given in Lemma 1 can be interpreted in terms of two possibilities:

1. If for typical λ,

   $$\operatorname{rank} \begin{bmatrix} \lambda I - A & B \\ -C & 0 \end{bmatrix} = n + m$$

   then the static bifurcation corresponds to an invariant zero (of the linearized dynamics) located at the origin. This is referred to as the *nondegenerate case.* Recall that the set of invariant zeros is composed of the following (see, for example, [2]):
   (a) input decoupling zeros (uncontrollable modes), λ satisfies

   $$\operatorname{rank} \begin{bmatrix} \lambda I - A & B \end{bmatrix} < n$$

   (b) output decoupling zeros (unobservable modes), λ satisfies

   $$\operatorname{rank} \begin{bmatrix} \lambda I - A \\ C \end{bmatrix} < n$$

   (c) transmission zeros (the remainder of the invariant zeros)
2. Otherwise, for typical λ,

   $$\operatorname{rank} \begin{bmatrix} \lambda I - A & B \\ -C & 0 \end{bmatrix} < n + m$$

   or equivalently,

   $$\det \left\{ C (\lambda I - A)^{-1} B \right\} = \det \left\{ G(\lambda) \right\} = 0$$

   This is referred to as the *degenerate case.* The degenerate case corresponds to the following possibilities:
   (a) insufficient independent controls, rank $B < p$
   (b) redundant outputs, rank $C < p$

In summary, we have the following necessary condition (from [12]).

Proposition 1. *The equilibrium point (x^*, u^*, μ^*) is a static bifurcation point of the (square) system (1) only if one of the following conditions obtains for its linearization (5):*

1. *there is a transmission zero at the origin,*
2. *there is an uncontrollable mode with zero eigenvalue,*
3. *there is an unobservable mode with zero eigenvalue,*
4. *it has insufficient independent controls,*
5. *it has redundant outputs.*

In view of this result it is interesting to investigate how degeneracies occur in linear, parameter-dependent control systems and what limitations they impose on linear regulator design. Such questions have been considered in [4] and [5].

2.2 Examples: a first look

In the following examples we illustrate some of the situations enumerated in Proposition 1.

Example 1 (Compressor stall). Compressor stall has been extensively studied over the past several years [16,3,6], [11], [9]. Vane adjustments are made to regulate compressor outlet plenum pressure in jet engines. Ideally a maximum pressure is desirable. The compressor characteristic normally achieves a peak within the admissible range of vane positions. However, the corresponding equilibrium point is a static bifurcation point associated with the emergence of a non-axisymmetric flow pattern called rotating stall. In fact the bifurcation is a subcritical pitchfork. At the bifurcation point the linearized system has a mode with zero eigenvalue that is both uncontrollable and unobservable.

Example 2 (Automobile directional stability). Consider the following situation. An automobile is driven along a straight path with the steering wheel locked. The vehicle forward speed is regulated using the throttle. It is well known that certain vehicles have a critical speed at which the normally stable equilibrium condition becomes unstable (see, for example, Doebelin [8]). This is the vehicle directional stability limit. The critical speed depends on a number of factors including the center of mass location and the tire cornering coefficients. In this example we will see that the directional stability limit is associated with loss of controllability and observability of the linearized system, and corresponds to a supercritical pitchfork bifurcation.

One way to change the bifurcation behavior of a regulator problem is to modify the zero structure by changing the input and output choices. For instance in this case, we might change the single-input single-output problem to a two-input two-output problem by adding the steering angle as a control input and vehicle angular velocity as an output. This completely eliminates the directional stability bifurcation and explains why steering angle stabilization of automobile has proved so successful (e.g., [1] and [17]). We examine this regulator problem below in Section 4.

Example 3 (Aircraft flight path regulation). Consider an aircraft in straight and level flight. The pilot attempts to maintain a flight path angle of $\gamma = 0$ while reducing airspeed ν [12]. This is accomplished using the throttle and elevator. At some minimum critical speed the equilibrium cannot be sustained and the aircraft stalls. This is associated with a fold bifurcation. At the bifurcation point the two linearized system control input vectors are linearly dependent.

Example 4 (Automobile cornering). Consider a vehicle with the throttle and steering angle employed to maintain a specified speed, V_s, and angular velocity, ω. Thus, the vehicle maintains a circular path of radius $R = V_s/\omega$. With the speed held constant, increasing ω (decreasing R) eventually leads to a loss of stability corresponding to a fold bifurcation. At the bifurcation point the linearized system has a transmission zero at the origin. We examine this problem more fully in Section 4.

3 Basic Computations

3.1 Modelling

Our approach to model formulation is a variant of Lagrange's equations referred to as Poincaré's equations [13]. Modelling proceeds in the usual way by formulating the kinetic and potential energy and constructing the generalized forces. In general, Poincaré's equations take the form

$$\dot{q} = V(q)\,p \tag{6a}$$

$$M(q)\dot{p} = -C(q,p)\,p - F(q,p,u,\mu) \tag{6b}$$

where q is a vector of generalized coordinates, p a vector of quasi-velocities, u a vector of exogenous inputs, and μ is a vector of parameters. Equation (6a) represents the system *kinematics* and (6b) the system *dynamics*. Since the inertia matrix, $M(q)$, is invertible for all q these equations can be put in the form of (1) with $x = (q,p)$ and g and h appropriately defined. In [13] and elsewhere, we provide a set of symbolic computing tools that enable the efficient assembly of models of the type (6a) and (6b) for complex multibody mechanical systems. Models derived in symbolic form can be manipulated via computer algebra constructions in many useful ways, such as to perform coordinate transformations or model simplification. Simulation code can be automatically generated.

Equilibria correspond to $\dot{q} = 0$, $\dot{p} = 0$, and $h(q,p,\mu) = 0$. Typically, $\dim q \leq \dim p$ so that the kinematics require $p = 0$. We will assume that this is the case. Then the equilibrium equations reduce to

$$F(q,0,u,\mu) = 0,\ h(q,0,\mu) = 0 \tag{7}$$

To find equilibria we need to solve (7) for q, u, with μ given. Clearly, these equations are of the form of (2). There can be a significant simplification in using (7), rather than the full state equation equivalent.

3.2 Locating static bifurcation points

We need to determine the static bifurcation point precisely in order to examine the linear (and local nonlinear) system properties. In some cases this can be accomplished analytically, e.g., the compressor stall and directional stability examples noted above. But in more typical engineering systems numerical calculation is required. One important computational tool is the Newton-Raphson-Seydel (NRS) method [18]. We seek a solution (x^*, u^*) of Eq. (2) for a parameter value μ^* at which the Jacobian $[D_xF, D_uF]$ is singular. Consequently, the Newton-Raphson method breaks down. The NRS method is appropriate for one parameter ($\mu \in R$) problems. In this case, generic bifurcations are of codimension 1 (rank $[D_xF, D_uF] = n + m - 1$). Here we seek $x \in R^n, u \in R^m, \mu \in R, v \in R^{m+n}$ that satisfy

$$F(x, u, \mu) = 0 \tag{8a}$$

$$Jv = 0, \quad J = \left[\, D_xF(x, u, \mu) \; D_uF(x, u, \mu) \,\right] \tag{8b}$$

$$\|v\| = 1 \tag{8c}$$

Eq. (8b) along with the eigenvector nontriviality condition (8c) require singularity of the Jacobian – a necessary condition for bifurcation. There are many variants of this formulation and it has been applied to fairly large systems. Of course, a key ingredient for success is the identification of good initial values for μ and v as well as x and u. Often these are obtained by continuing (in μ) a Newton Raphsom computation until close to singularity (see [15], for example). To this end have found it useful to use a singular value formulation, e.g.,

$$F(x, u, \mu) = 0 \tag{9a}$$

$$JJ^Tv = 0, \quad J = \left[\, D_xF(x, u, \mu) \; D_uF(x, u, \mu) \,\right] \tag{9b}$$

$$\|v\| = 1 \tag{9c}$$

3.3 Nonlinear control system properties

Of course, once the bifurcation point is determined it is a simple matter to compute the matrices defined in (5). Thus, we can identify the features of the linearized system that underly the bifurcation. Since a static bifurcation is always associated with a defect in the zero dynamics, the most obvious remedy is to change the input output structure. But this might not be the only alternative. To this end, we examine nonlinear control properties locally around the bifurcation point.

If the bifurcation is associated with a breakdown in linear controllability or observability, it is still possible that the system is controllable or observable in the nonlinear sense. To explore this notion further, assume that the system is affine so that the system (1) takes the form:

$$\dot{x} = f(x) + \sum_{i=1}^{m} g_i(x)\, u_i \tag{10a}$$

$$z = h(x) \tag{10b}$$

Recall that such a system has associated with it (see, for example, [10]) , the *controllability distributions*

$$\Delta_C = \langle f, g_1, \ldots, g_m \,|\text{span}\,\{f, g_1, \ldots, g_m\}\rangle \tag{11a}$$

$$\Delta_{C_O} = \langle f, g_1, \ldots, g_m \,|\text{span}\,\{g_1, \ldots, g_m\}\rangle \tag{11b}$$

and

$$\Delta_L = \text{span}\left\{f, ad_f^k g_i, 1 \le i \le m, 0 \le k \le n-1\right\} \tag{12a}$$

$$\Delta_{L_0} = \text{span}\left\{ad_f^k g_i, 1 \le i \le m, 0 \le k \le n-1\right\} \tag{12b}$$

Criteria for controllability for (10a) and (10b) are formulated in terms of *rank conditions* as illustrated in the following diagram

$$\begin{array}{ccccc}
\textit{weak local controllability} & \Leftarrow & \dim \Delta_C(x_0) = n & \Leftarrow & \dim \Delta_L(x_0) = n \\
\Uparrow & & \Uparrow & & \Uparrow \\
\textit{local controllability} & \Leftarrow & \dim \Delta_{C_0}(x_0) = n & \Leftarrow & \dim \Delta_{L_0}(x_0) = n \\
\Uparrow & & & & \Updownarrow \\
\textit{linear controllability} & & \Leftrightarrow & & \dim \left[B \cdots A^{n-1}B \right]
\end{array}$$

The essential observation is that the rank conditions are 'sufficient' but not 'necessary.' Necessity follows only if the relevant distribution is nonsingular at the point x_0. So, for example, if the system is not linearly controllable it may still be locally controllable.

Similarly, the *observability codistributions* are

$$\Omega_O = \langle f, g_1, \ldots, g_m \, | \mathrm{span}\{dh_1, \ldots, dh_p\} \rangle \tag{13a}$$

$$\Omega_L = \mathrm{span}\left\{ L_f^k(dh_i), 1 \le i \le p, 0 \le k \le n-1 \right\} \tag{13b}$$

The various observability rank conditions can be summarized in the following diagram.

$$\begin{array}{ccc}
\dim \Omega_O(x_0) = n & \Rightarrow & \textit{locally observable} \\
\Uparrow & & \Uparrow \\
\dim \Omega_L(x_0) = n & \Rightarrow & \textit{zero input observable} \\
\Updownarrow & & \Uparrow \\
\dim \begin{bmatrix} C \\ \vdots \\ CA^{n-1} \end{bmatrix} = n & \Leftrightarrow & \textit{linearly observable}
\end{array}$$

Once again, it is possible that the system is locally observable or zero-input observable even though it is not linearly observable.

There are important implications to the fact that a system may be controllable (or observable) in a nonlinear sense but not linearly controllable (or observable). Roughly, failure of linear controllability or observability means that any associated controller or observer will be non-smooth. Some observer examples may be found in [14].

Recall, the systematic construction (e.g., [10]) that reduces the square system (10a) and (10b) to the normal form

$$\dot{z} = Az + E\left[\alpha(x) + \rho(x)u\right] \tag{14a}$$

$$y = Cz \tag{14b}$$

where $z \in R^r$, $r \le n$, $\rho(x)$ is a square $m \times m$ matrix, and A, E, C constitute a Brunovsky form triple of matrices with indices, $r_1, \ldots, r_m$, and $r = r_1 + \cdots + r_m$. If $\rho(x_0)$ is of rank m, then the system is said to have well defined relative degree. Moreover, there are coordinates $\xi \in R^{n-r}$ and a transformation $x \to (\xi, z)$ that transforms the system (10a) and (10b) into (14a) and (14b) along with

$$\dot{\xi} = F(\xi, z, u) \tag{15}$$

Equations (10a) and (10b) are easily stabilized by smooth feedback, so that $z, u \to 0$ as $t \to \infty$, and the output is eventually zeroed. However, there remains the motion of ξ, which does not affect the states z, nor the output, y. Consequently, the *zero (output) dynamics* are defined as

$$\dot{\xi} = F(\xi, 0, 0) = f_z(\xi) \tag{16}$$

In the following examples, we evaluate the appropriate rank conditions using the symbolic computing tools described in [13]. These tools also allow computation of the (nonlinear) zero dynamics, relative degree, various normal forms, etc. To apply this analysis in the following examples, we first construct an affine approximation around the bifurcation point of interest.

4 The Automobile

We will illustrate the above concepts by considering their application to two automotive control problems. In recent years, electronic control systems have proved to be key contributors to improved vehicle handling and safety. Anti-lock brakes, traction control systems and electronic stabilization systems [1] are important innovations of the past decade. Each of these control systems attempts to address a fundamentally nonlinear stability problem. Future advances will include electronic steering and braking that will encourage more extensive 'drive-by-wire' control systems including full integration of lateral stabilization, active suspension, and power train control systems. Understanding the underlying nonlinear control issues will be essential to developing systems that work together in harmony.

In particular, we elucidate two important behavioral properties in terms of bifurcation behavior:

1. Depending on vehicle parameters, there may be a critical speed, $V_s^* < \infty$, above which constant speed, straight line motion is unstable. This is the *straight line (or, directional) stability* limit. Moreover, for $V_s < V_s^*$ the angular velocity satisfies $\omega\delta > 0$, whereas, for $V_s > V_s^*$, $\omega\delta < 0$.
2. At any fixed speed $V_s < V_s^*$, there is a critical angular velocity $\omega^* > 0$ (resp., $\omega^* < 0$) and a corresponding steering angle, $\delta^* > 0$, (resp., $\delta^* < 0$) above (resp., below) which there does not exist a stable equilibrium state. This is the *cornering stability* limit.

The vehicle to be considered is illustrated in Fig. 1. The difficulty in modelling such a vehicle is the algebraic complexity that arises when four distinct wheels, camber, caster and other practical details are included. Symbolic computing minimizes the painful, error-prone calculations.

For the vehicle shown the coordinates are $q = [\theta, X, Y]^T$ and the quasi-velocities are $p = [\omega, v_x, v_y]^T$. Note that v_x and v_y are the center of mass velocity coordinates in the body frame. The generalized forces involve rear tire drive forces and each tire also produces a cornering force modelled by an equation:

$$F_{yi} = \kappa_i \tan^{-1}(A_i \alpha_i), \quad i = 1, \ldots, 4 \tag{17}$$

[1] e.g., BMW's Dynamic Stability Control, Mercedes-Benz' Electronic Stability Program, Cadillac's StabiliTrak

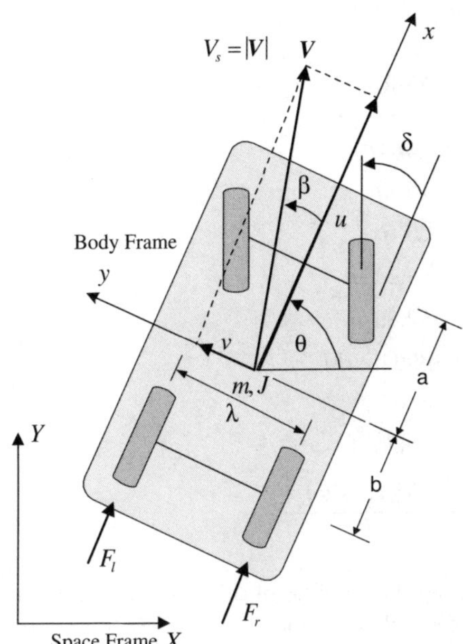

Fig.1. The automobile under study with reference frames and key parameters.

where κ_i and A_i are parameters that may differ among the four tires. α_i is the tire sideslip angle. It is convenient to introduce the vehicle sideslip angle, β, and transform the velocity coordinates $(v_x, v_y) \to (V_s, \beta)$ via the transformation relations:

$$v_x = V_s \cos\theta, \quad v_y = V_s \sin\theta \tag{18}$$

In these coordinates Eq. (6b) takes the form

$$\frac{d}{dt}\begin{bmatrix} \omega \\ V_s \\ \beta \end{bmatrix} = F\left(\omega, V_s, \beta, \delta, F_d\right) \tag{19}$$

Notice that (19) involves only the velocity coordinates ω, V_s, β and inputs F_d, δ. It does not involve any of the generalized coordinates q. It is, therefore, a closed system of differential equations. So we do not need the kinematic equations (6a). Once the velocity variables are determined the configuration coordinates can be obtained by quadratures.

Numerical calculations are based on the automobile data given in Table 1. The explicit set of equations used in the calculations are given in the Appendix.

4.1 Directional stability

A classic problem in automobile dynamics is the study of straight line directional stability. With $\delta = 0$, consider the equilibrium point corresponding to $\omega = 0, \beta = 0$ as the speed varies. Furthermore, we assume that both front tires are identical and

Table 1. Automobile Data

Symbol	Description	Value
$a+b$	wheelbase	111 *in*
r	tire radius	15 *in*
λ	track	60 *in*
a	front axle to center of gravity	58.14 *in*
b	rear axle to center of gravity	52.86 *in*
κ_f, κ_r	tire coefficient	6964.2 lbf/rad
A_f, A_r	tire coefficient	1
J	automobile inertia z	3,630 $lbf-sec^2$
m	automobile weight	155.28 *slug*

both rear tires are identical. We can determine the stability of this equilibrium point by examining the linear approximation for small deviations from $\omega = 0, \beta = 0$. To do this we compute the Jacobian

$$A\left(\bar{V}_s, \delta\right) = \begin{bmatrix} \frac{\partial f_1}{\partial \omega} & \frac{\partial f_1}{\partial \beta} \\ \frac{\partial f_2}{\partial \omega} & \frac{\partial f_2}{\partial \beta} \end{bmatrix}_{\omega=0,\beta=0}$$

and evaluate the eigenvalues, $\lambda_{1,2}$, of $A\left(\bar{V}_s, 0\right)$. We seek $\bar{V}_s$ such that $\mathrm{Re}\lambda = 0$. In this way, we find the critical speed

$$V_s^* = \sqrt{\frac{2\left(a+b\right)^2 \kappa_f \kappa_r A_f A_r}{m\left(a\kappa_f A_f - b\kappa_r A_r\right)}} \tag{20}$$

This formula is well known, e.g., in [8]. Notice that a critical speed, $0 < V_s^* < \infty$, exists if and only if $a\kappa_f A_f - b\kappa_r A_r > 0$. If this relationship is satisfied, then there is a V_s^* such that the origin is stable if $V_s < V_s^*$ and unstable if $V_s > V_s^*$.

We can learn more about the nature of this instability by examining the equilibrium point structure for varying V_s with fixed steering angle $\delta = 0$. The equilibrium surface for an automobile with parameters as defined in Table 1 is shown in Fig. 2. Notice that the bifurcation point $V_s = V_s^* = 132, \omega = 0, \beta = 0$ corresponds to a pitchfork bifurcation.

The system can be linearized at the critical point, $(\omega, V_s, \beta) = (0, 132, 0)$, to yield

$$\frac{d}{dt}\begin{bmatrix} \Delta\omega \\ \Delta V_s \\ \Delta\beta \end{bmatrix} = \begin{bmatrix} -1.247 & 0 & -1.691 \\ 0 & 0 & 0 \\ -1.002 & 0 & -1.359 \end{bmatrix} \begin{bmatrix} \Delta\omega \\ \Delta V_s \\ \Delta\beta \end{bmatrix} + \begin{bmatrix} 0 \\ 0.0129 \\ 0 \end{bmatrix} \Delta F_d$$

$$y = \begin{bmatrix} 0 & 1 & 0 \end{bmatrix} \begin{bmatrix} \Delta\omega \\ \Delta V_s \\ \Delta\beta \end{bmatrix}$$

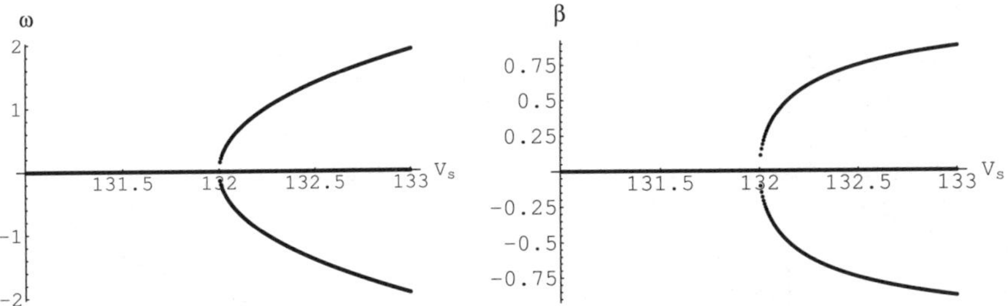

Fig.2. This figure shows the automobile equilibrium points for various speeds with steering angle fixed, $\delta = 0$. The critical speed is 132 fps.

It is easy to confirm that the system has a mode with eigenvalue, $\lambda = 0$ that is both uncontrollable and unobservable.

We also compute the controllability distributions and observability codistributions to find:

$$\dim \Delta_{L_0}(x) = 3, \quad \dim \Delta_{L_0}(x^*) = 1$$

and

$$\dim \Omega_L(x) = 3, \quad \dim \Omega_L(x^*) = 1$$

In each case we give first the generic rank of the distribution or codistribution and then the rank of the distribution or codistribution evaluated at the bifurcation point x^*. Thus, it is seen that both the controllability distribution and the observability codistribution are singular at the bifurcation point. The system may be (nonlinearly) controllable/observable at the bifurcation point, but we need to go further to establish this.

4.2 Cornering stability

Let us consider the behavior of a vehicle travelling with constant speed $\bar{V}_s$ and constant angular velocity $\bar{\omega}$. In view of Eq. (19), equilibrium points satisfy the algebraic equation

$$0 = F\left(\bar{\omega}, \bar{V}_s, \beta, \delta, F_d\right) \tag{22}$$

A typical equilibrium surface is shown in Fig. 3. The figure illustrates equilibria corresponding to constant speed and varying angular velocity. Notice that for angular velocity near zero there are three equilibrium points. The central branch consists of stable equilibria (at least for small $\bar{\omega}$). The other two are unstable. In this example, the sideslip angle, β, decreases with increasing $\bar{\omega}$. Eventually, the angular velocity approaches a critical value beyond which there is only one remaining equilibrium point and it is unstable. At the critical point, two equilibrium points merge and the disappear.

Unlike the previous example, we do not have a closed form equation to identify the bifurcation point. Instead, we use the NRS procedure. To do so, we set $\bar{V}_s = 100fps$ and treat $\bar{\omega}$ as a parameter. We find that the bifurcation occurs at:

$$(\omega, V_s, \beta, F_d, \delta) = (0.859899, 100, 0.0337081, 0.233148, 0.0615544)$$

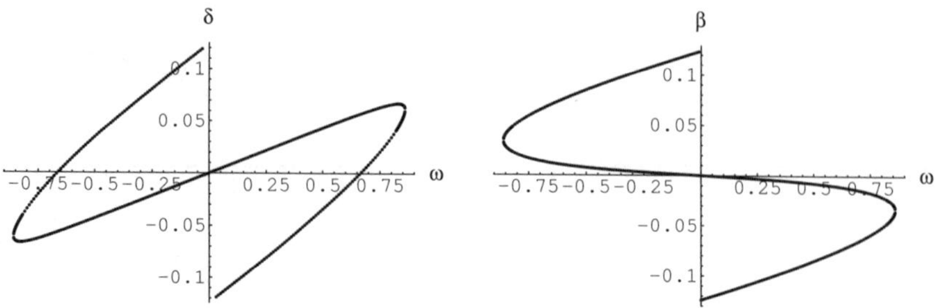

Fig.3. A typical equilibrium surface. This figure shows the principle component of the equilibrium manifold with $V_s = 100fps$. The surface characteristics vary considerably with tire parameters

The linearization at the bifurcation point is [2]

$$\frac{d}{dt}\begin{bmatrix}\omega\\ V_s\\ \beta\end{bmatrix} = \begin{bmatrix}-4.242 & 0.03647 & -86.35\\ 3.477 & -0.02991 & 157.9\\ -1.198 & -0.006895 & -4.115\end{bmatrix}\begin{bmatrix}\omega\\ V_s\\ \beta\end{bmatrix} + \begin{bmatrix}0 & 84.79\\ 25.74 & -122.1\\ 0.008681 & 4.054\end{bmatrix}\begin{bmatrix}F_d\\ \delta\end{bmatrix}$$

$$y = \begin{bmatrix}0\ 1\ 0\\ 1\ 0\ 0\end{bmatrix}\begin{bmatrix}\omega\\ V_s\\ \beta\end{bmatrix}$$

Now, it is easy to confirm that the linearized system is both observable and controllable, it has well defined relative degree, and it has a transmission zero at the origin.

In this case the bifurcation may be viewed as a classical saddle-node bifurcation in the (nonlinear) zero dynamics. The zero dynamics are well defined for all values of $\bar{\omega}$ on a neighborhood of its bifurcation value. They constitute a dynamical system (as opposed to a control system). Indeed, in the present case the one-dimensional zero dynamics are locally described by the differential equation:

$$\begin{aligned}\dot{\xi}_1 &= \left(-0.9944\Delta - 0.4808\Delta^2 - 0.0991\Delta^3\right)\\ &\quad + \left(22.21\Delta + 6.843\Delta^2\right)\xi_1 - (250.8 - 155.9\Delta)\,\xi_1^2 + 1165\xi_1^3\end{aligned} \tag{24}$$

Here, ξ_1 is the single zero dynamics state variable and Δ is a parameter that represents the deviation of $\bar{\omega}$ from its bifurcation value. Thus, when $\Delta = 0$, we have the zero dynamics at the bifurcation point. Fig. 4 shows the local equilibrium point structure of the zero dynamics.

[2] For readability we show the result with four significant figures. However, it is necessary to use the full numerical precision available with the Windows machine that was employed.

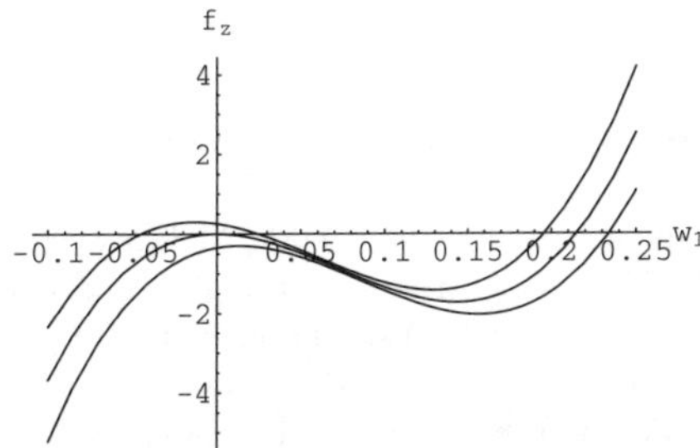

Fig.4. The right hand side of Eq. (24), $f_z(\xi_1, \Delta)$, is plotted for three different values of Δ, $\Delta = -0.3, 0.0, 0.3$. We see three equilibria for $\Delta = -0.3$, two for $\Delta = 0.0$, and one for $\Delta = 0.3$. Equilibria associated with negative slopes are stable and with positive slopes are unstable. Once again, $V_s = 100fps$.

5 Concluding Remarks

We have emphasized the importance of static bifurcations in the feedback regulation of nonlinear systems. Several practical situations ranging from compressor stall, automobile directional stability, aircraft stall, and automobile cornering stability have been described to illustrate the ubiquitous occurrence of these bifurcations. The relationship between static bifurcation and degeneracies in the linearized system zero dynamics has been described. The various types of zero dynamics defects have been connected to realistic situations. We argue that understanding the underlying cause of the bifurcation can suggest remedies.

We have noted that since these bifurcations are inextricably linked to zero dynamics defects, the most obvious approach to eliminating them is to change the system input-output structure. However, we have also indicated that when linear controllability/observability issues are involved, exploiting nonlinear controllability/observability around bifurcation points might afford other opportunities for remedy.

The automobile has been used as a vehicle to demonstrate the detailed computations. This example is useful because most readers will have sufficient experience with this system to appreciate the results. Also, the automotive industry has been implementing control devices intended to deal with bifurcation behavior – although that term is not often used.

Our computations integrate symbolic and numerical methods. As a result we are able to work efficiently with relatively complex models. Symbolic tools are used to assemble models, linearize them, implement nonlinear control computations, implement state transformations, assemble C-code for numerical implementation of NRS computations, and assemble C-code for simulation.

Acknowledgement.
The first two authors would like to acknowledge the support of this research by the NASA Langley Aeronautical Research Center under contract number NAG-1-01118.

References

1. Ackermann, J. (1997) Robust control prevents car skidding. IEEE Control Systems Magazine, 23–31
2. Antsaklis, P. J., Michel, A. N. (1997) Linear Systems. New York: McGraw-Hill
3. Banaszuk, A., Krener, A. J. (1998) Design of controllers for mg3 compressor models with general characteristics using graph backstepping. Automatica, **35**:1343–1368
4. Berg, J., Kwatny, H. G., (1995) A canonical parameterization of the kronecker form of a matrix pencil. Automatica **31**(5):669–680
5. Berg, J., Kwatny, H. G. (1996) Linear siso systems with extremely sensitive zero structure. IEEE Trans. Auto. Contr., **41**(7):1037–1040
6. Chen, X., Gu, G., Martin, P., Zhou K. (1998) Rotating stall control via bifurcation stabilization. Automatica **34**(4):437–443
7. Chow, S. N., Hale, J. K. (1982) Methods of Bifurcation Theory. New York: Springer-Verlag
8. Doebelin, E. O. (1980) System Modeling and Response: Theoretical and experimental approaches. New York: Wiley
9. Haddad, W. M., Leonessa, A., Chellaboina, V.-S., Fausz, J. L. (1999) Nonlinear robust disturbance rejection controllers for rotating stall and surge in axial flow compressors. IEEE Trans. Contr. Syst. Tech., **7**(3):391–398
10. Isidori, A. (1995) Nonlinear Control Systems. 3rd ed. London: Springer-Verlag
11. Krstic, M., Fontaine, D., Kokotovic, P. V., Paduano, J. D. (1998), Useful nonlinearities and global stabilization of bifurcations in a model of jet engine surge and stall. IEEE Trans. Auto. Contr., **43**(12):1739–1745
12. Kwatny, H. G., Bennett, W. H., Berg, J. M. (1991) Regulation of relaxed stability aircraft. IEEE Trans. Auto. Contr., **36**(11):1325–1323
13. Kwatny, H. G., Blankenship, G. L. (2000) Nonlinear Control and Analytical Mechanics: A Computational Approach. Control Engineering. Boston: Birkhäuser
14. Kwatny, H. G., Chang, B. C. (2002) Observer design tools for nonlinear flight regimes. In Proc. Amer. Contr. Conf., Anchorage, USA, 4203–4208
15. Kwatny, H. G., Fischl, R. F., Nwankpa, C. (1995) Local bifurcations in power systems: Theory, computation and application. Proceedings of the IEEE, **83**(11):1456–1483
16. Liaw, D. C., Abed, E. H. (1996) Active control of compressor stall inception: a bifurcation theoretic approach. Automatica **32**(1):109–115
17. Ono, E., Hosoe, S., Tuan, H. D., Doi, S. (1998) Bifurcation in vehicle dynamics and robust front wheel steering. IEEE Trans. Contr. Syst. Tech., **6**(3):412–420
18. Seydel, R. (1979) Numerical computation of branch points in nonlinear equations. Numerische Mathematik, **33**:339–352

A Automobile Model

In this paper we make several simplifying assumptions for the sake of efficiency of presentation. These include neglect of the tire mass and inertia about the axis, the assumption that castor and camber are zero, and the use of a relatively simple tire model Eq. (17).

The automobile dynamical equations (6b), after application of the velocity transformation, equations (18) take the form

$$M\left(V_s,\beta\right)\frac{d}{dt}\begin{bmatrix}\omega\\ V_s\\ \beta\end{bmatrix}=f\left(\omega,V_s,\beta,\delta,F_d\right) \tag{25}$$

where

$$\begin{aligned}f_1 = {} & \arctan\left[\left(\left(a\,\omega\,\cos[\delta]+V_s\sin[\beta-\delta]-\tfrac{1}{2}\,\lambda\,\omega\,\sin[\delta]\right)A_f\right)/\left(V_s\cos[\beta-\delta]+\tfrac{1}{2}\,\lambda\,\omega\,\cos[\delta]+\right.\right.\\ & \qquad \left.\left.a\,\omega\,\sin[\delta]\right)\right]\left(-a\,\cos[\delta]+\tfrac{1}{2}\,\lambda\,\sin[\delta]\right)\kappa_f-\\ & \arctan\left[\left(\left(a\,\omega\,\cos[\delta]+V_s\sin[\beta-\delta]+\tfrac{1}{2}\,\lambda\,\omega\,\sin[\delta]\right)A_f\right)/\right.\\ & \qquad \left.\left(V_s\cos[\beta-\delta]-\tfrac{1}{2}\,\lambda\,\omega\,\cos[\delta]+a\,\omega\,\sin[\delta]\right)\right]\\ & \left(a\,\cos[\delta]+\tfrac{1}{2}\,\lambda\,\sin[\delta]\right)\kappa_f+b\left(\arctan\left[\frac{\left(b\,\omega-V_s\sin[\beta]\right)A_r}{\frac{\lambda\,\omega}{2}-V_s\cos[\beta]}\right]+\arctan\left[\frac{\left(-b\,\omega+V_s\sin[\beta]\right)A_r}{\frac{\lambda\,\omega}{2}+V_s\cos[\beta]}\right]\right)\kappa_r\end{aligned}$$

$$\begin{aligned}f_2 = {} & 2\,F_d+V_s m\omega\sin[\beta]+\left(\arctan\left[\left(\left(a\,\omega\,\cos[\delta]+V_s\sin[\beta-\delta]-\tfrac{1}{2}\,\lambda\,\omega\,\sin[\delta]\right)A_f\right)/\right.\right.\\ & \qquad \left.\left(V_s\cos[\beta-\delta]+\tfrac{1}{2}\,\lambda\,\omega\,\cos[\delta]+a\,\omega\,\sin[\delta]\right)\right]+\\ & \quad \arctan\left[\left(\left(a\,\omega\,\cos[\delta]+V_s\sin[\beta-\delta]+\tfrac{1}{2}\,\lambda\,\omega\,\sin[\delta]\right)A_f\right)/\left(V_s\cos[\beta-\delta]-\right.\right.\\ & \qquad \left.\left.\left.\tfrac{1}{2}\,\lambda\,\omega\,\cos[\delta]+a\,\omega\,\sin[\delta]\right)\right]\right)\sin[\delta]\,\kappa_f\end{aligned}$$

$$\begin{aligned}f_3 = {} & -V_s\,m\,\omega\,\cos[\beta]-\left(\arctan\left[\left(\left(a\,\omega\,\cos[\delta]+V_s\sin[\beta-\delta]-\tfrac{1}{2}\,\lambda\,\omega\,\sin[\delta]\right)A_f\right)/\right.\right.\\ & \qquad \left.\left(V_s\cos[\beta-\delta]+\tfrac{1}{2}\,\lambda\,\omega\,\cos[\delta]+a\,\omega\,\sin[\delta]\right)\right]+\\ & \quad \arctan\left[\left(\left(a\,\omega\,\cos[\delta]+V_s\sin[\beta-\delta]+\tfrac{1}{2}\,\lambda\,\omega\,\sin[\delta]\right)A_f\right)/\left(V_s\cos[\beta-\delta]-\right.\right.\\ & \qquad \left.\left.\left.\tfrac{1}{2}\,\lambda\,\omega\,\cos[\delta]+a\,\omega\,\sin[\delta]\right)\right]\right)\cos[\delta]\,\kappa_f-\\ & \quad \left(\arctan\left[\frac{\left(b\,\omega-V_s\sin[\beta]\right)A_r}{\frac{\lambda\,\omega}{2}-V_s\cos[\beta]}\right]+\arctan\left[\frac{\left(-b\,\omega+V_s\sin[\beta]\right)A_r}{\frac{\lambda\,\omega}{2}+V_s\cos[\beta]}\right]\right)\kappa_r\end{aligned}$$

$$M=\begin{pmatrix}J & 0 & 0\\ 0 & m\,\cos[\beta] & -V_s\,m\,\sin[\beta]\\ 0 & m\,\sin[\beta] & V_s m\cos[\beta]\end{pmatrix}$$

Bifurcation and Chaos in Simple Nonlinear Feedback Control Systems

Wallace K. S. Tang

Department of Electronic Engineering
City University of Hong Kong, P. R. China
eekstang@cityu.edu.hk

Abstract. This chapter is to give the bifurcation analysis and the verification of chaotic dynamics in nonlinear feedback control systems based on numerical continuation techniques and the Shil'nikov theorem. The studied system is a low-order linear autonomous system with a simple nonlinear controller of the form $g(\nu) = \nu|\nu|$. The chaotic dynamics generated in this kind of systems are demonstrated by both computer simulation and circuitry implementation.

1 Introduction

Chaos is usually considered as an unfavourable phenomenon. However, it is very useful under certain situations and there is great interest in introducing chaotic behaviour in various applications. For example, human brain activities [13], network modelling [5], liquid mixing [11], secure communication [7] and so on.

This provides a strong motivation for the current research on a new task of making a non-chaotic dynamical system chaotic, which is known as "anticontrol of chaos" or "chaotification." Anticontrolling of chaos with the use of state-feedback [1] and small control perturbations or signals [14,15] has been proposed for discrete time systems. However, it would be a real challenge for anticontrolling of chaos in a non-chaotic continuous time systems, since a mathematical proof of the existence of chaos is usually very difficult if not impossible to be conducted.

In this chapter, by applying the local and global bifurcation, the ability of using nonlinear feedback controller for generating chaos in a non-chaotic linear autonomous continuous system is established. The designed nonlinear controller is with the form $g(\nu) = \nu|\nu|$. By means of one-parameter continuation on the periodic orbits and the homoclinic bifurcation analysis, the existence of chaos in the controlled system is verified.

The organization of this chapter is as follows: Some basic continuation techniques are reviewed in Sect. 2. Since our verification of the existence of chaos is based on the Shil'nikov theorem, the characteristic of the homoclinic bifurcation and the resonant conditions are discussed in Sect. 3. In Sects. 4 and 5, the second and third order linear systems are considered, respectively.

G. Chen, D.J. Hill, and X. Yu (Eds.): Bifurcation Control, LNCIS 293, pp. 83–98, 2003.

With the nonlinear feedback controllers, chaotic attractors can be generated in these low order systems. The bifurcation and chaotic behaviour of the controlled systems are studied in details. Conclusions are finally drawn in Sect. 6.

2 Numerical Continuation

Our study on the existence of chaotic attractors is based on the theories of local and global bifurcation. The major tool is the numerical continuation techniques [2,3] that are implemented in the softwares AUTO'97 [4] and CONTENTS [8]. The continuation methods translate the bifurcation analysis of equilibria and periodic cycles into the solution of an implicit algebraic equation for computation. They have been used in the study of many nonlinear differential equations and chaotic systems [4,9,10]. The advantages [9] for using continuation methods are listed as below:

- Both stable and unstable solutions are obtainable. It is essential since the unstable periodic orbits, especially the saddle-focus is critical for our verification of the existence of chaos.
- The results are independent on the initial conditions and the choice of the Poincaré section, since the problem has been converted in a numerical one.
- Hysteretic phenomena can be detected by forward and backward continuation.

Some of the basic functions are described briefly below. More detailed explanations can be referred to [2,3].

2.1 Continuation on regular solutions

Consider the following equation

$$G(x) = 0\,, \tag{1}$$

where $x = (u, \lambda)$ with $u \in R^n$; $\lambda \in R$ and $G : R^{n+1} \rightarrow R^n$.

A solution x_0 of $G(x) = 0$ is regular if the Jacobian matrix $G_x^0 = G_x(x_0)$ has maximal rank. Then, near x_0, there exists a unique one-dimensional continuum of solutions $x(s)$ with $x(0) = x_0$.

Assuming that λ is the continuation parameter and (u_0, λ_0) is a solution of $G(u, \lambda) = 0$ with a direction vector u_0', where $u' = \frac{du}{d\lambda}$, a new solution (u_1, λ_1) can be obtained by Newton's method and the convergence can be assured if the incremental step is small enough.

2.2 Continuation on periodic solutions

Consider the ordinary differential equation

$$\dot{x}(t) = f(x(t), \lambda), \tag{2}$$

where $\dot{x} = \frac{dx}{dt}$; $x, f \in R^n$ and $\lambda \in R$.

By transforming $t \to \frac{t}{T}$, we have

$$\dot{x}(t) = Tf(x(t), \lambda), \tag{3}$$

where $x, f \in R^n$ and $T, \lambda \in R$.

Assuming (x_k, T_k, λ_k) is known, $(x_{k+1}, T_{k+1}, \lambda_{k+1})$ can be found if

- a solution of period 1, $x(0) = x(1)$, is computed.
- a phase condition is computed as

$$\int_0^1 x(\tau)^* \dot{x}_k(\tau) d\tau = 0, \tag{4}$$

where * is the matrix transpose operator.
- pseudo-arclength continuation is used to trace out a branch of periodic solutions:

$$\int_0^1 (x(\tau) - x_k(\tau))^* x_k'(\tau) d\tau + (T - T_k)T_k' + (\lambda - \lambda_k)\lambda_k' = \Delta s, \tag{5}$$

where Δs is the small distance from $(x_{k+1}, T_{k+1}, \lambda_{k+1})$ to (x_k, T_k, λ_k) in a direction perpendicular to the vector (x_k', T_k', λ_k').

3 Homoclinic Bifurcation and the Resonant Conditions

The computation of the homoclinic orbit can be performed by continuing a periodic orbit until a very large T is obtained. The appearance of the periodic orbits with the continuation parameter λ from a homoclinic loop can be classified in two distinct cases [6]:

- *Saddle loop and stable saddle-focus loop*: Periodic orbit bifurcates from the homoclinic loop, either inward or outward with the continuation parameter.
- *Unstable saddle-focus loop*: Many periodic orbits bifurcate in both directions from the critical parameter value λ_h, giving the homoclinic orbit. The branch "wiggles" around such a value.

For the discussion of homoclinic bifurcations, the information about the eigenvalues is important. The controlled system is assumed to be a third order dynamical system. Consider the characteristic polynomial of the system Jacobian as follows:

$$P(\lambda) = \lambda^3 + p_1\lambda^2 + p_2\lambda + p_3 \,. \tag{6}$$

Let $\lambda_1, \lambda_2, \lambda_3$ be the eigenvalues at the equilibrium point of the system, without loss of generality, λ_1 is assumed to be real. Three resonant conditions are important to characterize these eigenvalues at the equilibrium:

$$\begin{aligned} \text{neutral saddle: } \sigma_1 &= \lambda_1 + \lambda_2 = 0 \\ \text{critical saddle: } \sigma_2 &= \lambda_2 - \lambda_3 = 0 \\ \text{neutral saddle-focus: } \sigma_3 &= \lambda_2 + \lambda_3 + 2\lambda_1 = 0 \end{aligned}$$

- $\sigma_1 = 0$ is the condition for a neutral saddle case if λ_1 and λ_2 are real but opposite in sign. It can also be used for testing the occurrence of Andronov-Hopf bifurcation if $\lambda_1 = \lambda_2 = \pm iw$.
- $\sigma_2 = 0$ gives the boundary condition for determining whether the equilibrium is saddle or saddle-focus. $\lambda_{2,3}$ are real or complex conjugate, depending on the value of σ_2.
- $\sigma_3 = 0$ gives the condition on the boundary of Shil'nikov inequality if λ_1 is real and $\lambda_{2,3}$ are complex conjugate.

In the following two sections, the dynamics of the low-order nonlinear feedback control systems are to be studied. The nonlinear feedback controller of the form $\nu|\nu|$ is designed and adopted. The existence of chaos will be verified by the aforementioned bifurcation techniques.

4 Second-order System

Consider a two-dimensional linear autonomous system with a negative state feedback control $-u$,

$$\begin{cases} \dot{x} = ax + by \,, \\ \dot{y} = cx + dy - u \,, \end{cases} \tag{7}$$

where a, b, c, and d are constants. The nonlinear feedback controller u is expressed as:

$$\begin{cases} u = \dot{z} \,, \\ \dot{z} = \alpha(y - z) + \beta(y - z)|y - z| \,, \end{cases} \tag{8}$$

where α and β are constant control gains, and the only nonlinearity is a piecewise quadratic function of the form $g(\nu) = \nu|\nu|$ [16].

4.1 Bifurcation analysis on equilibria

In this section, the bifurcations of the equilibria of Eqs. (7)–(8) are studied. To simplify the analysis, we first normalize some coefficients so that $b = -1$, $c = 1$, and d becomes $-d$, thereby obtaining

$$\begin{cases} \dot{x} = ax - y\,, \\ \dot{y} = x - dy - u(y,z)\,, \\ \dot{z} = u(y,z)\,, \end{cases} \tag{9}$$

where $u(y,z) = \alpha(y-z) + \beta(y-z)|y-z|$. This equivalent form of the controlled system provides some convenience for the following analysis.

It can be easily seen that Eq. (9) is odd-symmetric and is invariant under the mapping $(x,y,z) \mapsto (-x,-y,-z)$. In what follows, we assume that $\alpha = -0.053$, $\beta = 0.062$, $a > 0$, $d > 0$ and $ad < 1$.

There are three equilibria in Eq. (9):

$$P^0 = (0,0,0) \quad \text{and} \quad P^{\pm} = (0,0,q^{\pm})\,,$$

where, in the present selected numerical data, $q^{\pm} = \pm\,\alpha/\beta = \pm\,0.8548$.

The system Jacobian of Eq. (9) is

$$J = \begin{bmatrix} a & -1 & 0 \\ 1 & -d-u_y & u_y \\ 0 & u_y & u_y \end{bmatrix}, \tag{10}$$

where $u_y \equiv \frac{\partial u}{\partial y} = \alpha + 2\beta|y-z|$. $u_y = \alpha$ and $u_y = -\alpha$ at P^0 and $P^{\pm}$, respectively.

Consider its characteristic polynomial:

$$P(\lambda) = \lambda^3 + p_1\lambda^2 + p_2\lambda + p_3\,, \tag{11}$$

where

$$\begin{aligned} p_1 &= d + 2u_y - a\,, \\ p_2 &= 1 - ad + du_y - 2au_y\,, \\ p_3 &= u_y(1-ad)\,. \end{aligned} \tag{12}$$

Clearly, at P^0, $p_3 < 0$ while $p_3 > 0$ at $P^{\pm}$. Since the signs of p_3, are different at P^0 and at $P^{\pm}$, the topological types of P^0 and $P^{\pm}$ are expected to be different. With a positive p_3, one of the eigenvalues of Eq. (11) must be real and negative at $P^{\pm}$. On the other hand, one of the eigenvalues is real and positive at P^0.

By considering the resonant conditions, the bifurcation diagram for the equilibria, P^0 and $P^{\pm}$, can be obtained, as shown in Fig. 1.

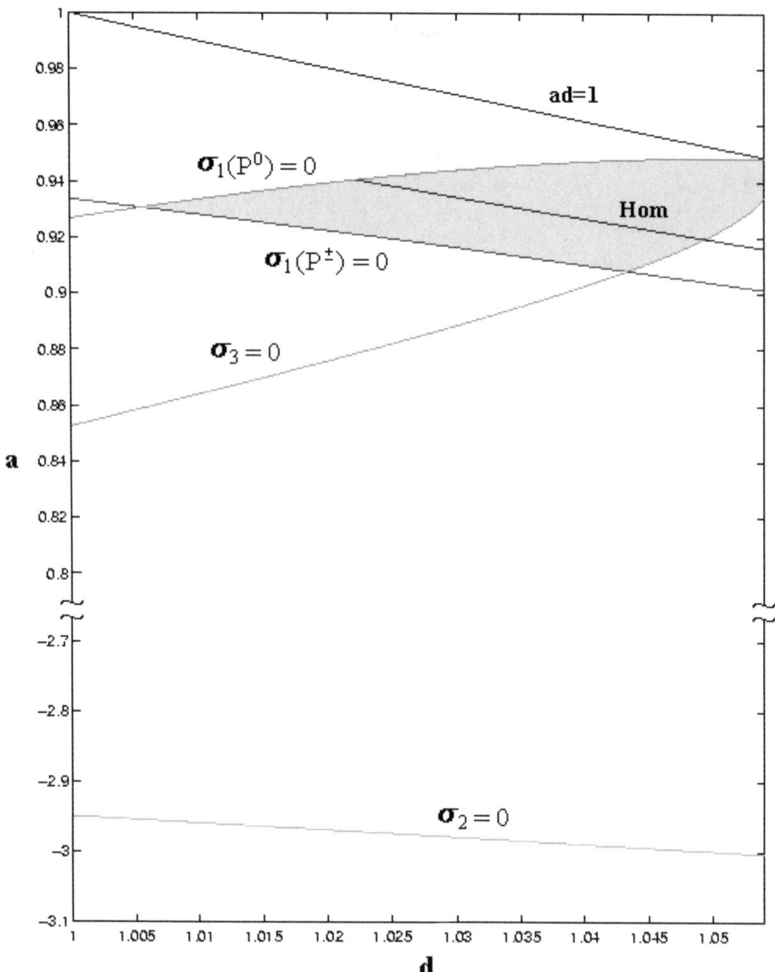

Fig.1. two-dimensional bifurcation diagram for P^0 and $P^{\pm}$, with $\alpha = -0.053$, $\beta = 0.062$.

It is known that an Andronov-Hopf bifurcation occurs when there are two purely imaginary complex conjugate eigenvalues. In our case, Hopf bifurcation occurs at P^0 and $P^{\pm}$. There are two purely imaginary complex conjugate eigenvalues along the Hopf curves in both cases. The curves are denoted as $\sigma_1(P^0) = 0$ and $\sigma_1(P^{\pm}) = 0$, respectively, in Fig. 1.

The equilibrium P^0 is unstable $(\lambda_1, Re\lambda_{2,3} > 0)$ when (a, d) is located above the curve $\sigma_1(P^0) = 0$. Below the curve $\sigma_1(P^0) = 0$, the equilibrium P^0 is considered as saddle-focus $(\lambda_1, Re\lambda_{2,3} < 0)$ since the values of (a, d) are within the regions above the curve $\sigma = 0$. Moreover, the equilibria $P^{\pm}$ are stable $(\lambda_1 < 0, Re\lambda_{2,3} < 0)$ if (a, d) is located below the curve $\sigma_1(P^{\pm}) = 0$.

The principal homoclinic bifurcation curve is computed by the continuation techniques. The curve is also depicted in Fig. 1, as denoted by 'Hom'. Only those homoclinic loops below the curve $\sigma_1(P^0) = 0$ are shown. When the homoclinic loop is above the resonant curves $\sigma_2 = 0$ and $\sigma_3 = 0$ (i.e. the values of (a, d) are within the shaded area), the equilibrium P^0 is saddle-focus and $\lambda_1 > |Re\lambda_{2,3}|$. Therefore, it can be concluded that chaotic attractor exists in the sense of Shil'nikov. For those homoclinic loops outside the shaded area, we have $\lambda_1 < |Re\lambda_{2,3}|$.

For example, considering the case with $d = 1.04$, homoclinic loop occurs when $a = 0.926721$. The eigenvalues at P^0 are $\lambda_1 = 0.0240$, $\lambda_{2,3} = -0.0156 \pm 0.2825i$. For the case with $d = 1.05$, the homoclinic loop occurs when $a = 0.919155$. The eigenvalues at P^0 are $\lambda_1 = 0.0238$, $\lambda_{2,3} = -0.0243 \pm 0.2779i$. The shape of the homoclinic orbit is depicted in Fig. 2. Since the system is odd-symmetric, two orbits are mapped onto each other under the symmetry.

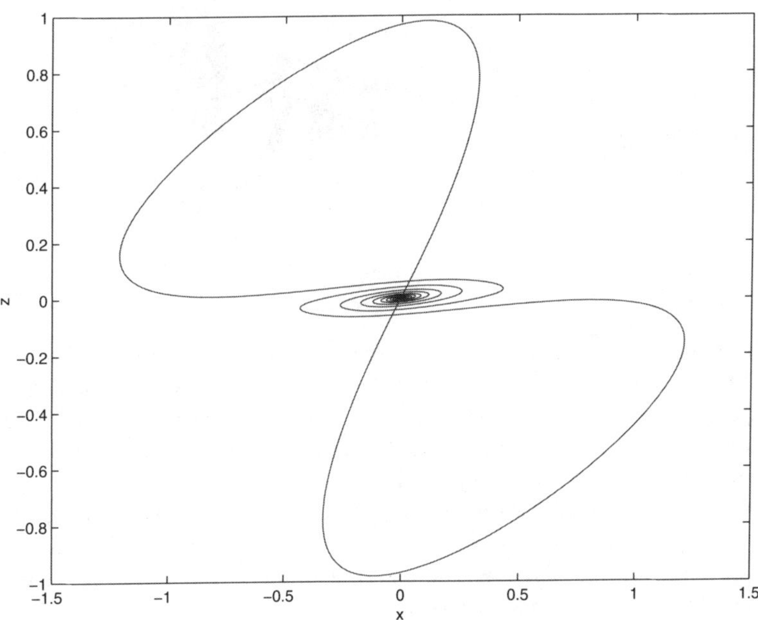

Fig.2. Homoclinic loop at P^0, with $a = 0.919155$, $d = 1.05$.

The bifurcation diagram for $d = 1.04$ with increasing a is depicted in Fig. 3. The phase portraits in Fig. 4, with increasing a, are used to visualize the details of the bifurcation. Initially, only the equilibria $P^{\pm}$ are stable and the system is rested at P^- (or P^+) with a small a. With the increase of the value of parameter a, the stability margin of these equilibria begins to deteriorate at the Hopf point ($a = 0.91023$) and a small periodic orbit appears. As a increases further, the system generates a cascaded period-doubling bifurcation for asymmetric periodic orbits (period-1 and period-2 limit cycles are shown in Fig. 4 (a) and (b), respectively).

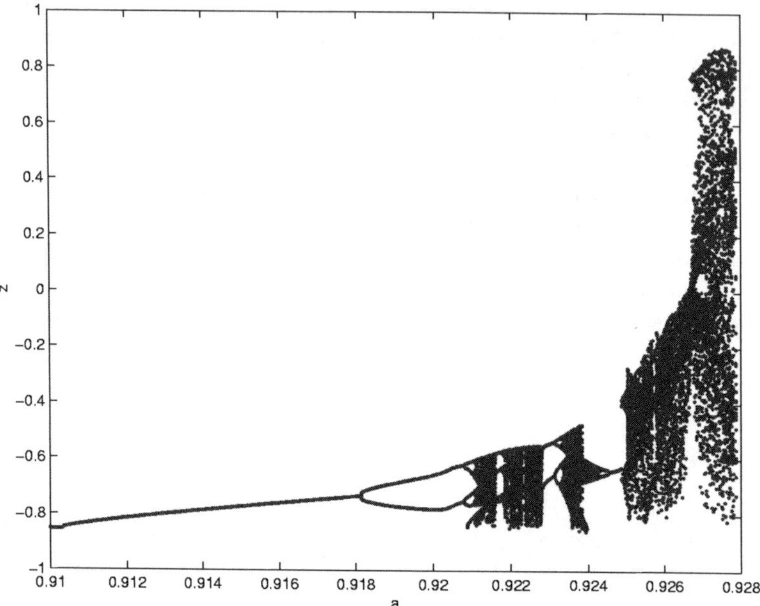

Fig.3. Bifurcation diagram of a, with $d = 1.04$.

At the end of the period-doubling cascade, two asymmetric chaotic attractors emerge and can be observed. For clarity, only the lower one is shown in Fig. 4 (c). These two asymmetric attractors are moving closer and closer toward each other, and eventually "glued" together, giving rise to a double-scroll attractor, as displayed in Fig. 4 (d).

The chaotic behavior of the system Eq. (9) can be realized by an electronic circuit [16]. Figures 5(a)–(d) shows the measured phase portraits in the x–z plane of different modes, showing the existing limit cycles, spiral attractor, and double-scroll attractor, respectively.

5 Third-order System

Consider a general third-order linear autonomous system expressed in its control form of

$$\begin{cases} \dot{x} = y\,, \\ \dot{y} = z\,, \\ \dot{z} = ax + by + cz\,, \end{cases} \tag{13}$$

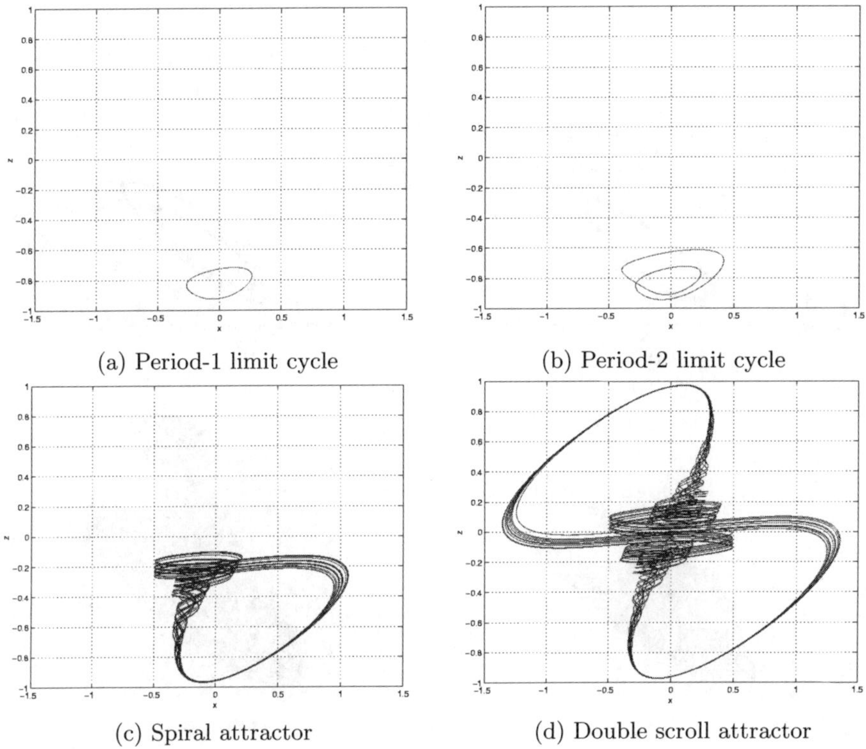

(a) Period-1 limit cycle (b) Period-2 limit cycle

(c) Spiral attractor (d) Double scroll attractor

Fig.4. Phase portraits, with (a) $a = 0.915$; (b) $a = 0.919$; (c) $a = 0.926$; (d) $a = 0.927$; with initial conditions $x(0) = y(0) = 0.1; z(0) = 0$.

where a, b, c are constants.

With the assumption that $a > 0$, $b < 0$ and $c < 0$, a nonlinear state feedback controller $g(x)$ [12] is applied to the system (13):

$$\begin{cases} \dot{x} = y\,, \\ \dot{y} = z\,, \\ \dot{z} = ax + by + cz + g(x)\,. \end{cases} \tag{14}$$

where $g(x) = \beta x|x|$ and $\beta < 0$ is a constant feedback gain. It should be noticed that the only nonlinearity is again in the form $g(\nu) = \nu|\nu|$. The block diagram of the controlled system is depicted in Fig. 6.

Similarly, Eq. (14) is odd-symmetric and invariant under the mapping $(x, y, z) \mapsto (-x, -y, -z)$. The three equilibria of Eq. (14) are:

$$P^0 = (0,0,0) \quad \text{and} \quad P^{\pm} = (q^{\pm}, 0, 0)\,,$$

where $q^{\pm} = \mp a/\beta$, a and β are in different signs. In our assumptions, $a > 0$ and $\beta < 0$ as a negative state feedback is adopted. If $a < 0$, the feedback controller can be modified as $\alpha x + \beta x|x|$ where $\alpha > |a| > 0$ and $\beta < 0$.

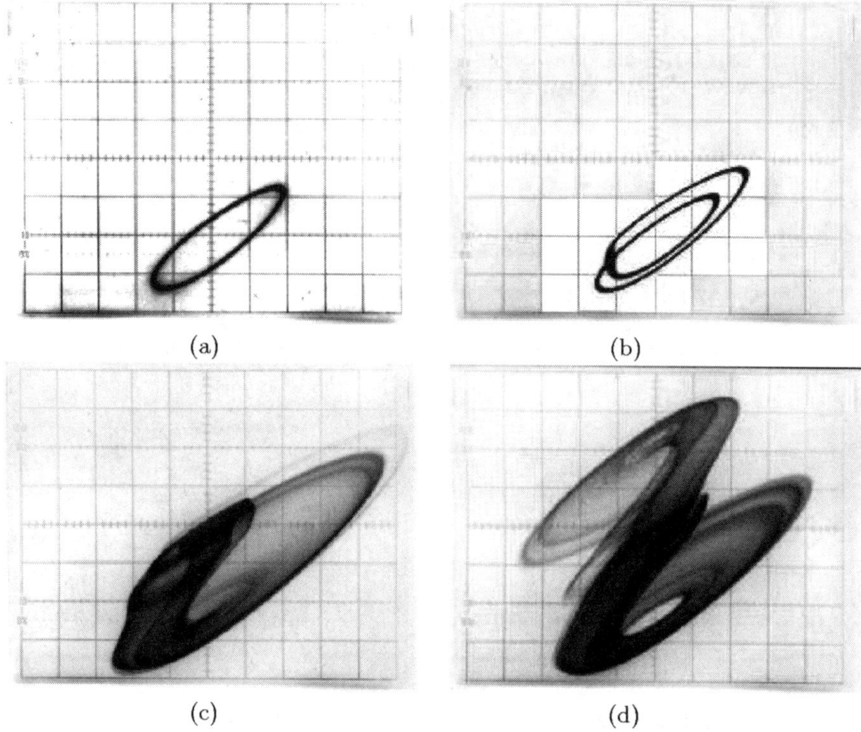

(a) (b) (c) (d)

Fig.5. (a) Period-1 limit cycle; (b) Period-2 limit cycle; (c) Spiral attractor; (d) Double-scroll attractor.

Consider the characteristic polynomial of the system Jacobian of Eq. (14):

$$P(\lambda) = \lambda^3 + p_1\lambda^2 + p_2\lambda + p_3 \,, \tag{15}$$

where

$$\begin{aligned} p_1 &= -c\,, \\ p_2 &= -b\,, \\ p_3 &= -a - 2\beta|x|\,, \end{aligned} \tag{16}$$

Assuming that $c = -1$ and $\beta = -1$, we can construct the bifurcation diagram of the equilibria, as shown in Fig. 7, using the same procedures described in previous section. The same notations as in Fig. 1 are used.

In this case, the equilibria $P^{\pm}$ undergo an Andronov-Hopf bifurcation on the curve $\sigma_1(P^{\pm}) = 0$ and $P^{\pm}$ are unstable above the curve, with one negative eigenvalue and two eigenvalues with positive real parts. They are stable if the values of (a, b) are below this curve.

For the equilibrium P^0, it is always unstable but it is a neutral saddle on the curve $\sigma_1(P^0) = 0$, having all the eigenvalues are real and $\lambda_1 = -\lambda_2$. Without loss of generality, it is assumed that $\lambda_1 > 0$ is real. From the diagram, it is

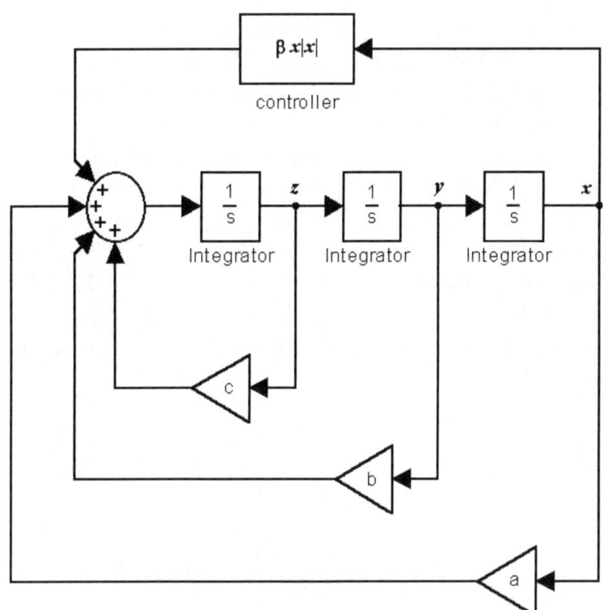

Fig.6. Third-order linear autonomous system with nonlinear feedback

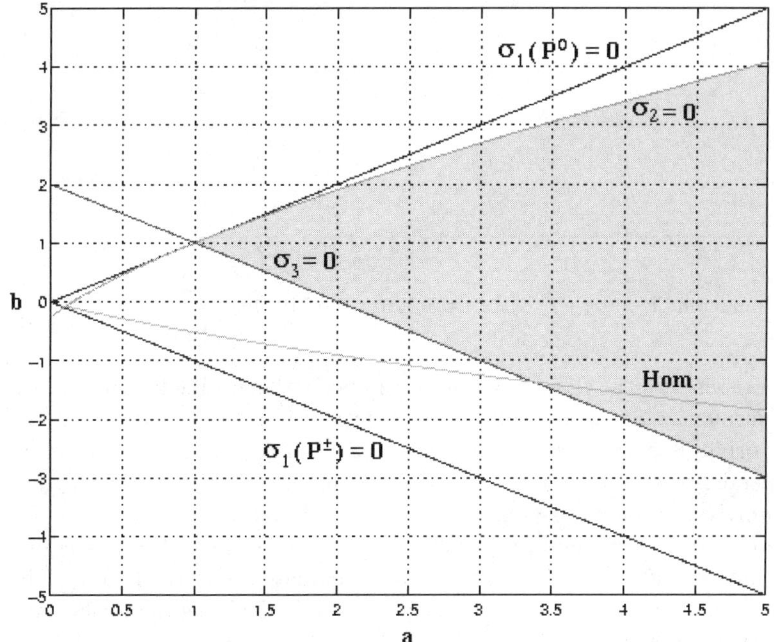

Fig.7. Bifurcation diagram for equilibria, with $c = -1$ and $\beta = -1$.

easily proved that the equilibrium P^0 is a saddle-focus ($\lambda_1 > 0, Re\ \lambda_{2,3} < 0$) and the eigenvalues satisfy the Shil'nikov criteria ($\lambda_1 > |Re\lambda_{2,3}|$), if the values of (a, b) are located in the shaded area. The principal homoclinic bifurcation curve is also computed and denoted by 'Hom.'

Consider the case with $a = 4$, the homoclinic loop occurs when $b = -1.56586$. The eigenvalues at P^0 are $\lambda_1 = 1.0636$, $\lambda_{2,3} = -1.0318 \pm 1.6420i$. Hence, there exists a homoclinic loop to the Shil'nikov saddle-focus equilibrium P^0. On the other hand, if $a = 1.5$, the homoclinic loop occurs when $b = -0.719149$. The eigenvalues at P^0 are $\lambda_1 = 0.7439$, $\lambda_{2,3} = -0.8719 \pm 1.1208i$. The shape of the homoclinic loop is depicted in Fig. 8. Since the system is odd-symmetric, two orbits are mapped onto each other under the symmetry.

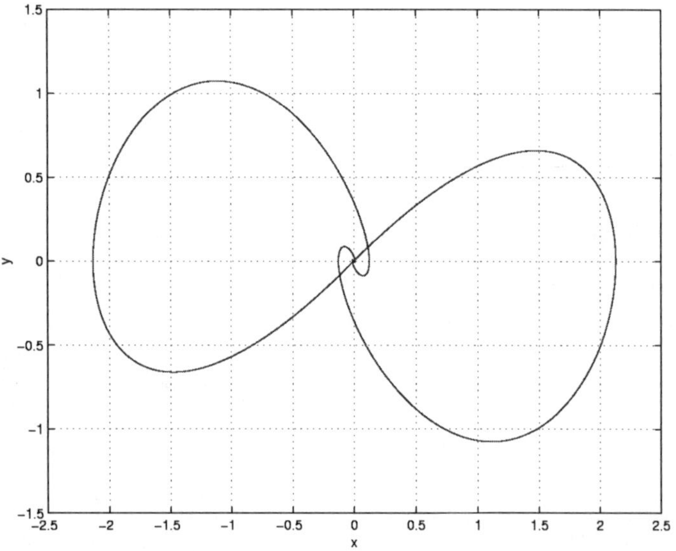

Fig.8. Homoclinic loop at P^0, with $a = 1.5$, $d = -0.719149$.

The existence of the chaotic attractor can be further verified with the help of the continuation on the periodic orbits. Consider $a = 4$, the continuation on periodic orbits is performed and the bifurcation diagram is shown in Fig. 9. The dependence of the period T on the parameter b is demonstrated. For clarity, only the principal asymmetric periodic orbits are depicted.

Referring to Fig. 9, the branch exhibits a growth of period in a wiggling manner, and eventually a homoclinic orbit occurs at the critical value. The wiggling behaviour of the branch indicates a Shil'nikov's saddle-focus type. The direct consequence of the wiggles on the path of periodic orbits is that several periodic orbits can exist simultaneously. Similar observation is obtained from the bifurcation diagram for $a = 4$ with increasing b as depicted in Fig. 10.

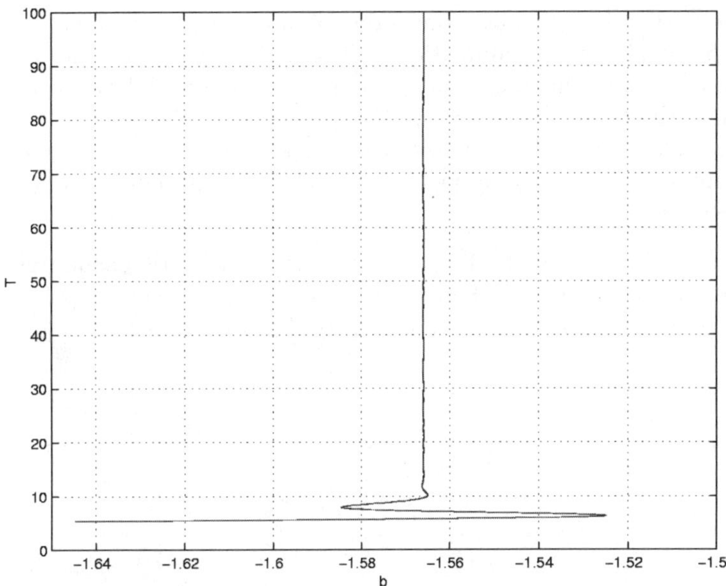

Fig.9. Bifurcation for periodic orbits on b

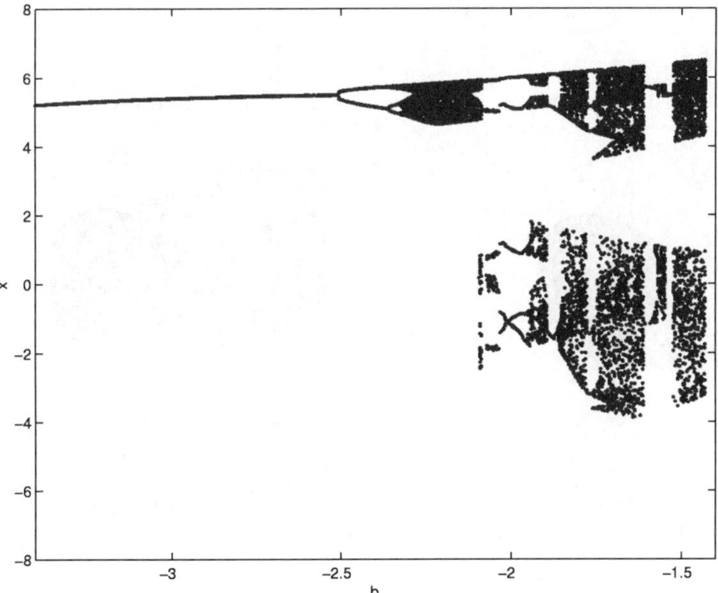

Fig.10. Bifurcation diagram of b, with $a = 4$.

The phase portraits, with increasing value of b, are depicted in Fig. 11 to visualize the details of the bifurcation. Initially, the system rests at P^+ (or P^-) with a small b, and the stability margin of these equilibria begins to deteriorate at the Hopf point ($b = -4$). As b increases further, the system generates a cascaded period-doubling bifurcation for asymmetric periodic orbits. An example of period-1 and period-2 limit cycles are shown in Fig. 11 (a) and (b), respectively.

At the end of the period-doubling cascade, two asymmetric chaotic attractors (one of them is shown in Fig. 11 (c)) emerge and "glued" together, giving rise to a double-scroll attractor, as displayed in Fig. 11 (d).

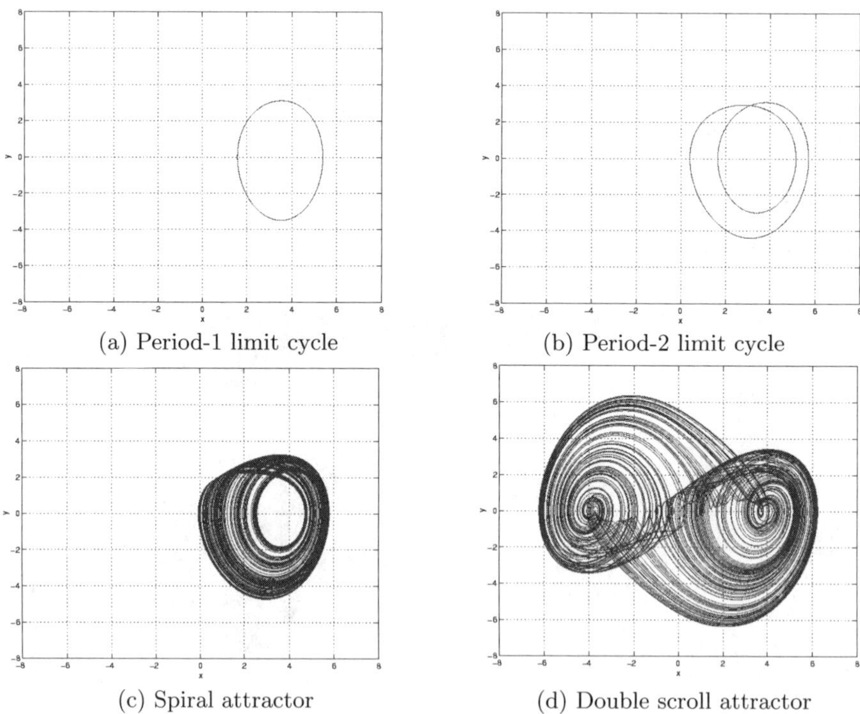

(a) Period-1 limit cycle (b) Period-2 limit cycle

(c) Spiral attractor (d) Double scroll attractor

Fig.11. Phase portraits, with (a) $b = -3$; (b) $b = -2.4$; (c) $b = -2.25$; (d) $b = -1.7$; with initial conditions $x(0) = 0.1; y(0) = 0.01; z(0) = 0.001$.

The chaotic behavior of the system Eq. (14) can also be realized by an electronic circuit [12]. The obtained phase portraits in the x–y for period-1, period-2 limit cycles, spiral attractor, and double-scroll attractor, are shown in Figs. 12 (a)–(d), respectively.

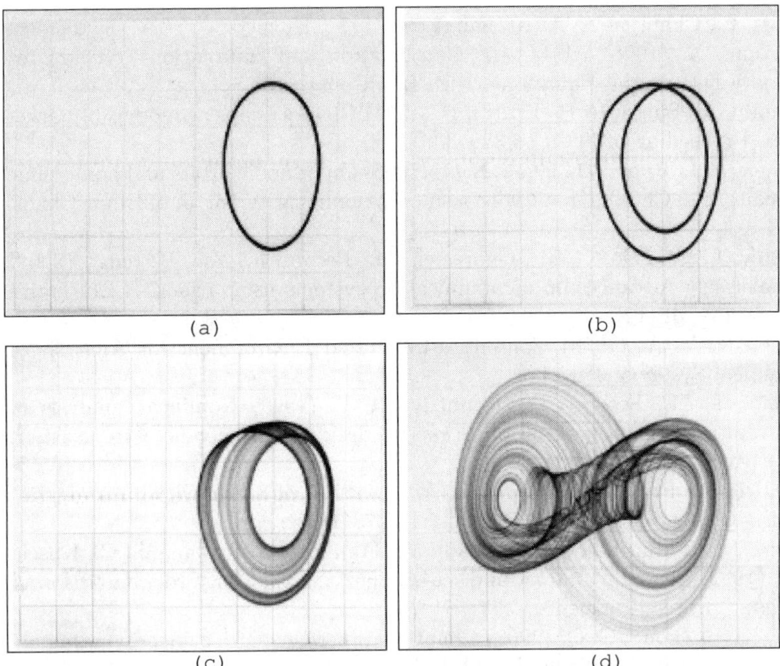

Fig.12. (a) Period-1 limit cycle; (b) Period-2 limit cycle; (c) Spiral attractor; (d) Double-scroll attractor.

6 Conclusions

In this chapter, the existence of chaotic attractors in low-order nonlinear feedback controlled continuous time system is verified by means of numerical continuation techniques and the Shil'nikov theorem. The only nonlinearity of the designed feedback controller is a piecewise-quadratic function in the form of $\nu|\nu|$. Bifurcation analysis has been carried out, and both the simulation and circuitry implementation have demonstrated the ability of the nonlinear feedback in generating chaos in such a system.

References

1. Chen, G., Lai, D. (1998) Feedback anticontrol of discrete chaos. Int. J. Bifur. Chaos, **8**:1585–1590
2. Doedel, E., Keller, H. B., Kernevez, J. P. (1991) Numerical analysis and control of bifurcation problems (I) Bifurcation in finite dimensions. Int. J. of Bifur. Chaos, **1**(3):493–520
3. Doedel, E., Keller. H. B., Kernevez, J. P. (1991) Numerical analysis and control of bifurcation problems (II) Bifurcation in infinite dimensions. Int. J. of Bifur. Chaos, **1**(4):745–772

4. Doedel, E., Champneys, A. R., Fairgrieve, T. F., Kuznetsov, Y. A., Sandstede, B., Wang, X. (1998) AUTO 97: Continuation and Bifurcation Software for Ordinary Differential Equations (with HomCont)
5. Erramilli, A., Singh, R. P., Pruthi, P. (1994) Chaotic maps as models of packet traffic. Proceeding of ITC, **14**:329–338
6. Khibnik, A. I., Roose, D., Chua, L. O. (1993) On periodic orbits and homoclinic bifurcaitons in Chua's circuit with a smooth nonlinearity. Int. J. of Bifur. Chaos, **3**:363–384
7. Kocarev, L., Maggio, G. M., Ogorzalek, M., Pecora, L., Yao, K. (eds.) (2001) Special Issue: Advances on communication systems using chaos. IEEE Trans. Circ. Syst.-I, **48**(12)
8. Kuznetsov, Y. A. (1998) Content – Integrated Environment for Analysis of Dynamical Systems: Tutorial
9. Maggio, G. M., Feo, O. D., Kennedy, M. P. (1999) Nonlinear analysis of the Colpitts oscillator and applications to design. IEEE Trans. Circ. Syst.–I, **46**(9):1118–1130
10. Ng, L., Rand, R. (2002) Bifurcations in a Mathieu equation with cubic nonlinearities. Chaos, Solitons and Fractals, **14**:173–181
11. Ottino, J. M., Muzzio, F. J., Tjahjadi, M., Franjione, J. G., Jana, S. C., Kusch, H. A. (1992) Chaos, symmetry and self-similarity: Exploiting order and disorder in mixing processes. Science, **257**:754–760
12. Tang, K. S., Zhong, G. Q. (2002) Chaotification of low-order linear continuous-time system using nonlinear feedback. Int. J. Bifur. Chaos, to appear
13. Schiff, S. J., Jerger, K., Duong, D. H., Chang, T., Spano, M. L., Ditto, W. L. (1994) Controlling chaos in the brain. Nature **370**:615–620
14. Wang, X. F., Chen, G. (2000) Chaotifying a stable LTI system by tiny feedback control. IEEE Trans. on Circ. Syst.-I, **47**(3):410–415
15. Yang, W., Ding, M., *et al.* (1995) Preserving chaos: Control strategies to preserve complex dynamics with potential relevance to biological disorders. Phys. Rev. E, **51**:102–110
16. Zhong, G. Q., Tang, K. S., Chen, G., Man, K. F. (2001) Bifurcation analysis and circuit implementation of a simple chaos generator. Latin American Applied Research, **31**:227–232

Bifurcation Dynamics in Control Systems

Pei Yu

Department of Applied Mathematics
The University of Western Ontario
London, Ontario, N6A 5B7, Canada
pyu@pyu1.apmaths.uwo.ca

Abstract. This chapter deals with bifurcation dynamics in control systems, which are described by ordinary differential equations, partial differential equations and delayed differential equations. In particular, bifurcations related to double Hopf, combination of double zero and Hopf, and chaos are studied in detail. Center manifold theory and normal form theory are applied to simplify the analysis. Explicit stability conditions are derived and routes of bifurcations leading to various complex dynamics are given. A system with time delayed feedback control is studied to show that time delay plays a important role in controlling and anti-controlling chaotic motions. Furthermore, a simple feedback controller is designed for anti-controlling Hopf bifurcation arising in the Lorenz system.

1 Introduction

Nonlinear dynamical system theory plays an important role in almost all the areas of science and engineering because real world systems are indeed nonlinear. The theory of dynamics is particularly useful in the study of complex behavior such as instability, bifurcation and chaos (e.g., see [1–7]), which are encountered in mechanics, aeronautics, electrical circuits, control systems, population problems, economics, financial systems, the stock market, ecological systems, etc. (e.g., see [8–23]). In general, studies in nonlinear system behavior may be divided into two main categories: local analysis and global analysis. For instance, post-critical behavior such as saddle-node bifurcation and Hopf bifurcation can be studied locally in the vicinity of a critical point, while heteroclinic and homoclinic orbits, and chaos are essentially global behavior and have to be studied globally. These two categories need to be treated with different theory and methodology.

For local dynamical analysis, usually the first step is to simplify a given system as much as possible, while keeping the dynamical behavior of the system unchanged. There exist many methodologies which have been proved good for dynamical systems, including center manifold theory, normal form theory, averaging method, multiple time scales, Lyapunov-Schmidt reduction, the method of succession functions, the intrinsic harmonic balancing technique, etc. These methods can be used to obtain "simplified" governing

G. Chen, D.J. Hill, and X. Yu (Eds.): Bifurcation Control, LNCIS 293, pp. 99–126, 2003.

equations in the vicinity of a point of interest. The "simplified" system is topologically equivalent to the original system, and thus the dynamic behavior of the original system can be studied on the basis of the "simplified" system, making the analysis much easier. Usually, normal form theory (e.g., see [1,2,5,24–29]) is applied together with center manifold [30] theory which is applied first to reduce a system to a low dimensional center manifold. Then the method of normal forms is employed to obtain further reduction of the system. However, there exist approaches which combine the two theories into one unified procedure (e.g., see [31–36]).

In general, a normal form is not uniquely defined and computing the explicit formula of a normal form in terms of the coefficients of the original system is not easy. In the past few years, symbolic computation of normal forms using computer algebra systems such as Maple, Mathematica, and Macsyma has received considerable attention. The method developed in [10] may be the first "automatic" symbolic program for systematically computing the normal forms of Hopf and generalized Hopf bifurcations. More recently, different methods and such "automatic" programs have been developed for considering other singularities [32–36].

Normal form theory may should be called conventional (or classical) normal form theory since it has been found that the CNF can be further simplified, though the basic idea of the CNF is to find a form "as simple as possible". The further reduction on the CNF leads to so called the simplest normal form, or unique normal form. A number of researchers have paid attention on the computation of the simplest normal form in the past few years (see [37–39] and references therein). However, we will not discuss the use of the simplest normal form in this chapter.

In the past two decades, there has been rapidly growing interest in bifurcation dynamics of control systems, including controlling and anti-controlling bifurcations and chaos (e.g., see [7,12,13,19,21,40–55]). There are a wide variety of promising potential applications of bifurcation and chaos control. In general, the aim of bifurcation control is to design a controller such that the bifurcation characteristics of a nonlinear system undergoing bifurcation can be modified to achieve some desirable dynamical behaviors, such as changing a Hopf bifurcation solution from subcritical to supercritical [13,53], eliminating chaotic motions [13], etc. Anti-control of chaos, on the other hand, is to purposely create chaos when it is beneficial. Many applications have been found, for example, in the areas of mechanical systems, fluid dynamics, biological systems and secure communications [50]. Chaos synchronization used in encryption is one good example of chaos application [56,57]. The basic idea of communication using chaos is to use a chaotic signal as a "mask" to scramble the message in the transmitter and then to fully recover the message from the receiver. In general engineering applications, one expects to design a system to be either chaotic or non-chaotic as one wishes.

A control system can be described by a map, a function or, in more general, a operator in either time domain or frequency domain. Differential equations are the most useful and widely applied tools in describing control systems. They may be ordinary differential equations (ODE), partial differential equations (PDE), delayed differential equations (DDE), or combination of differential equations and algebraic equations (DAE). For example, consider the control system given by the following ODE:

$$\dot{\boldsymbol{x}} = \boldsymbol{f}(\boldsymbol{x}, \boldsymbol{\mu}) + \boldsymbol{u}, \quad \boldsymbol{x},\ \boldsymbol{u} \in R^n, \quad \boldsymbol{\mu} \in R^m, \quad \boldsymbol{f} : R^n \to R^n, \tag{1}$$

where $\boldsymbol{x}$, $\boldsymbol{u}$ and $\boldsymbol{\mu}$ are state variable, control variable and system parameter, respectively, and the dot indicates the differentiation with respect to time t. $\boldsymbol{\mu}$ may be considered as control parameters. Usually, $\boldsymbol{\mu}$ is not explicitly shown in a control system. In this chapter, $\boldsymbol{\mu}$ is explicitly shown for the convenience of bifurcation analysis. The control function $\boldsymbol{u}$ can be, in general, any kind of function of the parameter $\boldsymbol{\mu}$ as well as time t, which renders system (1) non-autonomous. However, when a control law is determined system (1) may be transformed to autonomous. For instance, suppose the feedback, given by

$$\boldsymbol{u} = \boldsymbol{u}(\boldsymbol{x},\ \boldsymbol{\mu}), \tag{2}$$

is chosen, then system (1) becomes autonomous, and the bifurcation theory for autonomous ODE can be applied with the $\boldsymbol{\mu}$ as control parameter.

When a system is given by a non-autonomous ODE, one can always formally transform it to an autonomous system by introducing additional state variables. For example, consider the following equation:

$$\ddot{x} + \mu\,\dot{x} + \omega^2 x + \alpha\, x^3 = F \cos t, \tag{3}$$

where μ, α and F are parameters, and F can be considered as a control parameter. By introducing $x_1 = x$, $x_2 = \dot{x}$, $x_3 = \sin t$ and $x_4 = \cos t$, one can transform Eq. (3) into an autonomous system:

$$\begin{aligned}
\dot{x}_1 &= x_2\,, \\
\dot{x}_2 &= -\,\omega^2\, x_1 - \,\mu\, x_2 - \alpha\, x_1^3 + F\, x_4\,, \\
\dot{x}_3 &= x_4\,, \\
\dot{x}_4 &= -\,x_3\,,
\end{aligned} \tag{4}$$

which is a typical double Hopf bifurcation problem.

If a control system is described by a PDE, one way to analyze bifurcation of the system is first to transform the PDE to ODE with certain technique and then to apply ODE theory. When DDE is used to describe a control system, one may apply center manifold theory to obtain the governing equations described by ODE and thus greatly simplify the dynamical analysis.

In the following sections, we shall present three systems described by ODE, PDE and DDE, respectively, and mainly focus on double Hopf bifurcations, heteroclinic orbits and chaotic motions. This will show the complexity of bifurcation phenomena and demonstrate the importance of control. Furthermore, the Lorenz equation is used to consider a simple feedback control which stabilizes a subcritical Hopf bifurcation and eliminates chaotic motions. The center manifold theory and normal form theory are used in the analysis for all the four systems.

2 Double Hopf Bifurcation (ODE)

Double Hopf bifurcation, which is characterized by two pairs of purely imaginary eigenvalues of the Jacobian of a system, has been studied by many researchers (e.g., see [2,3,8,10,14,26,27,32,34,35]). In this section, we consider a general nonlinear system described by ODE which exhibits double Hopf bifurcation. Assume that the system is governed by the following differential equation:

$$\dot{\boldsymbol{x}} = J_c\,\boldsymbol{x} + \boldsymbol{f}(\boldsymbol{x},\boldsymbol{\mu}), \quad \boldsymbol{x} \in R^n, \quad \boldsymbol{\mu} \in R^m, \quad \boldsymbol{f} : R^n \to R^n, \tag{5}$$

with $\boldsymbol{x} = \boldsymbol{0}$ as an equilibrium of the system for any values of the parameter $\boldsymbol{\mu}$, i.e., $\boldsymbol{f}(\boldsymbol{0},\boldsymbol{\mu}) = \boldsymbol{0}$. The nonlinear function $\boldsymbol{f}$ is assumed analytic, and J_c can be, without loss of generality, further assumed in the form of

$$J_c = \begin{bmatrix} 0 & \omega_{1c} & 0 & 0 & 0 \\ -\omega_{1c} & 0 & 0 & 0 & 0 \\ 0 & 0 & 0 & \omega_{2c} & 0 \\ 0 & 0 & -\omega_{2c} & 0 & 0 \\ 0 & 0 & 0 & 0 & A \end{bmatrix}, \tag{6}$$

in which A is an $(n-4)\times(n-4)$ stable matrix (i.e., all eigenvalues of A have negative real parts). If the ratio ω_{1c}/ω_{2c} is an irrational number, it is called non-resonance. For this non-resonant double Hopf bifurcation, the normal form of the system up to 3rd order can be obtained as (see [2,35])

$$\begin{aligned} \dot{r}_1 &= r_1 \left[\alpha_{11}\,\mu_1 + \alpha_{12}\,\mu_2 + a_{20}\,r_1^2 + a_{02}\,r_2^2 \right], \\ \dot{r}_2 &= r_2 \left[\alpha_{21}\,\mu_1 + \alpha_{22}\,\mu_2 + b_{20}\,r_1^2 + b_{02}\,r_2^2 \right], \end{aligned} \tag{7}$$

$$\begin{aligned} \dot{\theta}_1 &= \omega_{1c} + \beta_{11}\,\mu_1 + \beta_{12}\,\mu_2 + c_{20}\,r_1^2 + c_{02}\,r_2^2, \\ \dot{\theta}_2 &= \omega_{2c} + \beta_{21}\,\mu_1 + \beta_{22}\,\mu_2 + d_{20}\,r_1^2 + d_{02}\,r_2^2, \end{aligned} \tag{8}$$

where r_k and θ_k represent the amplitudes and phases of motion, respectively. The unfolding is given in the terms of perturbation parameters μ_1 and μ_2. The "automatic" Maple program for computing the normal form and associated nonlinear transformation of system (5) has been developed using a perturbation approach in [35] in which another Maple program is given to compute bifurcation solutions and their stability, and the critical boundaries. The routes of bifurcations from periodic solutions to 2- and 3-dimensional tori can also be explicitly found using the software.

The steady-state solutions and their stability conditions, and the route of bifurcations can be derived from Equation (7) as follows. Setting $\dot{r}_1 = \dot{r}_2 = 0$ yields the steady-state solutions:

(1) The initial equilibrium solution (E.S.):

$$r_1 = r_2 = 0 \quad (\text{i.e., } x_i = 0); \tag{9}$$

(2) Hopf bifurcation solution (H.B.(I) with frequency ω_1):

$$\begin{aligned} r_1^2 &= -\frac{1}{a_{20}}\left(\alpha_{11}\,\mu_1 + \alpha_{12}\,\mu_2\right), \\ r_2 &= 0\,, \\ \omega_1 &= \omega_{1c} + \beta_{11}\,\mu_1 + \beta_{12}\,\mu_2 + c_{20}\,r_1^2\,; \end{aligned} \tag{10}$$

(3) Hopf bifurcation solution (H.B.(II) with frequency ω_2):

$$\begin{aligned} r_1 &= 0, \\ r_2^2 &= -\frac{1}{b_{02}}\left(\alpha_{21}\,\mu_1 + \alpha_{22}\,\mu_2\right), \\ \omega_2 &= \omega_{2c} + \beta_{21}\,\mu_1 + \beta_{22}\,\mu_2 + d_{02}\,r_2^2\,; \end{aligned} \tag{11}$$

(4) Quasi-periodic solution (2-D Tori with frequencies ω_1, ω_2):

$$\begin{aligned} r_1^2 &= \frac{a_{02}\left(\alpha_{21}\,\mu_1 + \alpha_{22}\,\mu_2\right) - b_{02}\left(\alpha_{11}\,\mu_1 + \alpha_{12}\,\mu_2\right)}{a_{20}\,b_{02} - a_{02}\,b_{20}}, \\ r_2^2 &= \frac{b_{20}\left(\alpha_{11}\,\mu_1 + \alpha_{12}\,\mu_2\right) - a_{20}\left(\alpha_{21}\,\mu_1 + \alpha_{22}\,\mu_2\right)}{a_{20}\,b_{02} - a_{02}\,b_{20}}, \\ \omega_1 &= \omega_{1c} + \beta_{11}\,\mu_1 + \beta_{12}\,\mu_2 + c_{20}\,r_1^2 + c_{02}\,r_2^2\,, \\ \omega_2 &= \omega_{2c} + \beta_{21}\,\mu_1 + \beta_{22}\,\mu_2 + d_{20}\,r_1^2 + d_{02}\,r_2^2\,. \end{aligned} \tag{12}$$

Then one can use the Jacobian matrix of Eq. (8) to find the stability conditions of the above solutions as well as the bifurcation route. For example, evaluating the Jacobian on the equilibrium solution (9) results in the stable region for the E.S. as

$$\alpha_{11}\,\mu_1 + \alpha_{12}\,\mu_2 < 0 \qquad \text{and} \qquad \alpha_{21}\,\mu_1 + \alpha_{22}\,\mu_2 < 0. \tag{13}$$

The boundaries defined by Eq. (13) leads to two critical lines, one of them is given by

$$L_1\,: \quad \alpha_{11}\,\mu_1 + \alpha_{12}\,\mu_2 = 0 \qquad (\alpha_{21}\,\mu_1 + \alpha_{22}\,\mu_2 < 0), \tag{14}$$

Table 1. Bifurcation solutions and stability conditions

Solution	Stability Condition	Critical Line	Solution
E.S. (9)	$\alpha_{11}\,\mu_1+\alpha_{12}\,\mu_2<0$ $\alpha_{21}\,\mu_1+\alpha_{22}\,\mu_2<0$	$L_1:\ \alpha_{11}\,\mu_1+\alpha_{12}\,\mu_2=0$ $L_2:\ \alpha_{21}\,\mu_1+\alpha_{22}\,\mu_2=0$	H.B.(I) H.B.(II)
H.B.(I) (10)	$\alpha_{11}\,\mu_1+\alpha_{12}\,\mu_2>0$ $\alpha_{21}\,\mu_1+\alpha_{22}\,\mu_2$ $-\frac{b_{20}}{a_{20}}(\alpha_{11}\,\mu_1+\alpha_{12}\,\mu_2)<0$	$L_3:\ (\alpha_{21}-\frac{b_{20}}{a_{20}}\alpha_{11})\,\mu_1+$ $(\alpha_{22}-\frac{b_{20}}{a_{20}}\alpha_{12})\mu_2=0$	2-D Tori
H.B.(II) (11)	$\alpha_{21}\,\mu_1+\alpha_{22}\,\mu_2>0$ $\alpha_{11}\,\mu_1+\alpha_{12}\,\mu_2$ $-\frac{a_{02}}{b_{02}}(\alpha_{21}\,\mu_1+\alpha_{22}\,\mu_2)<0$	$L_4:\ (\alpha_{11}-\frac{a_{02}}{b_{02}}\alpha_{21})\,\mu_1+$ $(\alpha_{12}-\frac{a_{02}}{b_{02}}\alpha_{22})\mu_2=0$	2-D Tori
2–D Tori (12)	$a_{20}\,b_{02}-a_{02}\,b_{20}>0$ $a_{20}\,(a_{02}-b_{02})(\alpha_{21}\mu_1+\alpha_{22}\mu_2)-$ $b_{02}(a_{20}-b_{20})(\alpha_{11}\mu_1+\alpha_{12}\mu_2)<0$	$L_5:\ [a_{20}(a_{02}-b_{02})\alpha_{21}\ -$ $b_{02}(a_{22}-b_{20})\alpha_{11}]\,\mu_1+$ $[a_{20}(a_{02}-b_{02})\alpha_{22}\ -$ $b_{02}(a_{22}-b_{20})\alpha_{12}]\mu_2=0$	3-D Tori

where a family of limit cycles bifurcates from the E.S. with the approximate solution H.B.(I) given by Equation (10). The second critical line is

$$L_2:\quad \alpha_{21}\,\mu_1+\alpha_{22}\,\mu_2=0\qquad(\alpha_{11}\,\mu_1+\alpha_{12}\,\mu_2<0),\tag{15}$$

from which another family of limit cycles, given by H.B.(II) solution (11), may occur.

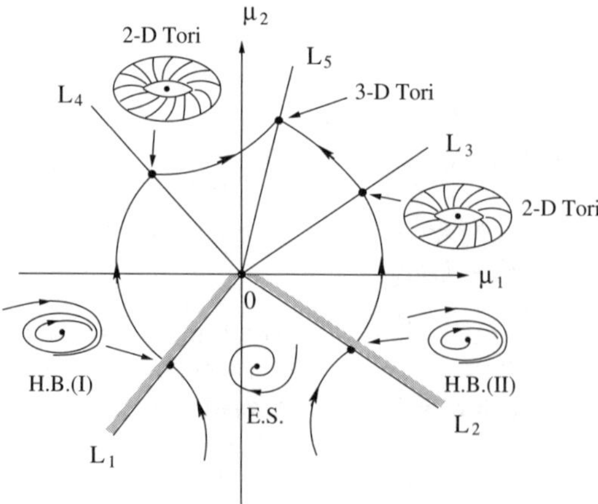

Fig.1. General bifurcation diagram

Similarly we can analyze the other steady-state solutions (10)–(12). The results are summarized in Table 1, and the bifurcation diagram is shown in Fig. 1. The details of the analysis can be found in [35].

In the following, two examples are given to show the application of the above results.

2.1 An electrical circuit

A nonlinear electrical circuit, shown in Fig. 2, consists of two capacitors C_1, C_2, two inductors L_1, L_2, a resistor R, and a conductance. L_1 and C_1 are connected in parallel, while L_2, C_2 and R in series. All the five elements, L_1, L_2, C_1, C_2 and R are assumed to be linear time-invariant elements, but C_1 and R may be varied as control parameters. The conductance, however, is a nonlinear element with the characteristic:

$$i_G = -\tfrac{1}{2}\, v_G + \alpha\, v_G^3, \tag{16}$$

where i_g and v_G represent the current and voltage of the conductance, respectively. α may be considered as another control parameter. The voltages across the capacitors and the currents in the inductors are chosen as the state variables (shown in Fig. 2), denoted by $(z_1, z_2, z_3, z_4) = (v_{C_1}, i_{L_1}, v_{C_2}, i_{L_2})$. Then by choosing $L_1 = \sqrt{2}$, $L_2 = 2 - \sqrt{2}$ and $C_2 = \sqrt{2} - 1$, one can obtain the state equations of the circuit as follows:

$$\begin{aligned}
\dot{z}_1 &= \eta_1 \left(\tfrac{1}{2}\, z_1 + z_2 - z_4 - \alpha\, z_1^3 \right), \\
\dot{z}_2 &= -\tfrac{\sqrt{2}}{2}\, z_1\,, \\
\dot{z}_3 &= (\sqrt{2} + 1)\, z_4\,, \\
\dot{z}_4 &= (2 - \sqrt{2})\, (z_1 - z_3 - \eta_2\, z_4),
\end{aligned} \tag{17}$$

where $\eta_1 = 1/C_1$ and $\eta_2 = R$ are treated as two independent control (perturbation) parameters.

It is easy to show that the Jacobian of Eq. (17) evaluated at the equilibrium $z_i = 0$ and at the critical point, defined by $\eta_{1c} = 2$ and $\eta_{2c} = 1 + \frac{\sqrt{2}}{2}$

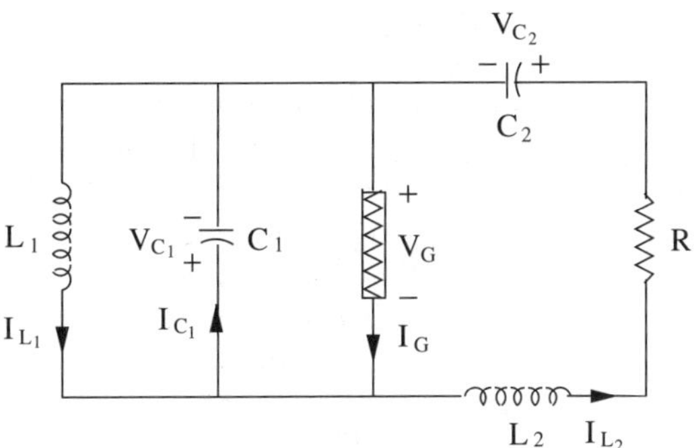

Fig.2. An electrical circuit

Table 2. Bifurcation solutions and stability conditions for the electrical circuit

Solution	Expression	Stability	Slope of C.L.
E.S.	$r_1 = 0$ $r_2 = 0$	$\mu_1 + (4\sqrt{2}-6)\,\mu_2 > 0$ $\mu_1 + (6\sqrt{2}-8)\,\mu_2 > 0$	$L_1:\ \frac{3+2\sqrt{2}}{2}$ $L_2:\ -\frac{4+3\sqrt{2}}{4}$
H.B. (I)	$r_1^2 = -\frac{1}{3}(\sqrt{2}+1)\left[\mu_1 + (6\sqrt{2}-8)\mu_2\right]$ $r_2 = 0$ $\omega_1 = 1 + \frac{1}{2}\sqrt{2}\left[\mu_1 + 2(\sqrt{2}-1)\mu_2\right]$	Stable	$L_3:\ -\frac{7+5\sqrt{2}}{2}$
H.B. (II)	$r_1 = 0$ $r_2^2 = \frac{1}{3}(2+\sqrt{2})\left[\mu_1 + (4\sqrt{2}-6)\mu_2\right]$ $\omega_2 = \sqrt{2} + \sqrt{2}\left[\mu_1 - (2-\sqrt{2})\,\mu_2\right]$	Stable	$L_4:\ \frac{16+11\sqrt{2}}{4}$
2-D Tori	$r_1^2 = \frac{\sqrt{2}}{18}\left[(5\sqrt{2}+6)\mu_1 - 4(\sqrt{2}-1)\mu_2\right]$ $r_2^2 = \frac{-\sqrt{2}}{9}\left[(3+2\sqrt{2})\mu_1 + 2(\sqrt{2}-1)\mu_2\right]$ $\omega_1 = 1 + \frac{\sqrt{2}}{2}\,\mu_1 + (2-\sqrt{2})\,\mu_2$ $\omega_2 = \sqrt{2} + \sqrt{2}\,\mu_1 - 2(\sqrt{2}-1)\,\mu_2$	Unstable	

has two pairs of pure imaginary eigenvalues: $\lambda_{1,2} = \pm\, i$ and $\lambda_{1,2} = \pm\,\sqrt{2}\, i$. For $\alpha = 1$, executing the Maple program [35] yields the normal form up to 3rd order as follows:

$$
\begin{aligned}
\dot{r}_1 &= r_1 \left[-\tfrac{1}{4}\,\mu_1 - \tfrac{1}{2}\,(3\,\sqrt{2}-4)\,\mu_2 - \tfrac{3}{4}\,(\sqrt{2}-1)\,(r_1^2 + 2\,r_2^2) \right], \\
\dot{r}_2 &= r_2 \left[\tfrac{1}{2}\,\mu_1 - (3 - 2\,\sqrt{2})\,\mu_2 - \tfrac{3}{4}\,(2-\sqrt{2})\,(2\,r_1^2 + r_2^2) \right],
\end{aligned} \tag{18}
$$

$$
\begin{aligned}
\dot{\theta}_1 &= 1 + \tfrac{1}{4}\,(\sqrt{2}-1)\,\mu_1 + \tfrac{1}{2}\,(\sqrt{2}-2)\,\mu_2 - \tfrac{3}{4}\,(r_1^2 + 2\,r_2^2), \\
\dot{\theta}_2 &= \sqrt{2} + \tfrac{1}{2}\,(\sqrt{2}-1)\,\mu_1 - (\sqrt{2}-1)\,\mu_2 + \tfrac{3}{4}\,\sqrt{2}\,(2\,r_1^2 + r_2^2),
\end{aligned} \tag{19}
$$

where $\boldsymbol{\mu} = (\mu_1,\, \mu_2) = (\eta_1 - \eta_{1c},\, \eta_2 - \eta_{2c})$. Further executing the Maple program for the bifurcation analysis [35] using Eqs. (18) and (19) gives the results summarized in Table 2, and the bifurcation diagram is shown in Fig. 3. If one changes the control parameter α in Eq. (16), one would find that as long as the α is chosen positive (which implies that the nonlinear function (16) has a typical nonlinearity), the bifurcation diagram does not change (see Fig. 3). However, when α is chosen negative, then all bifurcation solutions are unstable. Further, to obtain stable quasi-periodic solutions (2-D tori), one need to change other parameters L_1, L_2 and C_2 or need to design a feedback controller.

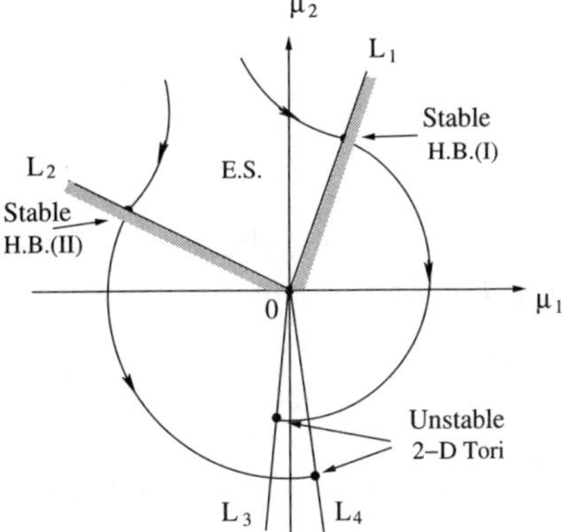

Fig.3. Bifurcation diagram for the electrical circuit

2.2 A mechanical system

The double pendulum system, shown in Fig. 4, has been studied by many researchers to demonstrate complex bifurcation phenomena. For example, a number of singularities associated with this system are considered in [10], including a double zero eigenvalue, a simple zero and a pair of purely imaginary eigenvalues, and two pairs of purely imaginary eigenvalues. In this section, we show the bifurcation property of the system in the vicinity of a non-resonant double Hopf critical point.

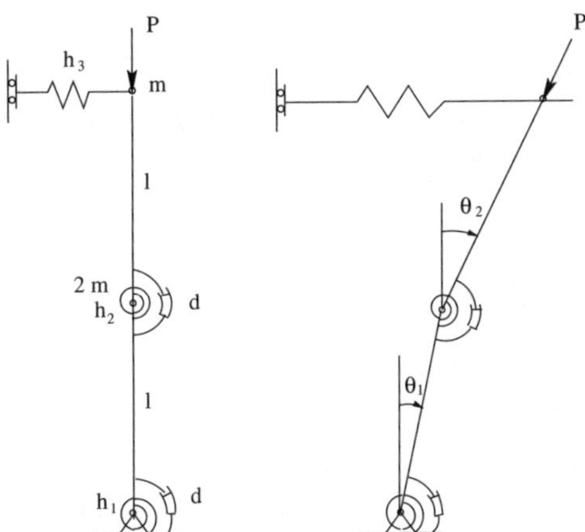

Fig.4. A double pendulum system

The system consists of two rigid weightless links of equal length l which carry two concentrated masses $2m$ and m, respectively. A follower force P is applied to this system. The system energy for the three linear springs h_1,

h_2 and h_3 is assumed to be given by [3]

$$V = \tfrac{1}{2} \left[(h_1+h_2+h_3 l^2)\,\theta_1^2 + 2(h_3 l^2 - h_2)\theta_1\theta_2 + (h_2+h_3 l^2)\theta_2^2 \right] - \tfrac{1}{6} h_3 l^2 (\theta_1+\theta_2)(\theta_1^3+\theta_2^3), \tag{20}$$

where θ_1 and θ_2 are generalized co-ordinates which specify the configuration of the system completely. The kinetic energy T of the system is expressed by

$$T = \tfrac{m\,l^2}{2\,\Omega^2} \left[3\,\theta_1'^2 + \theta_2'^2 + 2\,\theta_1'\,\theta_2' \cos(\theta_1-\theta_2) \right], \tag{21}$$

where Ω is an arbitrary value rendering the time variable non-dimensional. The generalized force corresponding to the generalized co-ordinates θ_1 and θ_2 may be written as

$$Q = P\,l\,\sin(\theta_1-\theta_2), \tag{22}$$

and the damping can be expressed by

$$D = \tfrac{1}{2} \left[d_1\,\theta_1'^2 + d_2\,(\theta_1'-\theta_2')^2 \right] + \tfrac{1}{4}\, d_4\,(\theta_1'-\theta_2')^4, \tag{23}$$

where d_1, d_2 represent the linear parts, while d_4 describes the non-linear part.

With the aid of the Lagrangian equations, in addition, choosing the state variables $z_1=\theta_1$, $z_2=\theta_1'$, $z_3=\theta_2$, $z_4=\theta_2'$ and rescaling the coefficients to be dimensionless coefficients as

$$\begin{aligned} & f_1 = \tfrac{h_1\Omega^2}{m\,l^2}, \quad f_2 = \tfrac{h_2\Omega^2}{m\,l^2}, \quad f_3 = \tfrac{h_3\Omega^2}{m}, \quad f_4 = \tfrac{P\Omega^2}{m\,l}, \\ & f_5 = \tfrac{d_4\Omega^4}{m\,l^2}, \quad \eta_1 = \tfrac{d_1\Omega^2}{m\,l^2}, \quad \eta_2 = \tfrac{d_2\Omega^2}{m\,l^2}, \end{aligned} \tag{24}$$

then at the critical point defined by $f_1=\frac{4}{7}$, $f_2=\frac{407}{56}$, $f_3=\frac{1}{56}$, $f_4=\frac{535}{28}$, $f_5=-1$, $\eta_1=\eta_2=0$, where the eigenvalues of the system has two pairs of purely imaginary eigenvalues: $\lambda_{1,2}=\pm i$, $\lambda_{3,4}=\pm\sqrt{2}\,i$, one finally finds the normal form of the system as follows:

$$\begin{aligned} r_1' &= r_1 \left[-\tfrac{11}{7}\,\mu_1 + \tfrac{75}{56}\,\mu_2 - \tfrac{625}{10976}\,r_1^2 - \tfrac{2025}{5488}\,r_2^2 \right], \\ r_2' &= r_2 \left[\tfrac{37}{28}\,\mu_1 - \tfrac{159}{56}\,\mu_2 + \tfrac{1325}{5488}\,r_1^2 + \tfrac{4293}{10976}\,r_2^2 \right]; \end{aligned} \tag{25}$$

$$\begin{aligned} \theta_1' &= 1 - \tfrac{6493901}{22127616}\,r_1^2 - \tfrac{1423069}{2458624}\,r_2^2, \\ \theta_2' &= \sqrt{2}\left[1 + \tfrac{14046397}{22127616}\,r_1^2 + \tfrac{2975501}{9834496}\,r_2^2 \right]. \end{aligned} \tag{26}$$

Executing the Maple program for bifurcation analysis on Eqs. (25) and (26) yields the bifurcation solutions and stability conditions given in Table 3, and the bifurcation diagram is shown in Fig. 5. This double pendulum example exhibits not only stable periodic solutions, but also stable quasi-periodic motion. The results also indicate that the 2-D torus loses stability at the critical line L_5 and bifurcates into quasi-periodic motion on a 3-D torus. Note that one may use the force P as a control parameter to change the bifurcations and stabilities.

Table 3. Bifurcation solutions and stability conditions for the double pendulum

Solution	Expression	Stability	Slope of C.L.
E.S.	$r_1 = 0$ $r_2 = 0$	$\mu_2 < \frac{88}{75}\,\mu_1$ $\mu_2 > \frac{74}{159}\,\mu_1$	$L_1: \frac{88}{75}$ $L_2: \frac{74}{159}$
H.B. (I)	$r_1^2 = \frac{10976}{625}\,(-\frac{11}{7}\,\mu_1 + \frac{75}{56}\,\mu_2)$ $r_2 = 0$ $\omega_1 = 1 + \frac{71432911}{8820000}\,\mu_1 - \frac{6493901}{940800}\,\mu_2$	Stable	$L_3: \frac{7478}{3975}$
H.B. (II)	$r_1 = 0$ $r_2^2 = -\frac{10976}{4293}\,(\frac{37}{28}\,\mu_1 - \frac{159}{56})\,\mu_2$ $\omega_2 = \sqrt{2}\,(1 - \frac{110093537}{107702784}\,\mu_1 + \frac{2975501}{1354752}\,\mu_2)$	Unstable	$L_4: -\frac{964}{3975}$
2-D Tori	$r_1^2 = \frac{188944}{99375}\,\mu_1 + \frac{196}{25}\,\mu_2$ $r_2^2 = -\frac{1465688}{31975}\,\mu_1 + \frac{196}{81}\,\mu_2$ $\omega_1 = 1 + \frac{104850832237}{50485680000}\,\mu_1 - \frac{47010917}{12700800}\,\mu_2$ $\omega_2 = \sqrt{2}\,(1 - \frac{34401875231}{201942720000}\,\mu_1 + \frac{580057817}{101606400}\,\mu_2$	Stable	$L_5: \frac{30031}{7950}$
3-D Tori	Bifurcating from the critical line L_5		

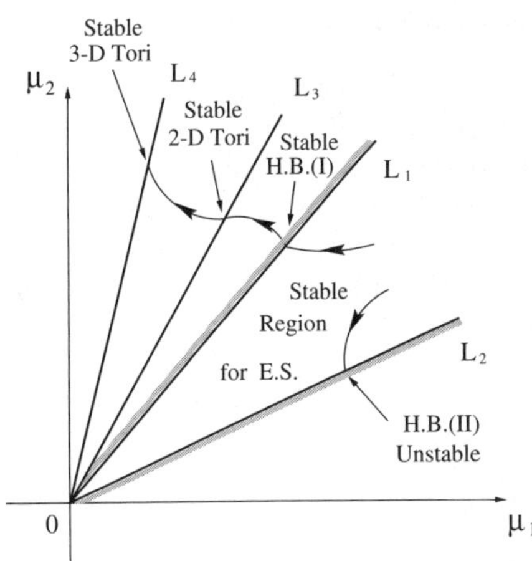

Fig.5. Bifurcation diagram for the double pendulum system

3 Heteroclinic Bifurcation (PDE)

In this section, we turn to consider PDE equation describing a rectangular thin plate subjected to transverse and in plane excitations simultaneously. The main attention here is focused on 1:1 semi-simple internal resonance as well as external resonances. The computation of the normal form is divided into two steps: First the method of multiple scales [27,31,35] is used to obtain the averaged equation from the original non-autonomous system. Then based on the autonomous system (averaged equations), the normal form theory is applied to find the explicit formulas of the normal form and nonlinear transformation. A codimension-3 case, associated with a double zero and a purely imaginary pair, is studied in detail.

The thin plate to be considered is a rectangular plate, having length a, width b and thickness h, respectively. It is simply supported at its four edges, and subjected to a transverse and an in-plane excitations simultaneously. A schematic illustration of the thin plate is given in Fig. 6.

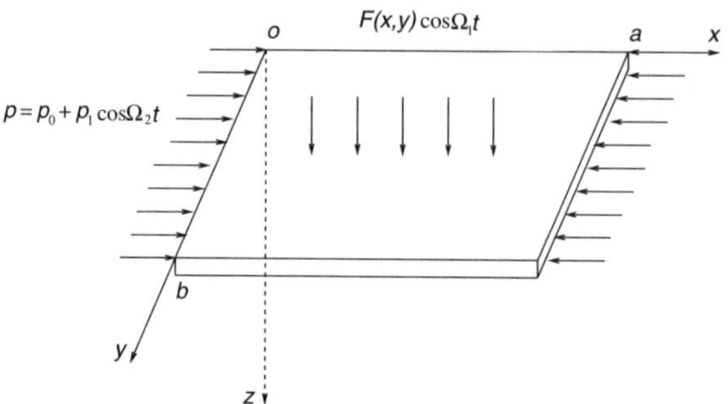

Fig.6. The structure of a thin plate and the coordinate system

A Cartesian coordinate system can be established such that the origin of the system is located at one corner of the rectangular thin plate (see Fig. 6). It is assumed that u, v and w represent the displacements of a point in the middle plane of the thin plate in the x, y and z directions, respectively. By applying the van Karman equation to a thin plate, we can establish the equations of motion for the rectangular thin plate as follows:

$$D\nabla^4 w+\rho h\frac{\partial^2 w}{\partial t^2}-\frac{\partial^2 w}{\partial x^2}\frac{\partial^2 \phi}{\partial y^2}-\frac{\partial^2 w}{\partial y^2}\frac{\partial^2 \phi}{\partial x^2}+2\frac{\partial^2 w}{\partial x\partial y}\frac{\partial^2 \phi}{\partial x\partial y}-\mu\frac{\partial w}{\partial t}=F(x,y)\cos\Omega_1 t,$$

$$\nabla^4\phi = E\,h\left[\left(\frac{\partial^2 w}{\partial x\partial y}\right)^2-\frac{\partial^2 w}{\partial x^2}\frac{\partial^2 w}{\partial y^2}\right], \tag{27}$$

where w is the transverse displacement, ρ is the density of thin plate, $D=\frac{E\,h^3}{12\,(1-\nu^2)}$ is the bending rigidity, E is Young's modulus, ν is the Possion's ratio, ϕ is the stress function, and μ is the damping coefficient, and p is the in-plane excitation forcing given by $p=p_0+p_1\cos\Omega_2 t$.

In this section, we mainly consider the nonlinear oscillations of the thin plate associated with the first two modes. After a length algebraic manipulation one obtains the following dimensionless equations of motion:

$$\begin{aligned}&\ddot{x}_1+\epsilon\,\mu\,\dot{x}_1+\left(\omega_1^2+2\,\epsilon f_1\cos\Omega_2 t\right)x_1+\epsilon\left(\alpha_1\,x_1^3+\alpha_2\,x_1x_2^2\right)=\epsilon\,F_1\cos\Omega_1 t,\\&\ddot{x}_2+\epsilon\,\mu\,\dot{x}_2+\left(\omega_2^2+2\,\epsilon f_2\cos\Omega_2 t\right)x_2+\epsilon\left(\beta_1\,x_2^3+\beta_2\,x_1^2x_2\right)=\epsilon\,F_2\cos\Omega_1 t\end{aligned} \tag{28}$$

in which α_k, β_k, etc. are parameters, ω_k $(k = 1,\, 2)$ are the two linear natural frequencies of the thin plate when the middle plane of the edges is subjected to the constant force p_0, and f_k $(k = 1,\, 2)$ are the amplitudes of parametric excitations.

To find the time averaged equations of Eq. (28), we focus on the case of 1:1 internal resonance and primary parametric resonance and assume the following relations [14]:

$$\omega_1^2 = \tfrac{1}{4}\,\Omega_2^2 + \epsilon\,\sigma_1\,, \quad \omega_2^2 = \tfrac{1}{4}\,\Omega_2^2 + \epsilon\,\sigma_2\,, \quad \Omega_1 = \Omega_2\,, \tag{29}$$

where σ_1 and σ_2 are the two detuning parameters. For convenience of the study, let $\Omega_1 = \Omega_2 = 2$. Then executing the Maple program developed in [14] yields four averaged equations. At the critical point defined by

$$\mu = f_2 = 0, \quad \sigma = -\,f_1\,, \tag{30}$$

the Jacobian of the averaged equations evaluated at the initial equilibrium solution has a double zero and a purely imaginary pair: $\lambda_{1,2} = 0$ and $\lambda_{3,4} = \pm\, i\,\frac{\sigma_2}{2}$. Further, letting $f_1 = 1$ yields the averaged equations:

$$\begin{aligned}
\tfrac{d\,x_1}{d\,T_1} &= x_2 - \tfrac{3}{2}\,\alpha_1\,x_2\,(x_1^2 + x_2^2) - \tfrac{1}{2}\,\alpha_2\,x_2\,(x_3^2 + 3\,x_4^2) - \alpha_2\;x_1\,x_3\,x_4\,,\\
\tfrac{d\,x_2}{d\,T_1} &= \tfrac{3}{2}\,\alpha_1\,x_1\,(x_1^2 + x_2^2) + \tfrac{1}{2}\,\alpha_2\,x_1\,(3\,x_3^2 + x_4^2) + \alpha_2\;x_2\,x_3\,x_4\,,\\
\tfrac{d\,x_3}{d\,T_1} &= \tfrac{1}{2}\,\sigma_2\,x_4 + \tfrac{3}{2}\,\beta_1\,x_4(x_3^2 + x_4^2) + \tfrac{1}{2}\,\beta_2\,x_4(3\,x_1^2 + x_2^2) + \beta_2\,x_1\,x_2\,x_3\,,\\
\tfrac{d\,x_4}{d\,T_1} &= -\,\tfrac{1}{2}\,\sigma_2 x_3 - \tfrac{3}{2}\,\beta_1 x_3(x_3^2 + x_4^2) - \tfrac{1}{2}\,\beta_2\,x_3(x_1^2 + 3x_2^2) - \beta_2\,x_1\,x_2\,x_4\,,
\end{aligned} \tag{31}$$

such that its Jacobian is in canonical form, where $T_1 = \epsilon\, t$.

Next, to find the normal form of system (31) associated with the singularity of a double zero and a pair of purely imaginary eigenvalues, we apply the results and Maple programs obtained in [14] to find

$$\begin{aligned}
\dot{y}_1 &= -\,\bar{\mu}\,y_1 + (1 + \bar{\sigma}_1)\,y_2\,,\\
\dot{y}_2 &= \bar{\sigma}_1\,y_1 - \bar{\mu}\,y_2 + \tfrac{3}{2}\,\alpha_1\,y_1^3 + \alpha_2\,y_1\,(y_3^2 + y_4^2),\\
\dot{y}_3 &= -\,\bar{\mu}\,y_3 + \bar{\sigma}_2\,y_4 + \tfrac{3}{2}\,\beta_1\,y_4\,(y_3^2 + y_4^2) + \beta_2\,y_1^2\,y_4\,,\\
\dot{y}_4 &= -\,\bar{\sigma}_2\,y_3 - \bar{\mu}\,y_4 - \tfrac{3}{2}\,\beta_1\,y_3\,(y_3^2 + y_4^2) - \beta_2\,y_1^2\,y_3\,,
\end{aligned} \tag{32}$$

where $\bar{\mu} = \frac{1}{2}\,\mu$ and $\bar{\sigma}_2 = \frac{1}{2}\,\sigma_2$. Further, let $y_3 = I\,\cos\gamma$ and $y_4 = I\,\sin\gamma$, then from Eq. (32) one obtains the following equations:

$$\begin{aligned}
\dot{y}_1 &= -\,\bar{\mu}\,y_1 + (1 + \bar{\sigma}_1)\,y_2\,,\\
\dot{y}_2 &= \bar{\sigma}_1\,y_1 - \bar{\mu}\,y_2 + \tfrac{3}{2}\,\alpha_1\,y_1^3 + \alpha_2\,y_1\,I^2,\\
\dot{I} &= 0,\\
\dot{\gamma} &= -\,\bar{\sigma}_2 - \bar{\mu} - \tfrac{3}{2}\,\beta_1\,I^2 - \beta_2\,y_1^2\,.
\end{aligned} \tag{33}$$

Under the simple linear transformation: $y_1 = (1+\bar{\sigma}_1)\,u_1\,,\ y_2 = \bar{\mu}\,u_1 + u_2$, Eq. (33) becomes

$$\begin{aligned}\dot{u}_1 &= u_2\,,\\ \dot{u}_2 &= -\,\mu_1\,u_1 - \mu_2\,u_2 + \alpha_2\,I^2\,(1+\bar{\sigma}_1)\,u_1 + \tfrac{3}{2}\,\alpha_1\,u_1^3\,,\\ \dot{I} &= 0,\\ \dot{\gamma} &= -\,\bar{\sigma}_2 - \bar{\mu} - \tfrac{3}{2}\,\beta_1\,I^2 - \beta_2\,(1+\bar{\sigma}_1)^2\,u_1^2\,,\end{aligned} \tag{34}$$

where $\mu_1 = \bar{\mu}^2 - \bar{\sigma}_1\,(1+\bar{\sigma}_1)$ and $\mu_2 = 2\,\bar{\mu}$. It is noted from (34) that I is a constant since $\dot{I} = 0$, and thus the first and second equations of (34) are (independent of γ) decoupled from the other two equations. Therefore, we may first consider these two decoupled equations:

$$\begin{aligned}\dot{u}_1 &= u_2\,,\\ \dot{u}_2 &= -\,\mu_1\,u_1 - \mu_2\,u_2 + \alpha_2\,I^2\,(1+\bar{\sigma}_1)\,u_1 + \tfrac{3}{2}\,\alpha_1\,u_1^3.\end{aligned} \tag{35}$$

Since system (35) has a double zero eigenvalue at $u_1 = u_2 = 0$ for all the values $\mu_1 \geq 0$, the system can exhibit *heteroclinic* bifurcation due to $\alpha_1 > 0$. It is easy to verify that when $\mu_1 - \alpha_2\,I^2\,(1+\bar{\sigma}_1) < 0$, the only equilibrium solution of Eq. (35) is the trivial zero solution, $u_1 = u_2 = 0$, which is a saddle point. On the curve defined by

$$\bar{\mu}^2 = (1+\bar{\sigma}_1)\,(\bar{\sigma}_1 - \alpha_2\,I^2) \qquad \text{or} \qquad I_{1,2} = \pm\left[\frac{\bar{\mu}^2 - \bar{\sigma}_1\,(1+\bar{\sigma}_1)}{\alpha_2\,(1+\bar{\sigma}_1)}\right]^{1/2}, \tag{36}$$

the trivial zero solution may bifurcate into three solutions through a *pitchfork* bifurcation. The three solutions are given by

$$q_0 = (0,\,0) \quad \text{and} \quad q_{\pm}(I) = \left(\pm\left\{\frac{2}{3\,\alpha_1}\left[\bar{\mu}^2 - (\bar{\sigma}_1 + \alpha_2\,I^2)\,(1+\bar{\sigma}_1)\right]\right\}^{1/2},\ 0\right).$$

One can verify that the singular points $q_{\pm}(I)$ are also saddle points. On the line defined by $\mu_2 = 0$, limit cycles (Hopf bifurcation) may bifurcate from the trivial zero solution, which are stable for $\mu_2 < 0$.

To study the heteroclinic bifurcation, we introduce the following scale transformation $u_1 \to \epsilon\,u_1,\ u_2 \to \epsilon\,u_2,\ \mu_2 \to \epsilon\,\epsilon_2\ \ \alpha_1 \to \frac{2}{3}\,\epsilon^{-2}\,\alpha_1$ into Eq. (35) to obtain

$$\dot{u}_1 = u_2\,, \quad \dot{u}_2 = -\,\epsilon_1\,u_1 + \alpha_1\,u_1^3 - \epsilon\,\epsilon_2\,u_2\,, \tag{37}$$

where $\epsilon_1 = \mu_1 - \alpha_2\,I^2\,(1+\bar{\sigma}_1)$. When $\epsilon = 0$, system (37) is a Hamiltonian system with Hamiltonian $H(u_1,\,u_2) = \frac{1}{2}\,u_2^2 + \frac{1}{2}\,\epsilon_1\,u_1^2 - \frac{1}{4}\,\alpha_1\,u_1^4$. When $H = \frac{\epsilon_1^2}{4\,\alpha_1}$, there exists a heteroclinic loop Γ^0 consisting of the two hyperbolic saddles $q_{\pm}$ and a pair of heteroclinic orbits $u_{\pm}(T_1)$, which can be found as

$$\begin{aligned}u_1(T_1) &= \pm\sqrt{\frac{\epsilon_1}{\alpha_1}}\,\tanh\left(\frac{\sqrt{2\,\epsilon_1}}{2}\,T_1\right),\\ u_1(T_2) &= \pm\,\frac{\epsilon_1}{\sqrt{2\,\alpha_1}}\,\mathrm{sech}^2\left(\frac{\sqrt{2\,\epsilon_1}}{2}\,T_1\right).\end{aligned} \tag{38}$$

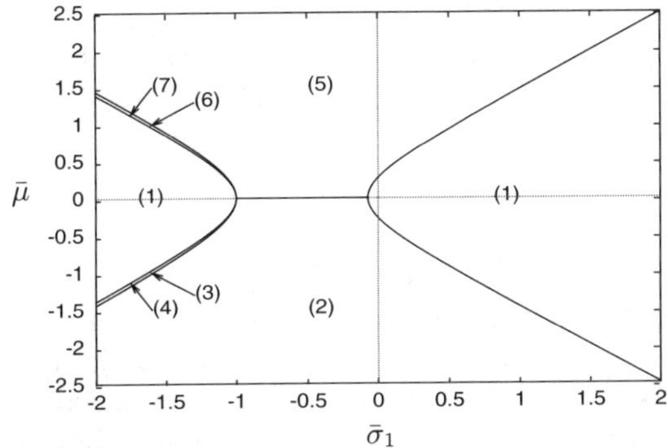

Fig.7. The bifurcation set of system (35): (1) saddle point, (2) stable limit cycle, (3) heteroclinic loop, (4) heteroclinic orbit, (5) unstable limit cycle, (6) heteroclinic loop, and (7) heteroclinic orbit

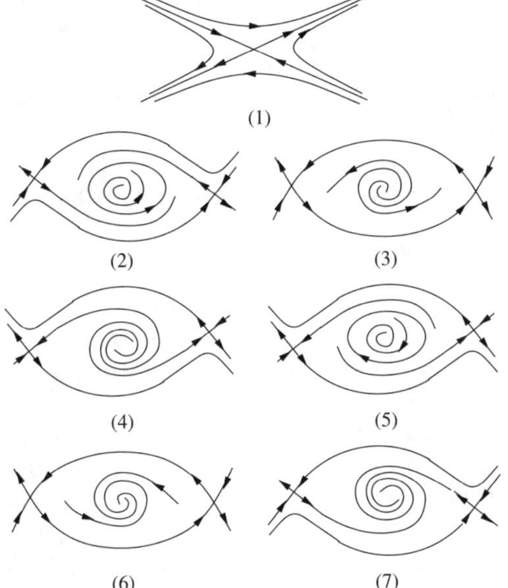

Fig.8. The phase portrait associated with the different bifurcation regions indicated in Fig 7

The Melnikov function for the heteroclinic orbits can be easily obtained by integration:

$$M(\epsilon_1, \epsilon_2, I) = \int_{-\infty}^{\infty} u_2(T_1)\,[-\epsilon_2\, u_2(T_1)]\; dT_1 = -\frac{2\sqrt{2}\,\epsilon_1^{3/2}\,\epsilon_2}{3\,\alpha_1}. \tag{39}$$

The necessary and sufficient condition to keep the heteroclinic loop preserved under a perturbation is $M(\epsilon_1, \epsilon_2, I) = 0$. Thus, Eq. (39) leads to either $\epsilon_2 = 0$ (i.e., μ_2 (or μ) $= 0$) which corresponds to the critical point, or $\epsilon_1 = 0$. Choosing $\epsilon_1 = 0$ yields

$$\bar{\mu}^2 = (1 + \bar{\sigma}_1)\,(\bar{\sigma}_1 - \alpha_2\, I^2). \tag{40}$$

Based on Eqs. (36) and (40), the bifurcation diagram is obtained as shown in Fig. 7, and the associated phase portraits are depicted in Fig. 8.

4 Chaos (DDE)

Having considered ODE and PDE systems in the previous sections, we now study a DDE system. In this section, a nonlinear oscillator with delayed time feedback is investigated and the main attention is focused on chaotic motions. It is well known that self-sustained oscillation plays a very important role in the study of nonlinear vibrating systems. If such a system contains time delayed component, it can even exhibit more interesting phenomena. The well-known example is that one dimensional autonomous differential equation with time delay can exhibit periodic solutions while a similar autonomous system without time delay requires at least two dimensions to have periodic motions. Further it has been shown that applying a time delay to a dynamical system may be one of the efficient approaches to control or anti-control complex dynamical motions such as chaos [50], since the time delay is easy to be controlled and realized in applications.

In this section, we consider a mathematical model including time delayed feedback control with a closed-loop, described by the following equation:

$$\ddot{x} + \omega_0^2\, x - \alpha_1\, \dot{x} + \alpha_3\, \dot{x}^3 = k\, \cos(\Omega t) + A\,(\dot{x}_\tau - \dot{x}) + B\,(\dot{x}_\tau - \dot{x})^3, \qquad (41)$$

where the parameters α_1, α_3 and k are assumed to be positive, $x_\tau = x(t - \tau)$, and τ is a time delay. The feedback is called negative if $A,\, B < 0$, and positive if $A,\, B > 0$. The case $\tau = 0$ corresponds to a typical nonlinear vibrating mechanical system.

Time delayed systems similar to system (41) but without external forcing (i.e., $k = 0$) have been considered by many researchers (e.g., see [58,59]). Recently, Reddy *et al.* [59] studied the effects of time delayed linear and nonlinear feedbacks on a so called Stuart-Landau system which may be obtained by avenging system (41) and setting $k = 0$. They showed rich dynamical behavior including several different types of bifurcations and chaos with linear or nonlinear feedback gains. Other similar studies using different feedbacks can be found in [58], which are though restricted to the study of limit cycles (periodic solutions). However, for the system with an external forcing, i.e., when $k \neq 0$ in Eq. (41), only a preliminary study has been given in [55].

In particular, here we shall investigate the effect of time delay on the dynamical solutions. The particular attention is focused on the dynamics of the system in the vicinity of the critical point where a double Hopf bifurcation may occur. A numerical time integration scheme is used to obtain complex dynamical solutions such as periodic, quasi-periodic and chaotic motions. It is shown that chaotic motions are very sensitive to the time delay τ.

First, we find the critical condition at which system (41) has a double Hopf bifurcation. To achieve this, the linear equation of system (41) corresponding to $k = 0$ leads to the characteristic equation:

$$\lambda^2 + (A - \alpha_1)\,\lambda + \omega_0^2 - A\,\lambda\,\exp(-\lambda\,\tau) = 0, \tag{42}$$

where ω_0 is a real positive constant. Letting $\lambda = i\,\omega$ and substituting it into (42) results in the following equations

$$\begin{aligned} \omega_0^2 - \omega^2 &= A\,\omega\,\sin(\omega\tau), \\ A\,\omega - \alpha_1\,\omega &= A\,\omega\,\cos(\omega\tau) \end{aligned} \tag{43}$$

from which one finds

$$\omega_\pm^2 = \left\{ 1 + \tfrac{(2A-\alpha_1)\,\alpha_1}{2\,\omega_0^2} \pm \sqrt{\left[1 + \tfrac{(2\,A-\alpha_1)\,\alpha_1}{2\,\omega_0^2}\right]^2 - 1} \right\} \omega_0^2\,, \tag{44}$$

which indicates that condition $\alpha_1(2A - \alpha_1) \geq 0$ must be satisfied since $\omega_\pm$ are real and positive. This condition actually implies that the critical values of α_1 and A, expressed in ω and τ, are located in the region in the α_1-A plane, bounded by the two straight lines $\alpha_1 = 0$ and $\alpha_1 = 2A$. Note that Eq. (43) contains three independent parameters α_1, A and τ since ω_0 is a constant, and ω can be determined in terms of the three parameters via Eq. (43). When $\tau = 0$, system (41) is reduced to an ordinary differential equation (without time delay), then it is obvious that $\alpha_1 = 0$ is the critical stability boundary. However, if $\tau > 0$, then the critical stability boundary is changed and the new boundaries can be found from Eq. (43). Therefore, it is important to study how the variation of the time delay τ affects the dynamical behavior of the system. It have been shown that [55]: (i) Hopf bifurcations may only occur on the critical stability boundaries; and (ii) it is possible to have double Hopf bifurcations (associated with two pairs of purely imaginary eigenvalues). Double Hopf bifurcation will be considered in this section.

To obtain the critical point at which a double Hopf bifurcation takes place, consider the following equations which are equivalent to Eq. (43):

$$\cos(\omega_+\tau_+) = 1 - \tfrac{\alpha_1}{A} \qquad \text{and} \qquad \cos(\omega_-\tau_-) = 1 - \tfrac{\alpha_1}{A}\,, \tag{45}$$

where τ_+ and τ_- are solved from Eq. (43), corresponding to ω_+ and ω_-, respectively. Equation (45) yields $\cos(\omega_+\tau_+) = \cos(\omega_-\tau_-)$, or $\omega_+\tau_+ = \pm\,\omega_-\tau_- + 2\,j\,\pi$, where j is an integer. On the double Hopf bifurcation, $\tau_+ = \tau_- = \tau_c$, thus $(\omega_+ + \omega_-)\,\tau_c = 2\,j\,\pi$. Further, let $\frac{\omega_-}{\omega_+} = \omega_r$ $(0 < \omega_r \leq 1)$, we can then find

$$\begin{aligned} \tau_c &= \frac{2\,j\,\pi\,\sqrt{\omega_r}}{(1 + \omega_r)\,\omega_0}\,, \\ \alpha_{1c} &= \pm\,\frac{\omega_0\,(1-\omega_r)}{\sqrt{\omega_r}}\,\tan\!\left(\frac{j\,\pi}{1 + \omega_r}\right). \\ A_c &= \pm\,\frac{\omega_0\,(1 - \omega_r)}{\sqrt{\omega_r}}\,\csc\left(\frac{2\,j\,\pi}{1 + \omega_r}\right) \end{aligned} \tag{46}$$

and $\omega_- = \omega_0 \sqrt{\omega_r}$ and $\omega_+ = \omega_0/\sqrt{\omega_r}$ for $0 < \omega_r < 1$. Therefore, at the critical point at which a double Hopf bifurcation occurs, all the critical values of the parameters α_{1c}, A_c, τ_c, $\omega_{1c} \equiv \omega_-$ and $\omega_{2c} \equiv \omega_+$ are expressed in terms of ω_r, ω_0 and k. Suppose the ω_0 is given and the integer k is fixed, then the critical values of all the parameters are uniquely determined by the ratio of the two frequencies, ω_r.

For the double Hopf bifurcation, three typical cases: 1:$\sqrt{2}$ non-resonance, and 1:2 and 1:3 resonant cases, have been studied in detail using a numerical approach [55]. In general, the system can exhibit very rich periodic and quasi-periodic solutions, and chaotic motions in the vicinity of the critical point. In particular, for certain values of τ, when $k = 0$, the system mainly exhibit periodic solutions, while $k \neq 0$ can yield both periodic and quasi-periodic motions. Moreover, it is found that when $k \neq 0$, the periodic solutions occur only for small values of k while the quasi-periodic motions for large values of k.

In the rest of the section, we shall consider a set of parameter values near the 1:$\sqrt{2}$ non-resonance to show that the system can indeed exhibit rich chaotic motions. In particular, we are interested in the sensitivity of the chaotic motions with respect to the time delay τ. The parameters other than τ are chosen as follows: $\omega_0 = 10$, $\alpha_3 = 0.3$, $\alpha_1 = 12.2$, $A = 6.5$, $k = 12$, $\Omega = 2$ and $B = 1$. We consider the variation of τ on the interval $[0.26, 0.28]$. The results are presented in Figs. 9, 10 and 11. Figure 9 depicts the Poincaré maps of the trajectories as τ is varied. It clearly shows a route of bifurcations from quasi-periodic motions to chaos when the time delay is increased from 0.26 to 0.268. At $\tau = 0.26$, the Poincaré map is a simple closed orbit, implying that the trajectory moves on a two-dimensional torus in the x–$\dot{x}$–t space (see Fig. 9(a)). The simple closed orbit becomes intersecting itself with one node at $\tau = 0.2643$ (Fig. 9(b)) and two nodes at $\tau = 0.266$ (Fig. 9(c)). These self-intersecting, closed orbits indicate that the motions are still quasi-periodic. Finally, when $\tau = 0.268$, the motion becomes chaotic since the Poincaré map shows a set of bounded, scattered dense points.

Figure 10 shows that different patterns of motions exist when τ is varied. The Poincaré map given in Figs. 10(a) ($\tau = 0.269$) and (e) ($\tau = 0.276$) indicate that the motions are chaotic, while those shown in Figs. 10(b) ($\tau = 0.2694$), (c) ($\tau = 0.27$) and (d) ($\tau = 0.275$) are quasi-periodic motions. It is clear that chaotic motions appear "periodically", transferring from one chaotic motion to another. It is interesting to observe that the motions given in Figs. 10(b) and (d) are almost "double" the motion shown in Fig. 10(c). This periodicity in the chaotic motions exists for other values of τ. In fact, chaotic motions have also been found for $\tau = 0.2665$, 0.288, 0.3004, etc., and there exist quasi-periodic and other types of complex motions between these chaotic motions. For example, for $\tau \in [0.285, 0.28536]$, all the Poincaré maps are simple closed orbits, suggesting that the motions for these values of τ are quasi-periodic. There also exist motions like that shown in Figs. 10(b)–(d).

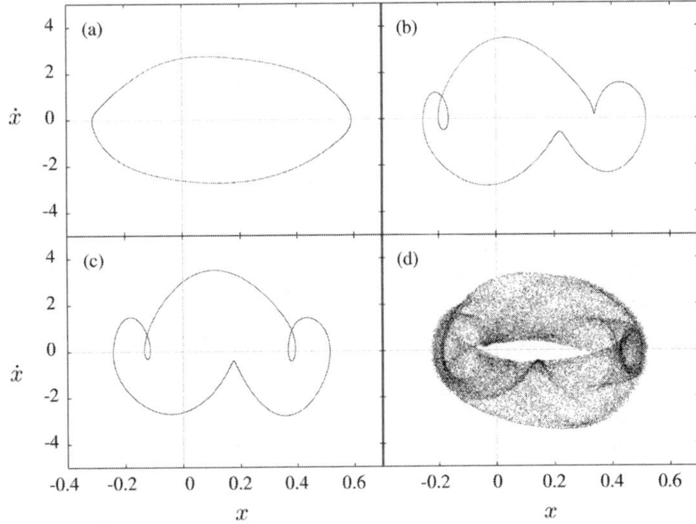

Fig.9. Route to chaos as τ equals: (a) 0.26; (b) 0.2643; (c) 0.266; (d) 0.268

It should be pointed out that the periodic appearing of the chaotic motions is not for single value, but for a short interval of τ. For instance, when $\tau \in [0.26635, 0.2665]$, all the motions are chaotic. When $\tau > 0.301$, the Poincaré maps show that all the motions are quasi-periodic.

It is also interesting to note that for τ chosen between 0.266 and 0.268 there is an interval $\tau \in [0.26626, 0.26631]$ in which intermittent chaos exists with the double nodded closed orbit as its frame, as shown in Fig. 11. The first intermittent chaos occurs at $\tau = 0.26626$ with less scattered points (see Fig. 11(a)), and the last intermittent chaos shown in Fig. 11(b) for $\tau = 0.26631$ has more scattered points.

The effects on the chaotic motion have also been considered for other parameters: the amplitude of the external vibration force, k, the frequency of the external force, Ω, and the nonlinear feedback gain, B [55]. It has been shown that the chaotic motion is not sensitive to these three parameters. However, it is found that the positive nonlinear feedback is a necessary condition to allow the system to exhibit chaotic motions. The extremely sensitivity of the chaotic motion on the time delay suggests that one may use the time delay to control or anti-control chaotic motions.

5 Anti-Control of Hopf Bifurcation

Finally, we would like to use the Lorenz equation to demonstrate how to use feedback control to change the stability of Hopf bifurcation solutions (limit

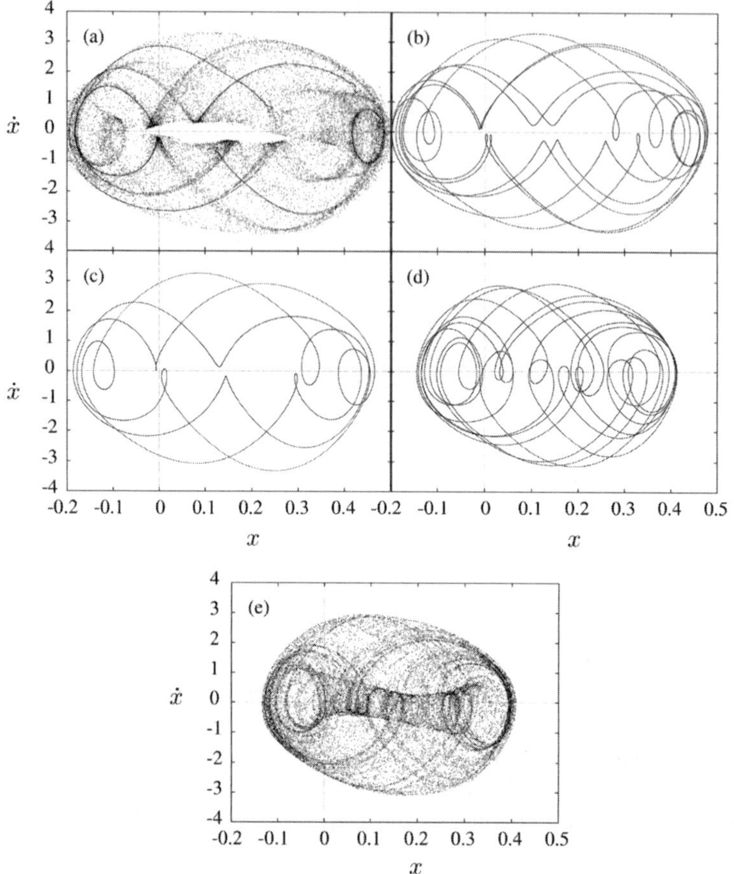

Fig.10. Route to chaos as τ equals: (a) 0.26; (b) 0.2643; (c) 0.266; (d) 0.268

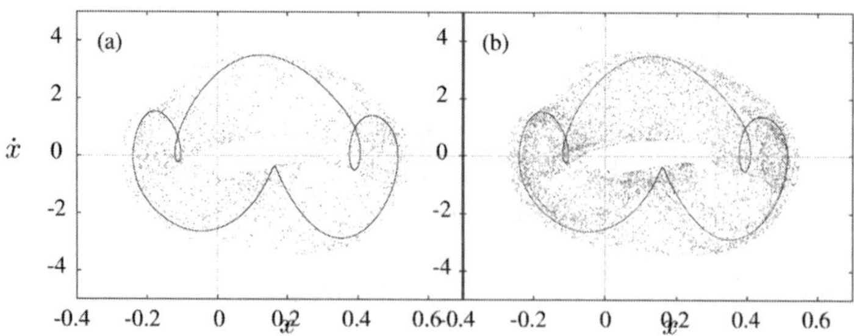

Fig.11. Intermittent chaos when (a) $\tau = 0.26626$; and (b) $\tau = 0.26631$

cycles) and eliminate chaotic motions. The equations are given by [13]

$$\begin{aligned} \dot{x} &= -p\,(x-y), \\ \dot{y} &= -x\,z - y, \\ \dot{z} &= x\,y - z - r, \end{aligned} \tag{47}$$

where p and r are positive parameters, which may be considered as control parameters. It can be shown that the system has three equilibrium solutions C_0, C_+ and C_-. The two equilibria C_+ and C_- lose their stabilities at a critical point $r_H = 16$, where subcritical Hopf bifurcation occur.

The anti-control of Hopf bifurcation of this system has been considered in [13,53]. It is shown that the system exhibit chaotic behavior when $r = 14$. Although general control strategy are proposed in the papers, the authors proposed a feedback control u, utilizing a washout filter, to obtain

$$
\begin{aligned}
\dot{x} &= -p\,(x-y), \\
\dot{y} &= -x\,z - y, \\
\dot{z} &= x\,y - z - r + u, \\
\dot{v} &= y - c\,v,
\end{aligned}
\tag{48}
$$

where v is the state of the washout filter used for control:

$$
u = -\,k_c\,(y - c\,v) - k_n\,(y - c\,v)^3, \tag{49}
$$

with constant gains k_c and k_n, and c is a constant chosen for the filter. They showed that with $c = 0.5$, $k_c = 2.5$ and $k_n = 0.009$, the Hopf bifurcation solution becomes supercritical, and the critical point is now about $r_H = 36$. The main disadvantages of this approach is that the controller increases the system dimension by one.

In this section, we present a simpler feedback control without increasing system dimension. To do this, first note that the system involves two parameters p and r, in general the critical condition for r should be a function of p. In fact, it is easy to find that the critical point is given by

$$
r_H = \frac{p\,(p+4)}{p-2}, \qquad (p > 2). \tag{50}
$$

If $r = 16$, then $p = 4$ or 8. We take $p = 8$ in the following analysis. We try to set the control law as

$$
u = -\,k_c\,y - k_1\,x^3 - k_2\,y^3 - k_3\,z^3, \tag{51}
$$

where the original three system states are used in the feedback control, with constant gains k_1, k_2 and k_3 to be determined. We could, in more general, add quadratic terms in the feedback control, and even further add control components to all the three equations of system (47). However, it can be shown that one 3rd-order term is enough, so even two k_i $(i = 1, 2)$ can be zero in Eq. (51).

As easily verified, the linear term will alter the stability condition for the equilibria C_+ and C_-. We want to increase the stable region (in terms of r) for the equilibria. However, the 3rd-order term also affect the equilibrium solutions and their stabilities. This results in solving a 3rd-degree polynomial and the analysis becomes involved. We thus restrict that k_i's be small so that they have very little influence on the equilibrium solutions and their stability. Consequently we may apply the method of normal forms to find the explicit stability conditions.

First, the three equilibrium solutions can be found, for arbitrary p and r, as follows:

$$\begin{aligned} &\mathrm{C}_0: \; x_e = y_e = 0, \quad z_e = -r, \\ &\mathrm{C}_+: \; x_e = y_e = \tfrac{1}{2}\left[k_c + \sqrt{k_c^2 + 4(r-1)}\right], \quad z_e = -1, \\ &\mathrm{C}_-: \; x_e = y_e = \tfrac{1}{2}\left[k_c - \sqrt{k_c^2 + 4(r-1)}\right], \quad z_e = -1. \end{aligned} \tag{52}$$

We only consider C_+ (C_- can be similarly analyzed). A straightforward calculation yields the characteristic polynomial:

$$\lambda^3 + (p+2)\lambda^2 + (p+r)\lambda + \tfrac{1}{2}p\left[k_c^2 + 4(r-1)\right] + \tfrac{1}{2}p\,k_c\sqrt{k_c^2 + 4(r-1)}\,. \tag{53}$$

The condition for the system to have Hopf bifurcation from the equilibrium C_+ is found as

$$(p+2)(p+r) - \tfrac{1}{2}p\left[k_c^2 + 4(r-1)\right] - \tfrac{1}{2}p\,k_c\sqrt{k_c^2 + 4(r-1)} = 0, \tag{54}$$

which, in turn, yields

$$r_\pm = \frac{p(p+4)}{p-2} + \frac{p\,k_c\left[k_c \pm \sqrt{(p^2-4)(p+1)+k_c^2}\right]}{(p-2)^2} \quad (p>2). \tag{55}$$

It is easy to check that the solution given in Eq. (55) with the positive sign, r_+, does not satisfy condition (54). Thus, we have only one solution:

$$r = r_- = \frac{p(p+4)}{p-2} + \frac{p\,k_c\left[k_c - \sqrt{(p^2-4)(p+1)+k_c^2}\right]}{(p-2)^2} \quad (p>2). \tag{56}$$

Further, in order for r to increase, so that the equilibrium C_+ will remain stable for larger values of r, we may utilize the parameter k_c. It requires that

$$k_c - \sqrt{(p^2-4)(p+1)+k_c^2} \geq 0, \qquad \text{i.e.,} \quad k_c \leq 2\sqrt{p+1}\,. \tag{57}$$

Now, with the choice $p = 8$, $k_c \leq 6$, and for determination we choose $k_c = -6$. Thus, we have the closed-loop system:

$$\begin{aligned} \dot{x} &= -8(x-y), \\ \dot{y} &= -xz - y, \\ \dot{z} &= xy - z - r + 6y - k_1x^3 - k_2y^3 - k_3z^3. \end{aligned} \tag{58}$$

The equilibrium C_+ becomes $x_e = y_e = 5$, $z_e = -1$, and the critical point is $r_H = 56$, which is much larger than $r_H = 16$ for the uncontrolled system. Notice that the eigenvalues of Jacobian evaluated at this equilibrium are:

$\lambda_{1,2} = \pm 8\,i$ and $\lambda_3 = -10$. Using the method of normal forms, we introduce the transformation,given by

$$\begin{aligned} x &= 5 + u + v + \tfrac{8}{5}\,w, \\ y &= 5 + v - \tfrac{9}{5}\,w, \\ z &= -1 - 4\,u + v + w, \end{aligned} \tag{59}$$

to Eq. (58), so as to obtain the following equations up to 3rd order:

$$\begin{aligned} \dot{u} &= 8\,v - \tfrac{94}{205}\,u^2 + \tfrac{32}{41}\,w^2 + \tfrac{212}{205}\,u\,v + \tfrac{336}{205}\,u\,w - \tfrac{848}{205}\,v\,w + k_3\,(\cdots), \\ \dot{v} &= -8\,u - \tfrac{421}{410}\,u^2 + \tfrac{124}{41}\,w^2 + \tfrac{104}{205}\,u\,v + \tfrac{687}{205}\,u\,w - \tfrac{416}{205}\,v\,w + k_3\,(\cdots), \\ \dot{w} &= -10\,w - \tfrac{47}{410}\,u^2 + \tfrac{8}{41}\,w^2 + \tfrac{53}{205}\,u\,v + \tfrac{84}{205}\,u\,w - \tfrac{212}{205}\,v\,w + k_3\,(\cdots), \end{aligned} \tag{60}$$

where $(\cdots)$ represents quadratic and cubic order terms.

To this end, one may apply the Maple program, developed for Hopf and generalized Hopf bifurcations [31], to obtain the normal form for system (60). It was found that k_1 does not help stabilize the bifurcation solution while k_2 is not important for stability. So, we simply select $k_1 = k_2 = 0$, thereby obtaining a fairly simple controller: $u = -k_c\,y - k_3\,z^3$. We thus have the following normal form given in polar coordinates:

$$\dot{R} = R\left[\tfrac{3}{378}\,\mu + \left(\tfrac{28449}{291920}\,k_3^2 - \tfrac{6864111}{2919200}\,k_3 + \tfrac{51771}{2919200}\right)R^2\right], \tag{61}$$

$$\dot{\Theta} = 8 + \tfrac{97}{1312}\,\mu - \left(\tfrac{8383053}{11676800}\,k_3 + \tfrac{1156143}{1167680}\,k_3^2 + \tfrac{581727}{11676800}\right)R^2, \tag{62}$$

where R and Θ represent the amplitude and phase of motion, respectively, $\mu = r - r_H = r - 56$, indicating the perturbation from the critical point. Steady-state solutions and their stabilities can be found from Eq. (61): The solution $R = 0$ represents the initial equilibrium solution C_+, which is stable when $\mu < 0$ (i.e., $r < r_H = 56$) and unstable when $\mu > 0$ ($r > 56$). The supercritical Hopf bifurcation solutions exist if

$$\tfrac{28449}{291920}\,k_3^2 - \tfrac{6864111}{2919200}\,k_3 + \tfrac{51771}{2919200} < 0, \tag{63}$$

or

$$0.0075 \approx \frac{762679}{63220} - \frac{71\sqrt{1037208369}}{189660} < k_3 < \frac{762679}{63220} + \frac{71\sqrt{1037208369}}{189660} \approx 24.12. \tag{64}$$

Since, at the beginning of the analysis, we assumed k_3 is small, we choose $k_3 = 0.01$. Then the control law is

$$u = 6\,y - 0.01\,z^3, \tag{65}$$

and the system with the feedback control is finally obtained as

$$\begin{aligned} \dot{x} &= -8\,(x - y), \\ \dot{y} &= -x\,z - y, \\ \dot{z} &= x\,y - z - r + 6\,y - 0.01\,z^3. \end{aligned} \tag{66}$$

Its normal form is obtained as

$$\dot{R} = R\,(0.007937\,\mu - 0.005769\,R^2)\,, \tag{67}$$

$$\dot{\Theta} = 8 + 0.073933\,\mu - 0.057098\,R^2\,, \tag{68}$$

and the supercritical Hopf bifurcation solutions are given by

$$R = 1.1729\,\sqrt{\mu} = 1.1729\,\sqrt{r-56}\,. \tag{69}$$

Some numerical simulation results are given in Figs. 12-14. Figure 12 shows trajectories of the original system (47) for $p = 8$. By linear analysis we know that the equilibrium C_+ is *locally* stable when $r < 16$. Figure 12(a) indicates that the trajectory converges to C_+ for certain initial conditions. However, (global) chaotic trajectories exist for some initial conditions. Figure 12(b) shows that with the same parameter values as used for Fig. 12(a), a chaotic motion is obtained for a different initial condition. When $r > 16$, for any initial conditions, only chaotic motions are found.

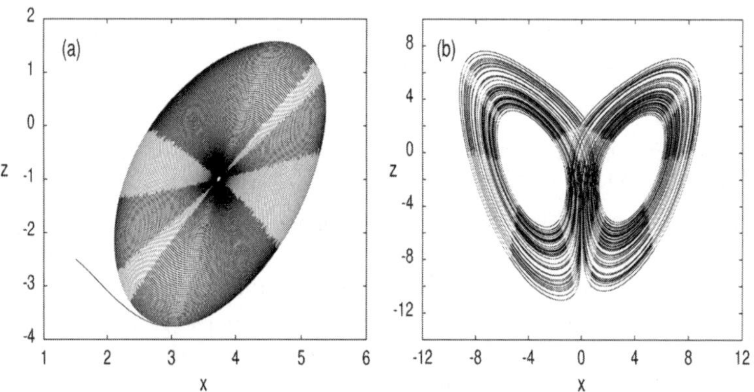

Fig.12. Trajectories of system (47) for $p = 8$ and $r = 15$ with the initial condition: (a) $(x_0,\, y_0,\, z_0) = (1.5,\, 2.5,\, -2.5)$; and (b) $(x_0,\, y_0,\, z_0) = (3.0,\, 0.1,\, 3.7)$

Typical results for the system under the proposed feedback control are shown in Figs. 13 and 14. Figure 13 depicts that the trajectories converge to the equilibrium C_+ for $1 < r < 56$, while Fig. 14 demonstrates the stable limit cycles of the system when $r > 56$. As verified by these numerical results, the designed feedback controller indeed changes the original subcritical Hopf bifurcation to a supercritical one, as claimed.

6 Conclusions

In this chapter we have presented an analytical method using center manifold and normal form theories to study bifurcation dynamics of control systems,

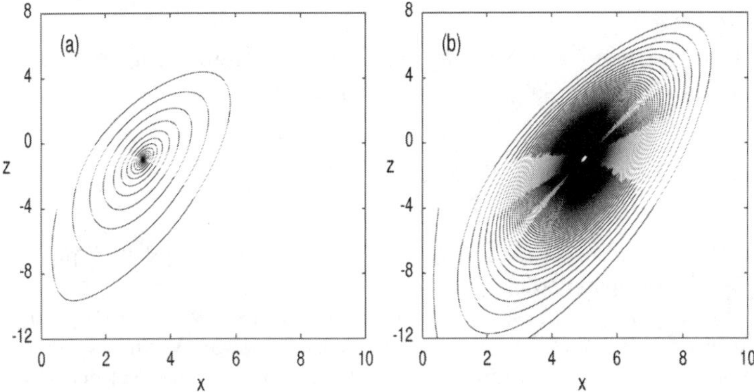

Fig.13. Stable equilibrium C_+ of system (66) with initial condition $(x_0,\ y_0,\ z_0) = (0.5,\ 0.1,\ -4.0)$ for (a) $r = 30$, and (b) $r = 56$

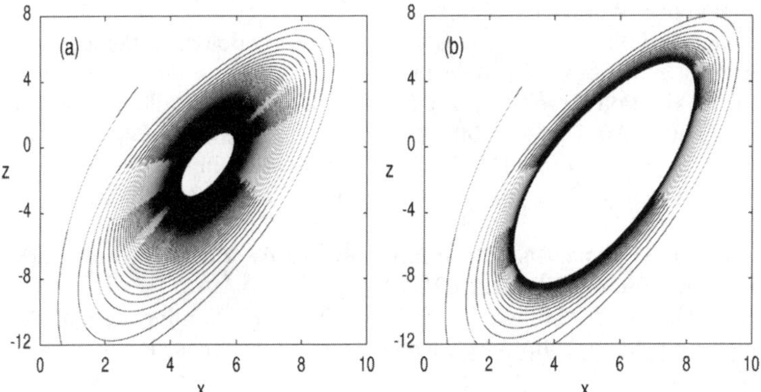

Fig.14. Stable limit cycles of system (66) with initial condition $(x_0,\ y_0,\ z_0) = (3.0,\ 0.1,\ 3.7)$ for (a) $r = 58$, and (b) $r = 65$

which are described by ODE, PDE or DDE. Particular attention is focused on bifurcation property in the vicinity of double Hopf critical point. It is shown that complex dynamical behavior such as limit cycle, quasi-periodic motion, heteroclinic bifurcation and chaos exist. Explicit bifurcation solutions and stability conditions are obtained for systems described by ODE. Heteroclinic bifurcation is obtained from a PDE system with twice applications of normal form theory. An oscillator with nonlinear feedback control is studied and the sensitivity analysis on time delay shows that using time delay may be an efficient way to realize control and anti-control of chaotic motions at one's will. Finally, the Lorenz equation is used to demonstrate a simple feedback control design through the use of normal form theory.

Acknowledgment.
The support received from the Natural Sciences and Engineering Research Council of Canada (NSERC) is greatly acknowledged.

References

1. Marsden, J. E., McCracken, M. (1976) The Hopf Bifurcation and Its Applications. New York: Springer-Verlag
2. Guckenheimer, J., Holmes, P. (1993) Nonlinear Oscillations, Dynamical Systems, and Bifurcations of Vector Fields. 4th ed. New York: Springer-Verlag
3. Huseyin, K. (1986) Multiple Parameter Stability Theory and Its Applications. Oxford: Oxford Univ. Press
4. Strogatz S. H. (1994) Nonlinear Dynamics and Chaos: with applications to physics, biology, chemistry, and engineering. Reading, MA: Addison-Wesley
5. Chow, S.-N., Li, C.-Z., Wang, D. (1994) Normal Forms and Bifurcation of Planar Vector Fields. Cambridge: Cambridge Univ. Press
6. Moiola, J. L., Chen, G. (1996) Hopf Bifurcation Analysis: A Frequency Domain Approach. Singapore: World Scientific
7. Chen, G., Dong, X. (1998) From Chaos to Order: Methodologies, Perspectives and Applications. Singapore: World Scientific
8. Yu, P., Shah, A. H., Pollplewell, N. (1992) Inertially coupled galloping of iced conductors. ASME J. Appl. Mech., **59**:140–145
9. Ono, E., Hosoe, S., Tuan, H. D., Doi, S. (1998) Bifurcation in vehicle dynamics and robust front wheel steering control. IEEE Trans. Contr. Syst. Tech., **6**:412–420
10. Yu, P., Bi, Q. (1998) Analysis of non-linear dynamics and bifurcations of a double pendulum. J. Sound Vib. **217**:691–736
11. Thompson, J. M. T., MacMillen, F. B. J. (eds.) (1998) Nonlinear flight dynamics of high-performance aircraft. Theme Issue, Phil. Trnas. Royal Society London, **A356**:2163–2333
12. Chen, Y. S., Leung, Y. T. (1998) Bifurcation and Chaos in Engineering. London: Springer
13. Chen, G., Moiola, J. L., Wang, H. O. (2000) Bifurcation control: Theories, methods, and applications. Int. J. Bifur. Chaos **10**:511–548
14. Yu, P., Zhang, W., Bi, Q. (2001) Vibration analysis on a thin plate with the aid of computation of normal forms. Int. J. Nonl. Mech., **36**:597–627
15. Hackl, K., Yang, C. Y., Cheng A. H. D. (1993) Stability, bifurcation and chaos of non-linear structures with control. Part 1. Autonomous case. Int. J. Nonl. Mech., **8**:441–454
16. Pinsky, M. A., Essary, B. (1994) Analysis and control of bifurcation phenomena in aircraft flight. J. Guidance, Contr. Dynam., **17**:591–598
17. Mees, A. I., Chua, L. O. (1986) The Hopf bifurcation theorem and its applications to nonlinear oscillations in circuits and systems. IEEE Trans. Circ. Syst., **26**:235–254
18. Hill, D. J., Hiskens, I. A., Wang, Y. (1993) Robust, adaptive or nonlinear control for modern power systems. In Proc. 32nd IEEE Conf. Decision Control, San Antonio, TX, 2335–2340

19. Chen, G., Moiola, J. L. (1994) An overview of bifurcation, chaos, and nonlinear dynamics in control systems. J. the Franklin Institute **331B**:819–858
20. Wang, H. O., Chen, D., Chen, G. (1998) Bifurcation control of pathological heart rhythms. In Proc. IEEE Conf. Contr. Appl., Trieste, Italy, 858–862
21. Chen, G. (ed.) (1999) Controlling Chaos and Bifurcations in Engineering Systems. Boca Raton, FL: CRC Press
22. Trippi, R. R. (1996) Chaos and Dynamics in Financial Markets: Theory, Evidence and Applications. New York: McGras-Hill/Irwin
23. Epstein, J. M. (1997) Nonlinear Dynamics, Mathematical Biology, and Social Science. New York: Addison-Wesley
24. Arnold, V. I. (1972) Lectures on bifurcations in versal families. Russ. Math. Surv., **27**:54–123
25. Takens, F. (1974) Singularities of vector fields. Publ. Math. IHES, **43**:47–100
26. Elphich, C., Tirapegui, E., Brachet, M. E., Coullet, P., Iooss, G. (1987) A simple global characterization for normal forms of singular vector fields. Physica D, **29**:95–127
27. Nayfeh, A. H. (1993) Methods of Normal Forms. New York: Wiley
28. Kang, W. (1998) Bifurcation and normal form of nonlinear control systems, Parts I and II. SIAM J. Contr. Optim. **36**:193–232
29. Yu, P., Huseyin, K. (1998) Parametrically excited nonlinear systems: A comparison of certain methods. Int. J. Nonl. Mech., **33**:967–978
30. Carr, J. (1981) Applications of Center Manifold Theory. New York: Springer-Verlag
31. Yu, P. (1998) Computation of normal forms via a perturbation technique. J. Sound Vib., **211**:19–38
32. Bi, Q., Yu, P. (1999) Symbolic software development for computing the normal forms of double Hopf bifurcation. J. Math. & Computer Modelling, **29**:49–70
33. Yu, P. (2000) A method for computing center manifold and normal forms. EQUADIFF 99, **2**:832–837, Singapore: World Scientific
34. Yu, P. (2001) Symbolic computation of normal forms for resonant double Hopf bifurcations using multiple time scales. J. Sound Vib. **247**:615–632
35. Yu, P. (2002) Analysis on double Hopf bifurcation using computer algebra with the aid of multiple scales. Nonl. Dynam., **27**:19–53
36. Yu, P. (2002) A simple and efficient method for computing center manifold and normal forms associated with semi-simple cases. Special Issue of Dynamics of Continuous, Discrete and Impulsive Systems for DCDIS'01, to appear
37. Yu, P. (1999) Simplest normal forms of Hopf and generalized Hopf bifurcations. Int. J. Bifur. Chaos, **9**:1917–1939
38. Yuan, Y., Yu, P. (2001) Computation of simplest normal forms of differential equations associated with a double-zero eigenvalues. Int. J. Bifur. Chaos, **11**:1307–1330
39. Yu, P. (2002) Computation of the simplest normal forms with perturbation parameters based on Lie transform and rescaling. J. Comput. and Appl. Math., to appear
40. Andronov, A. A., Leontovich, E. A., Gordon, I. I., Maier, A. G. (1967) Theory of Bifurcations of Dynamic Systems on a Plane. Moscow: Nauka; English Translation (1973) New York: Wiley
41. Yu, P., Huseyin, K. (1988) A perturbation analysis of interactive static and dynamic bifurcations. IEEE Trans. Auto. Contr. **33**:28–41

42. Srivastava, K. N., Srivastava, S. C. (1995) Application of Hopf bifurcation theory for determining critical value of a generator control or load parameter. Int. J. Electr. Power Energy Syst., **17**:347–354
43. Ueta, T., Kawakami, H., Morita, I. (1995) A study of the pendulum equation with a periodic impulse force – bifurcation and chaos. IEICE Trans. Fundam. Electr. Commun. Comput. Sci., **E78A**:1269–1275
44. Abed, E., Wang, H. O., Chen, R. C. (1994) Stabilization of period doubling bifurcations and implications for control of chaos. Physica D, **70**:154–164
45. Wang, H. O., Abed, E. H. (1995) Bifurcation control of a chaotic system. Automatica **31**:1213–1226
46. Yabuno, H. (1997) Bifurcation control of parametrically excited Duffing system by a combined linear-plus-nonlinear feedback control. Nonl. Dynam., **12**:263–274
47. Goman, M. G., Khramtsovsky, A. V. (1998) Application of continuation and bifurcation methods to the design of control systems. Phil. Trans. Roy. Soc. London, **A356**:2277–2295
48. Berns, D. W., Moiola, J. L., Chen, G. (1998) Predicting period-doubling bifurcations and multiple oscillations in nonlinear time-delayed feedback systems. IEEE Trans. Circ. Syst.-I, **45**:759–763
49. Calandrini, G., Paolini, E., Moiola, J. L., Chen, G. (1999) Controlling limit cycles and bifurcations. In Controlling Chaos and Bifurcations in Engineering Systems, Chen, G. (ed.). Boca Raton, FL: CRC Press, 200–227
50. Wang, X. F., Chen, G., Yu, X. (2000) Anticontrol of chaos in continuous-time systems via time-delay feedback. Chaos, **10**:771–779
51. Chen, G., Fang, J. Q., Hong, Y., Qin, H. S. (2000) Controlling bifurcations: Discrete-time systems. Discrete Dynamics in Nature and Society, **5**:29-33
52. Yap, K. C., Chen, G., Ueta, T. (2001) Controlling bifurcations of discrete maps. Latin American Appled Research, **31**:157–162
53. Chen, D., Wang, H. O., Chen, G. (2001) Anti-control of Hopf bifurcation. IEEE Trans. Circ. Syst.-I, **48**:661–672
54. Yu, P., Yuan, Y. (2001) Analysis on an oscillator with delayed feedback near double Hopf bifurcations. In Proc. Symp. on Vib. Contr. Mech. Syst., Nov. 11-16, 2001, New York
55. Yu, P., Yuan, Y., Xu, J. (2002) Study of double Hopf bifurcation and chaos for an oscillator with time delayed feedback. Int. J. Comm. in Nonl. Sci. Numer. Appl., **7**:69–91
56. Pecora, L. M., Carroll, T. L. (1990) Synchronization in chaotic systems. Phys. Rev. Lett., **64**:821–824
57. Cuomo, K. M., Oppenheim, A. V. (1993) Circuit implementation of synchronized chaos with applications to communications. Phys. Rev. Lett., **71**:65–68
58. Wulf, V., Ford, N. J. (2000) Numerical Hopf bifurcation for a class delay differential equations. J. Comput. Appl. Math., **115**:601–616
59. Reddy, D. V. R., Sen, A., Johnston, G. L. (2000) Dynamics of a limit cycle oscillator under time delayed linear and nonlinear feedbacks. Physica D, **144**:335–357

Analysis and Control of Limit Cycle Bifurcations

Michele Basso and Roberto Genesio

Università di Firenze
Dipartimento di Sistemi e Informatica
Via di S. Marta 3,
I-50139 Firenze, Italy
michele@dsi.ing.unifi.it

Abstract. The chapter addresses bifurcations of limit cycles for a general class of nonlinear control systems depending on parameters. A set of simple approximate analytical conditions characterizing all generic limit cycle bifurcations is determined via a first order harmonic balance analysis in a suitable frequency band. Moreover, due to the existing connection between limit cycle bifurcations and routes to chaos, the obtained predictions can also give a rough indication of possible regions of chaotic dynamics. Based on these analysis results, an approach to limit cycle bifurcation control is then proposed. The control design is based on a frequency interpretation of the bifurcation conditions obtained via harmonic balance approach. Examples of the control technique for a number of important bifurcations are reported.

1 Introduction

An important task in the study of nonlinear dynamical systems depending on parameters is the investigation of system bifurcations. A bifurcation is a qualitative change in the system dynamics which may occur when the parameters are quasi-statically varied [1]. They express the transition among different operating regions and their knowledge is essential either in the analysis of system dynamics, where different behaviors of a given process need to be predicted, and in control design, where a regulator must be employed to guarantee the required dynamics for the process. Indeed, in many practical applications a given behavior is required in a certain region of the parameter space. For example, in many biological systems the standard operating conditions are often chaotic or quasi-periodic [2], while in different applications a common objective is the stabilization of unstable periodic orbits embedded in a chaotic attractor [3]. In this context, a recent and interesting challenge in problems of nonlinear dynamics concerns control of bifurcations (see, e.g., [4–6] and references therein). Typical bifurcation control objectives include delaying the onset of an inherent bifurcation, stabilizing a bifurcated solution, changing the parameter value of an existing bifurcation point, modifying the shape or type of a bifurcation chain, introducing a new bifurcation at a preferable parameter value, monitoring the multiplicity, amplitude, and/or

G. Chen, D.J. Hill, and X. Yu (Eds.): Bifurcation Control, LNCIS 293, pp. 127–154, 2003.

frequency of some limit cycles emerging from bifurcation, optimizing the system performance near a bifurcation point.

In this chapter we focus on the analysis and control of limit cycle bifurcations in a rather general class of continuous dynamic systems. While the analysis of such situations is usually performed by numerical techniques on the related Poincaré maps, this works proposes a general input-output approach based on the harmonic balance (HB) principle. The outcome of this approach is the development of simple approximate analytical conditions characterizing all the generic bifurcations of codimension 1 [7]. Since these conditions can directly involve the controller parameters, they appear well suited to develop a technique for addressing several control problems.

Based on the harmonic balance prediction of such complex behaviors, we have developed a bifurcation control technique presenting analogies with the so called *delayed feedback control* (DFC) control strategy, introduced in recent years by Pyragas to address the problem of stabilizing unstable periodic orbits [8]. The key idea of the DFC approach is to use a feedback control signal formed by the difference between a measurable state of the system and the same state delayed by the period of the orbit of interest. The main advantage of this technique is the ability of keeping unchanged the limit cycle and the equilibria of the uncontrolled system, thus reducing the control effort. From the practical point of view, it can be applied using a continuous feedback loop to stabilize oscillations that are too fast to be handled by standard techniques based on measurements on a Poincaré section. Unfortunately, the choice of the controller parameters is by no means straightforward, in part due to the infinite dimensional nature of such a class of controllers. In fact, they can be set either experimentally or by using numerical techniques for stability analysis as, for example, the computation of the Lyapunov exponents, which is rarely an easy task [9,10]. As an alternative, we propose in a more extended context some design techniques for a class of linear time invariant finite dimensional controllers which can be interpreted as an approximation of the DFC structure. In this case, the control design is based on a frequency interpretation of the bifurcation conditions obtained via harmonic balance approach, exploiting the ideas of "washout filters" and *subharmonic control.*

The procedure presented herein allows one to analyze and control limit cycle bifurcations in a large class of nonlinear systems. Therefore, the results of the methods presented in this work can be considered in accordance with two points of view: the qualitative information makes clear the essential relations among the system parameters for the occurrence of certain phenomena, and could be used for synthesis or control problems, while the quantitative information can allow one to set up a numerical search of actual bifurcations in the parameter space. Moreover, due to the existing connection between limit cycle bifurcations and routes to chaos, the obtained predictions can also give a rough indication of possible regions of chaotic dynamics.

The chapter is organized as follows. Section 2 introduces the class of systems under study. Section 3 summarizes the main results of first order HB for limit cycle prediction, while Section 4 extends the same technique to perturbations for analyzing limit cycle bifurcations. The case of Lur'e systems is specialized in Section 5, whereas stability issues are considered in Section 6. An approach to the bifurcation control design is presented in Section 7. Finally, Section 8 considers the application of the design techniques for controlling a number of typical bifurcations.

2 The System

Consider the system $\mathcal{S}$ depicted in Fig. 1. This system consists in the *feedback interconnection* of a linear time-invariant subsystem and a nonlinear subsystem. At this stage, we do not take into account any parameter dependence

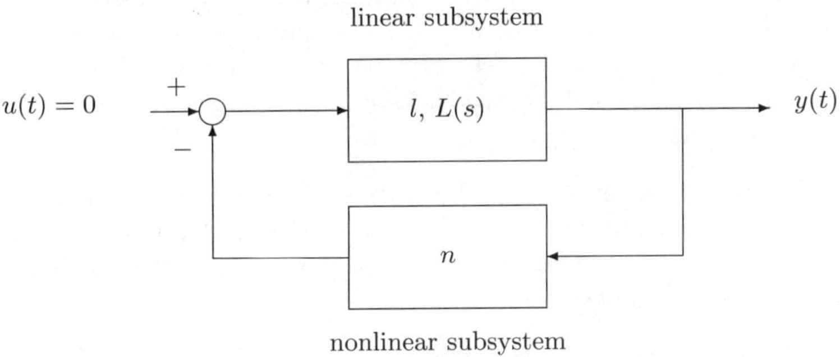

Fig.1. Basic system $\mathcal{S}$.

in system $\mathcal{S}$ and, since our interest is in the steady-state behavior of $\mathcal{S}$, we address the problem of the initial conditions of the two subsystems by simply assuming that they are consistent with our steady-state analysis. In Fig. 1, l denotes the operator of the linear subsystem and $L = L(s)$ denotes its transfer function. The nonlinear subsystem is given by a scalar time-invariant operator n which does not need to be memoryless (further clarification is given below). Thus, the overall system is governed by the operator equation

$$l \circ (n \circ y) + y = 0, \tag{1}$$

where $\circ$ denotes the operator application and $y = y(t)$ denotes the scalar output signal of system $\mathcal{S}$.

The class of systems $\mathcal{S}$ resembles that studied in the classical Lur'e problem (see, for example, [11–13]), but the restrictions are now less stringent.

Then, some of the results stated in this work for the system $\mathcal{S}$ simplify considerably if appropriate conditions are imposed on the linear and/or nonlinear subsystems. The corresponding specialized results are also given in this paper for completeness.

Now, we further detail the class of nonlinearities n that is here allowed, assuming that these are *explicit nonlinearities* [14,15]. This means that we can directly compute the output corresponding to any given input. For implicit nonlinearities, the output corresponding to a general input is defined only implicitly, as through the solution of an auxiliary system of equations.

Static and dynamic nonlinearities can be considered. We define a nonlinearity n to be static if for any input to n, the resulting output at any time t does not depend on input values at times outside an infinitesimal interval containing t. The acceptance of a dependence on input values infinitesimally close to t guarantees that multi-valued nonlinearities (i.e., relations) are included in this category [15]. Any nonlinearity which is not static is called dynamic. A *memoryless nonlinearity* is one which is single-valued and static.

Remark 1. A general class of explicit nonlinearities n consists of interconnections of linear dynamic and nonlinear static subsystems for which no nonlinear subsystem appears in a feedback path. If this latter requirement is violated, then the corresponding steady-state output cannot be explicitly written and it might not be unique.

Remark 2. In the case of explicit nonlinearities n possessing memory, the fact that the steady-state output might only result from the given input after a transient is immaterial for the following application based on the harmonic balance approach (see next Section).

Remark 3. The HB approach allows system $\mathcal{S}$ to be also externally forced. Moreover, the subsystems could be multivariable. In the latter case, there is a significantly increased computational burden associated with the extension.

3 Background on the Harmonic Balance (HB) Prediction of Limit Cycles

In this section we summarize the main results concerning the prediction of equilibria and limit cycles via (first order) harmonic balance (see, for example, [11–14,16–19]).

In the system $\mathcal{S}$ equilibria and limit cycles are intended as solutions where the output variable $y(t)$ is constant and periodic, respectively. Indeed, the restrictions we have placed on the structure of the subsystems of $\mathcal{S}$ are not sufficient to ensure that output equilibrium points and limit cycles always correspond to their classically defined counterparts for state space descriptions of $\mathcal{S}$. Nonetheless, it is convenient to use in the following these terms in the above sense. To derive these solutions, assume that any periodic signal of period $2\pi/\omega$ can be represented as

$$y_1(t) \doteq A + B\cos\omega t, \qquad B \geq 0,\ \omega > 0. \tag{2}$$

In principle, we could use a general k-th truncated form $y_k(t)$ so performing a closer approximate analysis of periodic solutions, but loosing the characteristics of a qualitative analysis to which the paper is oriented. Concerning the accuracy of description (2) and consequent results, see the discussion below.

Let the signal $y_1(t)$ of equation (2) enter in the nonlinear system n, and consider its steady state periodic output signal which is denoted by $(n \circ y_1)(t)$ and written in a series form as

$$(n \circ y_1)(t) = N_0 A + \mathbf{Re}\left[N_1 e^{j\omega t}\right] B + \dots. \tag{3}$$

Here, N_0 and N_1 represent the bias gain and the first harmonic gain, respectively, of system n and have the usual expression of the Fourier coefficients as

$$N_0 = N_0(A, B, \omega) \doteq \frac{1}{2\pi A} \int_{-\pi}^{\pi} (n \circ y_1)(t)\ d\omega t, \tag{4}$$

$$N_1 = N_1(A, B, \omega) \doteq \frac{1}{\pi B} \int_{-\pi}^{\pi} (n \circ y_1)(t) e^{-j\omega t}\ d\omega t. \tag{5}$$

We also assume without loss of generality that $(n \circ y_1)(t) = 0$ for $y_1(t) = 0$.

The well-known first order *harmonic balance* (HB) involves equating the 0 and ω frequency terms of $y_1(t)$ with the corresponding terms, so neglecting the higher harmonics, generated by l driven by $(n \circ y_1)(t)$ of equation (3). This leads to the algebraic equations, the first one real and the second complex

$$L(0) N_0(A, B, \omega) A + A = 0, \tag{6}$$

$$L(j\omega) N_1(A, B, \omega) B + B = 0, \tag{7}$$

which resemble to equation (1) and have to be solved for A, B, and ω.

The procedure, usually known as the *describing function method*, can yield the following kind of solutions which indicate specific system features:

- $A = A_0$, $B = 0$: equilibria;
- $A = 0$, $B = B_1 \neq 0$, $\omega = \omega_1 \neq 0$: symmetric limit cycles;
- $A = A_1 \neq 0$, $B = B_1 \neq 0$, $\omega = \omega_1 \neq 0$: asymmetric limit cycles.

A signal $y_1(t)$ of equation (2) where A, B and ω satisfy (6) and (7) is defined a predicted limit cycle of first order or, more simply, a *predicted limit cycle* (PLC).

It is to observe that the derivation of the simple equations (6) and (7) relies on the main hypotheses made for the system $\mathcal{S}$ in Section 2. In particular, the scalar feedback interconnection allows one to describe the periodic solution via parameters A, B and ω only, while the explicit structure of the nonlinearity n leads to a direct computation of the steady-state signal $n \circ y$ for any periodic signal y, specifically for the harmonic signal y_1 given by (2), and therefore via (4) and (5) to the gains N_0 and N_1.

With regard to the accuracy of the method, the results obtained on equilibria are exact, while those concerning limit cycles are approximate, due to the first harmonic analysis carried out on the system dynamics. The reliability of a PLC is based on a strong attenuation (*filtering hypothesis*) of the higher frequency components $2\omega, 3\omega, \ldots$ along the loop of Fig. 1 (see, for example, [11–13,16,17,19]). In this sense a computable measure of the error can be the distortion index defined for any PLC as the norm ratio

$$\Delta(A_1, B_1, \omega_1) = \frac{\|\tilde{y}_1(t) - y_1(t)\|_2}{\|y_1(t)\|_2}, \tag{8}$$

where $\tilde{y}_1(t)$ is the open loop periodic output of the system when $y_1(t)$ is injected into n. The possibility of quantifying a sufficiently small distortion can be connected to the considerations of Remark 7 below. In presence of complicated oscillations which require more harmonics to be described, the method can still be applied as a numerical technique (in the same context see, for example, [20–23]), but it looses its typical structural insight and the efficacy of the related developments following in the paper.

Some other comments concerning the presented HB method are in order.

Remark 4. The case of symmetric oscillations ($A = 0$) is typically observed when the nonlinearity n is odd, i.e., such that $n \circ y = -n \circ (-y)$.

Remark 5. In the important case of *memoryless nonlinear subsystem* n, when this reduces to a static single-valued nonlinearity and $\mathcal{S}$ become a so-called Lur'e system, it easily results for (4) and (5), with $(n \circ y_1)(t) = n[y_1(t)]$, that N_0 and N_1 are independent of ω and real. Therefore, the imaginary part of equation (7) simply becomes

$$\mathbf{Im}\,[L(j\omega)] = 0, \tag{9}$$

so determining the limit cycle frequency by a decoupled equation which depends only on the linear subsystem l.

Remark 6. The presence of forcing signals $u(t)$, constant or sinusoidal, acting on the system $\mathcal{S}$ can be taken into account by suitably introducing them in the balance of equations (6) and (7). In the sinusoidal case the frequency of oscillation is fixed by the forcing terms, but an unknown phase must be considered in $y_1(t)$ of equation (2), so that the system of equations (6) and (7) always has three unknowns. Forcing signals on n enter in the computation of N_0 and N_1 with the just outlined modifications in the sinusoidal case.

Remark 7. Apart from its positive empirical evidence over many years, the existence of periodic solutions derived by the describing function method can be supported by rigorous arguments. Under suitable conditions one can guarantee that the system really has a limit cycle in a computable neighborhood of the PLC [11–13,16,19].

Remark 8. For B tending to zero the solutions of equations (6) and (7) indicate, under some restrictions on the subsystem n, for example to be memoryless and smooth, the Hopf bifurcation conditions of the system $\mathcal{S}$.

Finally, it is to observe that the linear elements, $L(0)$ and $L(j\omega)$, and the nonlinear elements, N_0 and N_1, in equations (6) and (7) are functions of the parameters of basic system $\mathcal{S}$ given in Fig. 1. Thus, the predicted solutions derived by these HB conditions result in terms of such parameters, in particular of the real quantity μ which will be considered in the following as the *bifurcation parameter* of interest.

4 HB for Bifurcations of Limit Cycles

Now, we present an approach to the analysis of limit cycle bifurcations based on the results of the previous section and their extensions (see also [7,15] and, more numerically oriented, [20–23]).

Assume that the system $\mathcal{S}$ under study possesses a PLC of parameters A_1, B_1 and ω_1 defining the output $y_1(t)$ as given by equation (2). Since any local bifurcation of a limit cycle can be characterized studying its dynamics in the neighborhood, the idea is that of giving a small perturbation on $y_1(t)$ and finding conditions under which such a perturbation can be sustained by the system $\mathcal{S}$.

The structure of the perturbation depends on the particular bifurcation we want to detect and a new application of HB techniques is then performed, taking into account any signal variation with respect to the nominal PLC arising in the system $\mathcal{S}$, which can be evaluated in virtue of the small amplitude of such a perturbation.

In order to give a complete picture of the *bifurcations of codimension* 1 of the considered limit cycle, it is convenient to separate the case where the perturbed signal has the same structure of the PLC $y_1(t)$ from the case where the perturbed signal requires necessarily a modification of this structure. The need for this separation, makes it clear a distinction between non singular and singular perturbations. In the latter case, the application of HB requires the writing of additive equations. Next we present the application of the method for the different limit cycle bifurcations.

4.1 Cyclic fold bifurcation

In terms of perturbation this means that another limit cycle is present in the vicinity of the original one. Referring to a stable orbit, the typical scenario is that an unstable orbit coexists very close to it and at the bifurcation these orbits collide and disappear (see, for example, [1,24–26]). Indeed, the same situation of coexistence of limit cycles around the bifurcation can be present where the nonlinearities of the system $\mathcal{S}$ satisfy symmetry or other regularity properties, and different phenomena (for example symmetry breaking) take place at the bifurcation point. In view of the control purposes explained in the following we are not interested to such cases and we only consider the first situation, as expressly indicated by the bifurcation name.

When the limit cycle is described by $y_1(t)$ as given by equation (2), its perturbation of the same structure leads to the signal

$$y(t) = (A_1 + \Delta A) + (B_1 + \Delta B)\cos(\omega_1 + \Delta\omega)t, \tag{10}$$

where ΔA, ΔB and $\Delta\omega$ are small variations in bias, amplitude and frequency of the nominal oscillation. Taking into account the HB equations (6) and (7), the related conditions which ensure that such variations are sustained result

$$\left[L^{-1}(0) + N_0 + A_1\frac{\partial N_0}{\partial A}\right]\Delta A + \left(A_1\frac{\partial N_0}{\partial B}\right)\Delta B + \left(A_1\frac{\partial N_0}{\partial \omega}\right)\Delta\omega = 0, \tag{11}$$

$$\left(\frac{\partial N_1}{\partial A}\right)\Delta A + \left(\frac{\partial N_1}{\partial B}\right)\Delta B + \left\{\frac{\partial}{\partial\omega}\left[L^{-1}(j\omega) + N_1\right]\right\}\Delta\omega = 0, \tag{12}$$

where the inverse transfer function of l is used for simplicity and any term must be evaluated at A_1, B_1 and ω_1. In order that equations (11) and (12), the first one real and the second complex, have a non trivial solution in ΔA, ΔB and $\Delta\omega$, their determinant is equated to zero obtaining the relation

$$\arg\left\{\frac{\partial N_1}{\partial B} - A_1\frac{\partial N_1}{\partial A}\frac{\partial N_0}{\partial B}\left[L^{-1}(0) + N_0 + A_1\frac{\partial N_0}{\partial A}\right]^{-1}\right\} =$$

$$= \arg\left\{\frac{\partial}{\partial\omega}\left[L^{-1}(j\omega) + N_1\right] - A_1\frac{\partial N_1}{\partial A}\frac{\partial N_0}{\partial\omega}\left[L^{-1}(0) + N_0 + A_1\frac{\partial N_0}{\partial A}\right]^{-1}\right\}. \tag{13}$$

A number of special cases can be detailed due to the generality of system $\mathcal{S}$, also considering the situation where $L^{-1}(0) + N_0 + A_1\frac{\partial N_0}{\partial A} = 0$, but we can say that equation (13) essentially takes into account the cases of interest, according to the above considerations made on the bifurcation under study.

4.2 Flip and secondary Hopf (Neimark-Sacker) bifurcations

Now, consider a perturbed output signal of system $\mathcal{S}$ having the form [7]

$$y(t) = A_1 + B_1\cos\omega_1 t + \epsilon\cos(\omega' t + \phi) + \eta\cos[(\omega_1 - \omega')t + \psi], \tag{14}$$

where ϵ and η denote small numbers and the frequency $\omega' \in (0, \omega_1/2]$.

The structure assumed for the perturbation of the nominal PLC $y_1(t)$ comes from the need of having a model which can describe the situations arising in the vicinity of the considered bifurcation phenomena [1,24–26]. In fact, around the secondary Hopf bifurcation a different limit cycle having in addition a new frequency ω'(possibly *incommensurate* with respect to ω_1) occurs close to $y_1(t)$. On the other hand, the presence of the nonlinear subsystem n operating at the frequency ω_1 can surely produce *intermodulation terms* at the $\omega_1 - \omega'$ frequency which are included in equation (14).

For the small size of the perturbation, the steady state output of n due to the above signal $y(t)$ results in the general form [7]

$$\begin{aligned}(n \circ y)(t) &= (n \circ y_1)(t)+\\ &+\epsilon \mathbf{Re}\left\{P(\omega')e^{j(\omega' t+\phi)} + Q(\omega_1-\omega')e^{-j(\phi+\psi)}e^{j[(\omega_1-\omega')t+\psi]}\right\}+\\ &+\eta \mathbf{Re}\left\{Q(\omega')e^{-j(\phi+\psi)}e^{j(\omega' t+\phi)} + P(\omega_1-\omega')e^{j[(\omega_1-\omega')t+\psi]}\right\}+\dots\end{aligned} \tag{15}$$

Apart from the nominal term given by formula (3), this expression only emphasizes the components due to the perturbation and appearing in the band $[0, \omega_1]$, i.e. at the frequencies ω' and $\omega_1 - \omega'$. The complex gains P and Q obviously depend on A_1, B_1, and ω_1. They can be computed from the knowledge of the explicit subsystem n of Fig. 1, usually by substituting the memoryless nonlinearities (see Remark 1) by their derivatives at $y_1(t)$ (or by different computations in nonsmooth cases, see Remark 10 below).

Now, the HB concerning the frequencies ω' and $\omega_1 - \omega'$ is considered for the system $\mathcal{S}$. The corresponding small terms result to be decoupled by those of the PLC $y_1(t)$ satisfying the main equations (6) and (7). The components of the perturbation in (14) are equated with the output of the linear subsystem l (see Fig. 1) due to the input (15) at the same frequencies. The following system of complex equations descends [7]

$$\left[L^{-1}(j\omega') + P(\omega')\right]\epsilon + \left[Q(\omega')e^{-j(\phi+\psi)}\right]\eta = 0\,, \tag{16}$$

$$\left[Q(\omega_1-\omega')e^{-j(\phi+\psi)}\right]\epsilon + \left[L^{-1}(j(\omega_1-\omega')) + P(\omega_1-\omega')\right]\eta = 0\,, \tag{17}$$

where $\omega' \in (0, \omega_1/2]$.

Remark 9. As explained in Section 3 for the determination of PLCs, the HB conditions are approximate in nature and now equations (16)-(17) neglect the frequency component outside the band $[0, \omega_1]$. Then, to expect reliable results from such conditions it is required a more severe system filtering than the one indicated in Section 3 for a correct limit cycle prediction.

In order to have a non-zero solution of the homogeneous system of equations (16)-(17), with real values of the unknown amplitudes ϵ and η, further conditions can be easily derived. By equating to zero the determinant of equations (16)-(17) it follows [7]

$$\left|\left[L^{-1}(j\omega') + P(\omega')\right]\left[L^{-1}(j(\omega_1 - \omega')) + P(\omega_1 - \omega')\right]\right| = \left|Q(\omega')Q(\omega_1 - \omega')\right| , \tag{18}$$

$$\arg\left[\frac{Q(\omega')}{L^{-1}(j\omega') + P(\omega')}\right] = \arg\left[\frac{Q(\omega_1 - \omega')}{L^{-1}(j(\omega_1 - \omega')) + P(\omega_1 - \omega')}\right] = \phi + \psi , \tag{19}$$

where $\omega' \in (0, \omega_1/2]$, as a set of conditions corresponding to the existence of small subharmonics on the PLC and giving their phase. Observe that for $\omega' = \omega_1/2$ equation (19) is automatically satisfied. Therefore, in this approach the flip bifurcation follows as a particular case of the Neimark-Sacker bifurcation.

The three possible scenarios which have been considered concern the possible way how the *Floquet multipliers* of a limit cycle cross the unit circle, giving rise to local bifurcations, by varying one suitable system parameter [1,24–26]. If we move from a stable periodic solution, the cyclic fold occurs when a multiplier leaves the unit circle through +1, the flip when a multiplier leaves the circle through -1, and finally, at the secondary Hopf, two complex conjugate Floquet multipliers leave the unit circle away from the real axis. In terms of solution due to the described bifurcations, the original limit cycle: i) disappears, ii) becomes a *period-two cycle* and iii) becomes (in general) a *quasiperiodic orbit* (a torus), respectively.

The derived conditions (13) and (18)-(19) summarize the main result of the section. They represent an additive equation, with respect to those determining the PLC $y_1(t)$, and the different structure of these conditions follows from the need of representing a perturbation similar to the nominal PLC or a perturbation of different form, respectively.

The bifurcation equations must be solved in terms of the parameter μ to identify, according to the method approximation, the considered phenomena. Observe that generally (18) and (19) are two equations, but the unknown variable ω' can be eliminated between them. The form of these conditions follows from the generality of system $\mathcal{S}$: the role of subsystem l is expressed through the related inverse transfer function L^{-1}, while the nonlinear subsystem n influences N_0, N_1, P and Q.

The above results will be now applied to the important class of *Lur'e systems* in order to obtain more specific bifurcation conditions which also have a greater insight.

5 Limit Cycle Bifurcations in Lur'e Systems

Assume that the nonlinear subsystem n of the system $\mathcal{S}$ under study (Fig. 1) reduces to a single-valued function n for which are valid the considerations of Remark 5.

The investigation of the dynamics around a given PLC $y_1(t)$ proposed in Section 4 leads to determine the gains P and Q appearing in equation (15) and following. Denoting by $n'[y_1(t)]$ *the derivative* of n at $y = y_1(t)$ we obtain in a straightforward way the two real terms

$$P \doteq N_0'(A, B) = \frac{1}{2\pi} \int_{-\pi}^{\pi} n'[y_1(t)] \, d\omega_1 t, \tag{20}$$

$$Q \doteq \frac{1}{2} N_1'(A, B) = \frac{1}{2\pi} \int_{-\pi}^{\pi} n'[y_1(t)] \cos \omega_1 t \, d\omega_1 t. \tag{21}$$

Here, N_0' and N_1' denote the Fourier coefficients of order 0 and 1 of the derivative of n, and are connected to the derivatives of the harmonic gains N_0 and N_1. In fact, taking into account (4),(5) and (20),(21), the following relations can be easily derived

$$N_0' = N_0 + A\frac{\partial N_0}{\partial A} = N_1 + B\frac{\partial N_1}{\partial B}, \tag{22}$$

$$N_1' = 2A\frac{\partial N_0}{\partial B} = B\frac{\partial N_1}{\partial A}, \tag{23}$$

which allows one to directly express P and Q as functions of N_0 and N_1 determining the PLC.

Remark 10. In the case of nonsmooth nonlinearity n the evaluation of P and Q requires the direct computation (for small ϵ and η) of gains corresponding to the related frequencies. The ratio between ω' and ω_1 can be approximated in rational form in order to make use of the Fourier coefficient formulas.

By using the properties of memoryless nonlinearities (see Remark 5) and relations (20) and (21) we can restate the bifurcation conditions of the previous Section in the following form.

Cyclic fold bifurcation

$$\frac{d}{d\omega}\left\{\mathbf{Im}\left[L^{-1}(j\omega)\right]\right\}_{\omega_1} = 0, \tag{24}$$

Flip bifurcation

$$\left|L^{-1}\left(j\frac{\omega_1}{2}\right) + N_0'\right|^2 = \frac{1}{4} N_1'^2, \tag{25}$$

Secondary Hopf bifurcation

$$\left|\left[L^{-1}(j\omega') + N_0'\right]\left[L^{-1}(j(\omega_1 - \omega')) + N_0'\right]\right| = \frac{1}{4}\, N_1'^2, \tag{26}$$

$$\arg\left[L^{-1}(j\omega') + N_0'\right] = \arg\left[L^{-1}(j(\omega_1 - \omega')) + N_0'\right]. \tag{27}$$

Again, it is to be recalled that these are *additive equations* with respect to (6) and (7) and their solution is intended in terms of the bifurcation parameter μ. Observe that the first condition only depends by the linear subsystem appearing in $\mathcal{S}$ and, according to (9), it requires that its transfer function be *tangent* to the real axis.

Remark 11. Indeed, the general cyclic fold condition (13) can also leads, for Lur'e systems, to a different solution only depending by the nonlinearity and corresponding to the condition $\frac{\partial N_0}{\partial A}\frac{\partial N_1}{\partial B} = \frac{\partial N_0}{\partial B}\frac{\partial N_1}{\partial A}$, where the two limit cycles coexisting close to the bifurcation have the same frequency. This special situation is not considered in view of the proposed bifurcation control.

The flip and Neimark-Sacker bifurcation conditions appear to be more complicate, involving at a same time the characteristics of linear and nonlinear subsystems of $\mathcal{S}$: in the latter case also the new frequency ω' of the solution is introduced.

Summing up, the previous analysis gives a *structural view* of the main bifurcations of limit cycles in a feedback Lur'e system, and the related results obviously appear more readable than in the general case of Section 4. The simplified conditions which have been derived for the existence of bifurcations concern subharmonic perturbations and they result as more valid as the system satisfies the filtering hypothesis outside the frequency band $[0, \omega_1]$. These conditions are expressed in terms of the main elements of the system under study and can be an useful tool, not fully numerical, for a first order prediction of very complicated behaviors. For example, as indicators of possible transitions towards chaotic motions [27,28]. Moreover, the method gives the location of bifurcating limit cycles in the phase space, so that the obtained results can be useful for simulating the predicted behaviors.

Indeed, an important point must yet be faced. It concerns the *stability properties* of the PLCs and the related properties of the arising perturbations. So that, for example, one can distinguish between *supercritical and subcritical cases* (in the flip and Neimark-Sacker bifurcations) [1,24–26]. This will be considered in the next Section, as a necessary check of the solutions derived by the presented method.

6 Stability of Predicted Limit Cycles and Their Bifurcations

The solutions of equations (6) and (7) concerning equilibria and limit cycles have been obtained by using steady-state relations among the blocks of system $\mathcal{S}$. In fact, the possible initial conditions of dynamic subsystems have been neglected, intending that they must be consistent with the results of the analysis.

To investigate the *local stability* of solutions, we consider the system transients due to small perturbations of such solutions, and precisely when

- A_0 is varied to $A_0 + \Delta A(t)$ for the equilibria ($B = 0$);
- B_1 is quasistatically varied to $B_1 + \Delta B$ for the limit cycles, in the assumption that other perturbations give rise to motions decaying to zero.

In the first case the consequent system behavior is obtained by *linearizing* the subsystem n at $y = A_0$ and considering the corresponding dynamics of $\mathcal{S}$. Apart from critical cases, a standard linear criterion (for example the Nyquist criterion) yields the local stability features of any equilibrium point.

For the PLCs of expression (2), according to the *Loeb criterion* [14,17], [18], we assume that the harmonic balance equation (7) still holds for slight perturbations, so that a damped oscillation of complex frequency $j\omega_1+(\Delta\sigma+j\Delta\omega)$ corresponds to a modified amplitude $B_1 + \Delta B$. Since the dependence of the bias is derived from equation (6) in the simplified static form $A = A(B, \omega)$, the equation concerning the PLC dynamics becomes

$$\begin{aligned} &1 + L[j\omega_1 + (\Delta\sigma + j\Delta\omega)]\cdot \\ &\cdot N_1[A(B_1 + \Delta B, \omega_1 + \Delta\omega), B_1 + \Delta B, \omega_1 + \Delta\omega - j\Delta\sigma] = 0. \end{aligned} \tag{28}$$

Considering the variations around the nominal solution defined by A_1, B_1 and ω_1, the Loeb criterion requires for the stability of the PLC that $\Lambda \doteq \Delta\sigma/\Delta B$ is negative. In other words, any perturbation which brings the initial condition of the system "inside" the PLC ($\Delta B < 0$) is expected to correspond to an output growth as a function of time ($\Delta\sigma > 0$), and vice-versa for a perturbation "outside" the PLC. The condition, applied to equation (28), gives rise to the stability formula

$$\Lambda = -\mathbf{Im}\left\{\overline{L(j\omega_1)\frac{dN_1}{dB}[A(B,\omega),B,\omega]}\cdot\frac{d}{d\omega}L(j\omega)N_1[A(B,\omega),B,\omega]\right\} < 0, \tag{29}$$

where the overbar indicates complex conjugation and the derivatives are evaluated at the nominal point.

As for the prediction results of Section 3, the conclusions derived by means of preceding criteria are exact for the equilibria and approximate in nature for the periodic solutions. The approximation is less good than that concerning the existence of PLCs, for the additional simplifying assumptions made in the approach. On the other hand, the Loeb criterion results very often correct and in most cases, when N_0 and N_1 are frequency independent, is intuitively appealing since it represents a kind of extension of Nyquist stability criterion.

Remark 12. In the case of Lur'e systems, where N_1 is real and frequency independent and $L(j\omega_1)$ is consequently real, the Loeb stability condition (29) simply reduces to

$$\Lambda = \frac{dN_1}{dB}[A(B), B]_{|B_1} \cdot \mathbf{Im}\left\{\frac{d}{d\omega}L(j\omega)_{|\omega_1}\right\} < 0. \tag{30}$$

Remark 13. In applying conditions (29) and (30) it is necessary take into account the underlying above assumption that the perturbation ΔB is the only one considered as critical for the PLC stability.

The results obtained for the PLC stability, beyond their direct meaning, can also be used to check the kind of bifurcation recognized by the conditions of Section 4 and Section 5. In fact, considering now the dependence of the system by the parameter μ, let us assume that the bifurcation point occurs for $\mu = \mu_c$. The essential idea consists in applying the above criterion for the PLC stability on the side of bifurcation point where only an isolated periodic solution is present. This allows one to consider the cases supercritical and subcritical by studying a situation where the Loeb stability criterion appears surely more reliable than if it is applied when more solutions coexist [7,15].

Such analysis requires first the application of HB at the different frequencies involved, which are 0, ω_1, ω' and $\omega_1 - \omega'$ for the perturbed signal (14). Now, we must consider the first neglected terms in the expression of $(n \circ y)(t)$, namely the quadratic terms in ϵ and η. This means that the derived equations of HB will form now a unique algebraic system whose solution allows one to evaluate also the above small perturbation amplitudes as functions of the system parameter μ. From this result, by looking for example at the amplitude ϵ and taking into account the well-known typical bifurcation diagrams of *supercritical and subcritical cases* [1,24–26], the application of the above idea leads to the following conclusion. Consider values of μ sufficiently close to μ_c and such that

$$\left.\frac{d\epsilon^2}{d\mu}\right|_{\mu=\mu_c} (\mu - \mu_c) < 0, \tag{31}$$

and apply for the corresponding PLC the stability formula (29). Then

- $\Lambda(\mu) < 0$, i.e. a stable PLC, indicates a supercritical bifurcation at $\mu = \mu_c$;
- $\Lambda(\mu) > 0$, i.e. an unstable PLC, indicates a subcritical bifurcation at $\mu = \mu_c$.

Observe that the application of this criterion, which requires the computation of Λ at values of μ satisfying (31), has only sense for bifurcations satisfying the general equations (18) and (19) (and for equations (25) and (26)-(27)) and not for cyclic fold bifurcations as defined by (13) and (24).

Moreover, the derivative in (31) can be directly evaluated by linearizing the complete HB equation system without an explicit computation of $\epsilon^2 = \epsilon^2(\mu)$ (see, for example, [15] in the flip bifurcation case).

7 Control of Limit Cycle Bifurcations

In this section we intend to apply the previous results on the analysis and prediction of limit cycle bifurcations in order to control such phenomena. The objective of this study is to control the above bifurcations, in the sense of suitably modifying their occurrence with respect to the representative parameter μ, which is assumed to undergo quasistatical variations.

In fact, it is well-known that at any bifurcation there is a qualitative change in the dynamics features of the system, so that different behaviors (aperiodic, periodic, chaotic, etc.) and stability conditions are introduced in a continuous or discontinuous way. Depending on the system under study such new behaviors may result in desirable or dangerous dynamics. Therefore, the control considered in the following is oriented to *move the bifurcation point* of interest, by anticipating or delaying it, and even to suppress such phenomenon, according to specific requirements on the system.

With regard to the bifurcations here studied, which refer to limit cycles, we remark that very often they indicate the beginning of routes to chaos of the system (see, for example, [24,25,29]). So, in the supercritical case, the flip bifurcation opens the *period doubling* scenario which can give rise to the well-known Feigenbaum cascade, while the Neimark-Sacker bifurcation can lead to one of the *quasiperiodic routes* to chaos. Also, the cyclic fold bifurcation and the two bifurcations mentioned above in the subcritical case, can originate one of the *intermittency* mechanisms which occur when, at the same time, other system features are present. Therefore, the bifurcation control studied in this context is strongly related to *controlling chaos* in many dynamical systems (see, for example, [4,30–32].

Before introducing more formally the bifurcation control problem, it is important to remark that we want to perform this operation essentially maintaining the general dynamic behaviors of the system, with the required modification around the bifurcation point of interest, but preserving the qualitative dynamics exhibited in other ranges of the bifurcation parameter μ, where the system is supposed to operate in a satisfactory way.

7.1 Problem formulation

We assume that a *linear feedback controller* is employed in system $\mathcal{S}$ of Fig. 1, described by the operator k and having transfer function denoted by $K(s)$. The overall controlled system $\mathcal{S}^*$ is depicted in Fig. 2 and it is governed by the operator equation

$$l \circ (n \circ y + k \circ y) + y = 0, \tag{32}$$

presenting the same structure of equation (1) which holds for the original system $\mathcal{S}$.

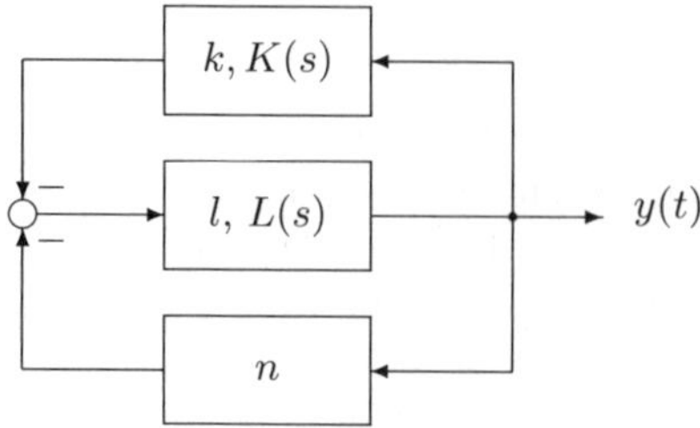

Fig.2. The controlled system $\mathcal{S}^*$.

Suppose now that the following scenario, which commonly occurs in many applications, applies to the uncontrolled system. The system $\mathcal{S}$ presents a Hopf bifurcation at $\mu = \mu_H$, resulting in a stable limit cycle in the parameter range $\mu_H < \mu < \mu_c$, which undergoes a limit cycle bifurcation (cyclic fold, flip or secondary Hopf) at $\mu = \mu_c$.

The primary objective is to obtain a feedback controller k such that for the controlled system $\mathcal{S}^*$ the above limit cycle bifurcation is moved to some value μ_c^*, different from μ_c and greater than μ_H. This is to be achieved without altering the existence of aperiodic and periodic regimes before the value $\mu = \mu_c^*$.

In the framework of HB where the analysis has been performed, the problem can be stated as follows [5,15,33]. Suppose we are given a system $\mathcal{S}$ which possesses a family of PLCs, stable in the range $\mu_H < \mu < \mu_c$, and has a predicted bifurcation of limit cycle at $\mu = \mu_c$, according to the conditions given in the previous sections. Determine a feedback controller k satisfying the following conditions:

i) $\mathcal{S}^*$ has a predicted bifurcation of limit cycle at $\mu = \mu_c^*$;
ii) $\mathcal{S}^*$ has the same PLC of $\mathcal{S}$ at $\mu = \mu_p$;
iii) $\mathcal{S}^*$ has a family of stable PLCs for $\mu < \mu_c^*$;
iv) $\mathcal{S}^*$ has the same equilibrium points of $\mathcal{S}$.

In particular, apart the main requirement i), condition ii) is imposed to maintain in the system $\mathcal{S}^*$ a representative element of the periodic behavior derived for the uncontrolled system $\mathcal{S}$, selected at the generic $\mu = \mu_p$. In other words, in reshaping the bifurcation diagram of $\mathcal{S}^*$ related to the parameter μ we intend to preserve, in terms of amplitude and frequency which define the corresponding PLC, the point at $\mu = \mu_p$ just found for $\mathcal{S}$, by assuming that around such a point the behavior of the new diagram has slight modifications with respect to the one of the uncontrolled system. Therefore, the same scenario is expected for the controlled system $\mathcal{S}^*$, presenting a Hopf bifurcation followed by a family of stable limit cycles (close to the original corresponding features) and then the interesting limit cycle bifurcation at the new imposed value $\mu = \mu_c^*$. The value of μ_p appearing in condition ii) can be selected in different ways, in order to simplify the controller design procedure and also depending on the considered bifurcation, as it will be shown in the following. In principle, significant values of the bifurcation diagram can be: $\mu_p = \mu_H$, $\mu_p = \mu_c$ and $\mu_p = \mu_c^*$.

Finally, condition iv) has been imposed because of the common requirement in feedback control of engineering systems that constant operating points be unaltered by the applied control.

7.2 Controller design

The control strategy is motivated by previous works which employ "washout filters" for preserving the original equilibria in the controlled system, with a related extension of the idea to preserve also limit cycles of a given frequency [5,15,33,34]. To reach this objective the transfer function $K(s)$ of the controller k is assumed to satisfy the constraints

$$K(0) = 0, \tag{33}$$

$$K\big[j\omega_1(\mu_p)\big] = 0, \tag{34}$$

which guarantee that the steady state action of such controller vanishes at the frequencies 0 and ω_1, also resulting in a small control energy in the neighborhood of such behaviors.

According to the assumptions made in applying the techniques proposed in the previous sections - the so-called "filtering hypothesis" - it follows that the above controller essentially preserves the limit cycles having such fundamental frequencies, since their higher harmonics result to be negligible. Apart from the fulfillment of constraints (33) and (34), it is also suitable to impose that the controller transfer function is rational, proper (or strictly proper) and stable, depending by a free parameter vector ρ which is to be defined in the design. The standard way to meet conditions (33) and (34) is to include the factors s and $(s^2 + \omega_1^2)$ in $K(s)$, while other possible incorporations can consist, for example, in a gain ρ_1 and a denominator factor $(s + \rho_2)^3$, with $\rho_2 > 0$. A general representation of the controller transfer function can be in the form $K(s, \rho)$, considering that its structure is fixed, according to (33) and (34) and to the other requirements imposed.

Coming back to the bifurcation control problem, as previously stated, we apply the analysis method of Section 3 through Section 6 to the controlled system $\mathcal{S}^*$ of Fig. 2, in order to satisfy the condition i)-iv) of such problem in terms of the free parameters ρ of the controller. We remark that the similar structure of $\mathcal{S}$ and $\mathcal{S}^*$ allows one to consider all the HB equations previously written for deriving PLCs and their bifurcations in the uncontrolled system as valid for the controlled system, with the substitution of $L(s)/(1+L(s)K(s))$ in place of $L(s)$ or, more simply, of $L^{-1}(s)+K(s)$ in place of $L^{-1}(s)$. Then, we propose the general procedure for the design assuming that the PLC whose characteristics have to be preserved is the one corresponding at $\mu_p = \mu_c^*$, just where the new limit cycle bifurcation is imposed. This implies that, due to the constraints (33) and (34) on k, the HB equations are the same for $\mathcal{S}$ and $\mathcal{S}^*$, so giving the same A_1, B_1 and ω_1 for the two systems at such a point of the bifurcation curve. Moreover, the bifurcation conditions ((13) or (18)-(19), in the general case) only require the above substitutions, while all the existing terms remain unchanged, and must be solved for the value ρ satisfying the equation for $\mu = \mu_c^*$.

To explain the procedure, consider for example the case of *flip bifurcation* for Lur'e systems. The condition equation (25) becomes for the controlled system

$$|L^{-1}(j\omega_1/2) + K(j\omega_1/2, \rho) + N_0'|^2 = \frac{1}{4}\, N_1'^2, \tag{35}$$

where the variable A_1, B_1 and ω_1 affecting L, N_0' and N_1' are the solutions of the original PLC equations (6) and (7). That is, in addition to the application of (6) and (7), the solution of bifurcation equation (35) is obtained in the uncontrolled case for $\mu = \mu_c$ and $K = 0$, while for the controlled case is obtained for $\mu = \mu_c^*$ and for a suitable value ρ affecting $K(j\omega_1/2, \rho)$. Then, the bifurcating PLC is always the solution of (6) and (7): at $\mu = \mu_c$ for the uncontrolled system, and at $\mu = \mu_c^*$ for the controlled system. In particular, in the latter case the PLCs of the two systems coincides but without any bifurcation for that of the uncontrolled system.

Indeed, due to the selection of μ_p the described technique only employs the additive bifurcation condition, without modifying the main limit cycle equations derived by the HB principle. This has been possible because we have considered that a PLC of $\mathcal{S}$ has been found just at the value of μ, denoted by μ_c^*, which is imposed for the bifurcation of the controlled system $\mathcal{S}^*$. In the considered scenario, such a situation usually applies for the flip and secondary Hopf bifurcations, where the derived PLCs extends over a range from $\mu = \mu_H$ to values of μ which are beyond the bifurcation point predicted for the system at μ_c. The same situation exists for the cyclic fold bifurcation when the objective of the control is to anticipate the bifurcation point, so that the PLC to be preserved is one on the stable branch before the tangency.

On the contrary, the proposed selection $\mu_p = \mu_c^*$ cannot be applied for delaying a cyclic fold bifurcation, because after the tangency no PLC is derived for the uncontrolled system $\mathcal{S}$. In such case a different technique can be followed. We impose $\mu_p = \mu_c$, and this corresponds to preserving the PLC at the uncontrolled bifurcation, but we use the bifurcation condition of the controlled system to delay its solution with respect to μ_c.

Unfortunately, in order to derive the new bifurcation point it is now necessary to solve the whole system given by (6), (7) and (13), with the substitution of $L^{-1}(s) + K(s)$ instead of $L^{-1}(s)$. So, the procedure results much more complicate than in other cases, where the only equation including the controller was that related to the bifurcation, while those defining the PLC remain unchanged. On the other hand, by continuity it can be expected that slight modifications of the linear subsystem l, by use of the controller k, results in corresponding modifications of the bifurcation value, and a suitable selection of the controller leads to the required delay of such a bifurcation in terms of the parameter μ.

Finally, in any design procedure that we employ it remains to check the requirement iii) of the bifurcation control problem: the stability of the *PLC family* preceding the bifurcation point. This can be done by the techniques presented in Section 7 on the basis of the Loeb criterion. As a principle, we can again remark that slight variations of the transfer function of the linear subsystem will intuitively preserve the stability features of the uncontrolled system, while greater variations can violate such situations and will require the above check.

7.3 A simplified technique: the subharmonic bifurcation control

A simplified approach to the control of limit cycle bifurcations is now presented. It results in a solution inspired by the controller action on the specific perturbations arising from a periodic solution close to its bifurcation, as shown in Section 4. On the other hand, the controller k preserves the above features of a washout filter.

With respect to the problem formulation of the previous section, we relax point i) which only indicates the requirement of delaying the bifurcation value $\mu = \mu_c$ of the PLC and does not give an a-priori prediction of the new bifurcation value μ_c^*. We fix in point ii), as invariant PLC for the uncontrolled and controlled system, that appearing at $\mu = \mu_p = \mu_c$, while the points iii) and iv) remain unchanged.

The basic idea consists in considering the linear part of the controlled system, having transfer function $L(s)/(1 + L(s)K(s))$, as a filter which can reduce the onset of *subharmonic terms* in the nonlinear loop of system $\mathcal{S}^*$. Since such subharmonic terms, which are related to the occurrence of limit cycles bifurcations, are the components of frequency $\omega_1/2$ for the flip and generally $\omega' \in (0, \omega_1/2]$ for the secondary Hopf, we impose a high gain of the controller at such frequencies and then we reduce the above filter gain [35–37]. Therefore, the transfer function of the controller k, satisfying the constraints (33) and (34), can have the form

$$K(s, \rho) = \rho_1 \frac{s(s^2 + \omega_1^2)}{(s^2 + 2\rho_2\omega' s + \omega'^2)(s + \rho_3)}. \tag{36}$$

For small values of the damping coefficient ρ_2, the controller has a resonance close to ω' (or $\omega_1/2$) which tends to stop the circulation of such frequency along the controlled system $\mathcal{S}^*$, resulting in a delay of the considered limit cycle bifurcation. In this case, the controller design only requires the knowledge of the PLC features at the uncontrolled system bifurcation, in particular the frequencies ω_1 and ω' (for the Neimark-Sacker case). Since the selection of ρ_2 (small and positive) and ρ_3 (positive) is not critical, the objective of moving the bifurcation point is reached by moving more or less the gain ρ_1, expecting that for greater value of such parameter there is a greater delay of the bifurcation point from the uncontrolled value μ_c.

In conclusion, such an approach is more *qualitative* than the one presented in Section 7. It is devoted to *bifurcation delay* and requires less information, having solution by a trial and error procedure on the controller parameter ρ, in particular the gain ρ_1 which allows one to modulate the imposed delay. Indeed, this technique has been presented with regard to flip and Neimark-Sacker bifurcations, where signal perturbations far from frequencies 0 and ω_1 induce such phenomena. For the case of cyclic fold, and referring to the need of delaying it, we must consider that the frequency of the perturbation is now close to ω_1, which remain unchanged for (33), Then, the problem can be faced by putting the frequency resonance ω' of the controller just close to ω_1 and adjusting the parameter ρ in order to destroy the uncontrolled bifurcation and recreate it successively for the controlled system $\mathcal{S}^*$.

Finally, as in the other procedures, the design must be completed by checking the stability of the PLC family before the bifurcation point obtained. For controllers k which satisfy the required specifications inducing small modifications in the original subsystem l, it can be expected that the stability features of $\mathcal{S}$ are preserved in $\mathcal{S}^*$.

8 Application Examples

8.1 Control of a period doubling bifurcation

We consider a system $\mathcal{S}$ as in Fig. 1, where the linear subsystem has transfer function

$$L(s) = \frac{1}{s^3 + \mu s^2 + 1.2s + 1}, \tag{37}$$

and the bifurcation parameter μ is restricted to be negative. The nonlinear subsystem is the memoryless nonlinearity

$$n(y) = -y^2. \tag{38}$$

Note that system $\mathcal{S}$ has been shown to possess a rich variety of complex dynamics [15,27]. It is easy to verify that the system possesses a stable equilibrium point in $y = 0$ for all $\mu < -5/6$ whereas it shows a supercritical Hopf bifurcation at $\mu = -5/6$.

We now compute the first order PLCs. Equations (4) and (5) give the nonlinear gains

$$N_0 = -A - \frac{B^2}{2A}, \tag{39}$$
$$N_1 = -2A. \tag{40}$$

Substitution of equations (39)-(40) and of $L(j\omega)$ from (37) into (6) and (7) leads to the following HB limit cycle equations:

$$A\left[1 - A - \frac{B^2}{2A}\right] = 0, \tag{41}$$
$$1 - \frac{2}{1 + \mu\omega^2 + j\omega(1.2 - \omega^2)} A = 0. \tag{42}$$

Solving (41) and (42) results in the following one-parameter family of PLCs for $\mu \in (-5/6, 0)$:

$$A(\mu) = \frac{1 + 1.2\mu}{2}, \tag{43}$$
$$B(\mu) = \sqrt{\frac{1 - (1.2\mu)^2}{2}}, \tag{44}$$
$$\omega(\mu) = \sqrt{1.2}\ . \tag{45}$$

Next, we study the presence of period doubling (flip) bifurcations for the uncontrolled system. Applying equations (22) and (23) to the given nonlinearity, we get

$$N_0' = -2A, \qquad N_1' = -2B.$$

Equation (25) is now used to determine the above bifurcation. This equation gives

$$\left| -\frac{3}{4}1.2\mu + j\frac{3.6\sqrt{1.2}}{8} \right| = \sqrt{\frac{1 - (1.2\mu)^2}{2}}\ . \tag{46}$$

Solving equation (46) provides a predicted period doubling bifurcation at $\mu_c = -0.41$, $\omega_1 = \sqrt{1.2}$.

To complete the analysis it remains to evaluate the stability of the one-parameter family of PLCs (43)-(45) for $\mu \in (-5/6, -0.41)$. These PLCs are indeed predicted to be stable by exploiting the Loeb Criterion presented in Section 6.

Numerical simulations show that the true behavior of the system agrees qualitatively with the predictions above. The bifurcation diagram of system $\mathcal{S}$ is given in Fig. 3 (light dots). The nominal stable limit cycle exists in the interval $(-5/6, -0.48)$, and it undergoes a period doubling bifurcation at $\mu = -0.48$. This bifurcation is the first of a sequence of period doubling bifurcations leading to chaotic motion at $\mu = -0.39$.

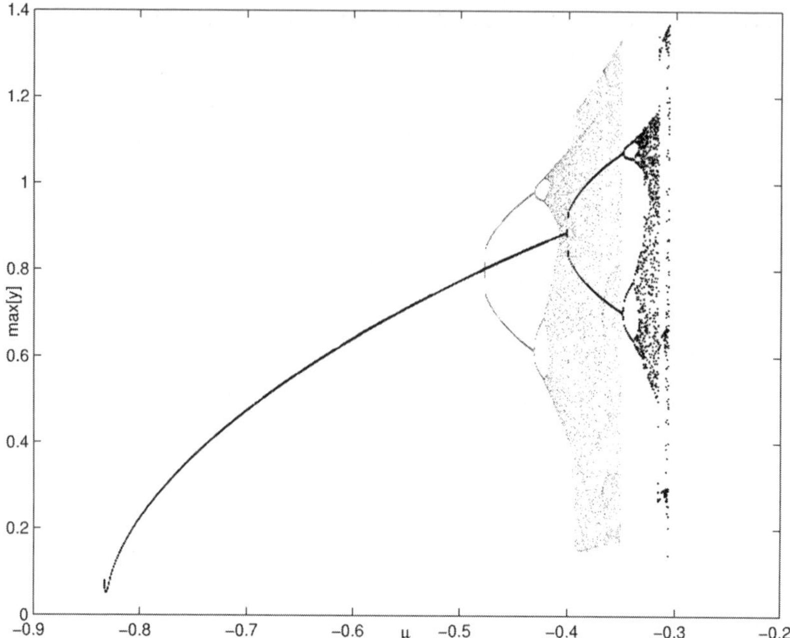

Fig.3. Uncontrolled (light dots) / Controlled (dark dots) bifurcation diagrams.

The objective of the control design is as in Section 7, where specifically we desire a delay of the period doubling bifurcation from $\mu_c = -0.41$ to $\mu_c^* = -0.30$, with a general structure for the feedback linear controller as follows

$$K(s) = \rho_1 \frac{s(s^2 + \omega_1^2)}{(s + \rho_2)^3} \ . \tag{47}$$

Since we found in equation (45) that $\omega(\mu)$ is constant, it follows that $K(j\omega(\mu)) \equiv 0$ for all μ. Using this in the harmonic balance equations, we find that in such an example systems $\mathcal{S}$ and $\mathcal{S}^*$ have the same one-parameter family of PLCs reported in equations (43)-(45).

In order to select the correct parameter ρ, we impose the flip bifurcation condition (35) for the controlled system $\mathcal{S}^*$. The derived equation must be solved with respect to the controller parameter ρ which generally has infinite solutions. Nevertheless, we select the solution which provide the lowest controller amplitude at the frequency $\omega_1/2$. Such solution is simply found by imposing that the phase of $K(j\omega_1/2)$ must be equal to that of the complex vector $L^{-1}(j\omega_1/2) + N_0'$. In this example, the above condition is met for $\rho_2 = 3.25$, while the remaining parameter $\rho_1 = 0.672$ is uniquely selected to satisfy the bifurcation condition (35).

Numerical simulations show that actual parameter value at which system $\mathcal{S}^*$ undergoes a period doubling bifurcation is $\mu = -0.39$. The bifurcation diagram of the controlled system is given in Fig. 3 (dark dots) superimposed with the uncontrolled one (light dots). Comparing the bifurcation diagrams, it is observed that chaotic motion is also delayed in system $\mathcal{S}^*$, as expected.

8.2 Control of a Neimark-Sacker bifurcation

Consider the system of Fig. 1, where [35]

$$L(s) = \frac{(s+1)^2}{s^3(s^2+0.3\mu s+\mu^2)}\,, \quad n(y) = 25.9y - 11.4y^3,$$

and focus on possible symmetric PLCs ($A = 0$). The conditions for the onset of a secondary Hopf bifurcation (Neimark-Sacker bifurcation) are reported in (26)-(27) and give the unique solution

$$\mu_c = 4, \qquad \omega_1 = 3.66, \qquad \omega' = 1.09.$$

This prediction is compared with the simulation results showing good accuracy of the proposed analysis. Indeed, Fig. 4 shows the attractor of the uncontrolled system in a parameter region where the limit cycle ($\mu = 4.05$) becomes a torus ($\mu = 3.95$).

For the problem at hand, subharmonic control is a convenient way to operate to delay a secondary Hopf bifurcation. Figure 5 shows the time transition between the uncontrolled motion and the controlled one for both cases $\mu = 4.05$ and $\mu = 3.95$, when the feedback linear filter

$$K(s) = 0.3\frac{s(0.075s^2+1)}{(0.83s^2+0.17s+1)\,(0.125s+1)}\,,$$

from expression (36) is connected to the system at $t = 0$.

8.3 Control of a cyclic fold bifurcation

Consider system $\mathcal{S}$ depicted in Fig. 1, where the linear subsystem l with transfer function [38]

$$L(s) = \frac{(s+\mu)^2}{(s+1)^3} \tag{48}$$

is feedback interconnected with a relay defined by

$$n[y(t)] = \mathrm{sgn}[y(t)]. \tag{49}$$

In order to study the existence of periodic solutions for the system we compute the bias and harmonic gains (4) and (5) of the relay as

$$N_0(A,B) = \frac{2}{\pi A}\varphi, \tag{50}$$

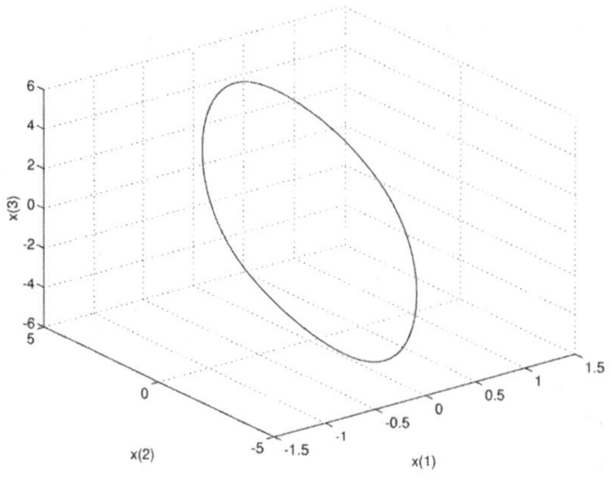

(a)

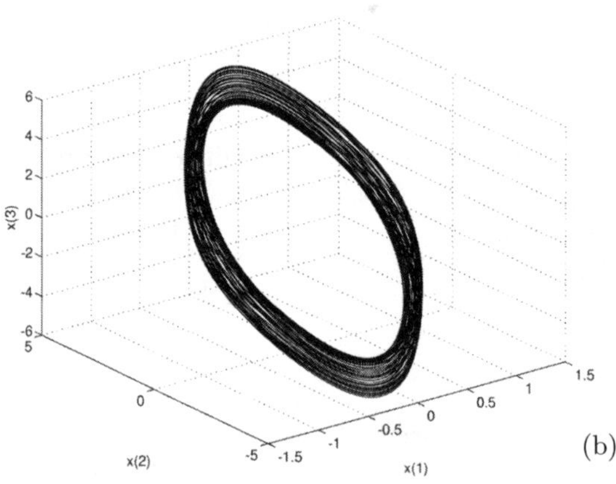

(b)

Fig.4. Uncontrolled trajectory: (a) $\mu = 4.05$; (b) $\mu = 3.95$.

$$N_1(A,B) = \frac{4}{\pi B}\cos\varphi, \tag{51}$$

where

$$\varphi = \arcsin\frac{A}{B} \in \left(-\frac{\pi}{2}, \frac{\pi}{2}\right). \tag{52}$$

By means of (50) and (51), the first order HB given by the general conditions (6) and (7), yields

$$A = -\frac{2}{\pi}L(0)\varphi, \tag{53}$$

$$B = -\frac{4}{\pi}L(j\omega)\cos\varphi. \tag{54}$$

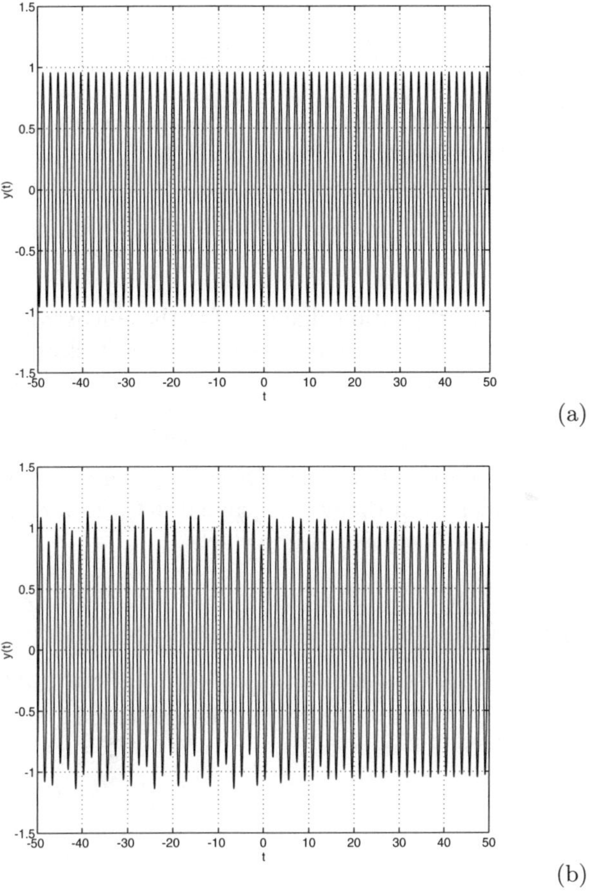

Fig.5. Uncontrolled / Controlled dynamics transition ($t \geq 0$): (a) $\mu = 4.05$; (b) $\mu = 3.95$.

The above equations can be restated as

$$\mathrm{Im}[L(j\omega)] = 0, \tag{55}$$

$$\sin 2\varphi = \frac{L(0)}{L(j\omega)}\varphi, \tag{56}$$

to be solved in the unknowns ω and φ. Reducing to the simpler case of symmetric PLCs ($A = 0$), this implies $\varphi = 0$ and, therefore

$$B = -\frac{4}{\pi}\, L(j\omega) = -\frac{4}{\pi}\, \frac{2\mu}{3 - \omega(\mu)^2}\ , \tag{57}$$

where

$$\omega(\mu)^2 = \frac{\mu^2 - 6\mu + 3 \pm \sqrt{(9-\mu)(1-\mu)^3}}{2} \,. \tag{58}$$

By exploiting the bifurcation conditions of Section 5 for memoryless nonlinearities, we observe that a cyclic fold bifurcation is predicted via equation (24) for

$$\mu_c = 9 \,, \quad \omega_1 = 3.87 \,, \quad B_1 = \frac{6}{\pi} \,. \tag{59}$$

Figure 6 shows the predicted bifurcation diagram for the uncontrolled system (dash-dotted curve). The related stability analysis makes it clear the presence of a stable branch which collides with the unstable one (dotted curve) at $\mu_c = 9$. Simulations confirm such a behavior, except that the actual bifurcation parameter value is slightly anticipated at $\mu = 9.25$.

The objective of the control design is to delay (μ is here considered as decreasing) the cyclic fold bifurcation in the way explained in Section 8. To this aim the following subharmonic controller is introduced

$$K(s) = -0.2 \frac{s(s^2 + \omega_1^2)}{(s+0.1)(s^2 + 0.774s + \omega_1^2)} \,, \tag{60}$$

whose parameters have been chosen in order to keep unchanged the stable PLC at the parameter value $\mu = 9$. Figure 6 shows that the controlled system $\mathcal{S}^*$ succeed in delaying the predicted bifurcation from $\mu_c = 9$ to $\mu_c^* = 8$ (solid curve).

References

1. Arnold, V. I. (1983) Geometrical Methods in the Theory of Ordinary Differential Equations. New York: Springer-Verlag
2. Glass, L., Mackey, M. (1988) From Clock to Chaos. Princeton: Princeton Univ. Press
3. Ott, E., Sauer, T., Yorke, J. (1994) Coping with Chaos. New York: Wiley
4. Abed, E. A., Wang, H. O. (1995) Feedback control of bifurcations and chaos in dynamical systems, In Recent Developments in Stochastic and Nonlinear Dynamics: Applications to Mechanical Systems, Sri Namachchivaya, N., Kliemann, W. (eds.).Boca Raton, FL: CRC Press
5. Wang, H. O., Abed, E. H. (1995) Bifurcation control of a chaotic system. Automatica, **31**:1213–1226
6. Chen, G., Moiola, J. L., Wang, H. O. (2000) Bifurcation control: Theories, methods, and applications. Int. J. Bifur. Chaos, **10**:511–548
7. Basso, M., Genesio, R., Tesi, A. (1997) A frequency method for predicting limit cycle bifurcations. Nonl. Dynam., **13**:339–360
8. Pyragas, K. (1992) Continuous control of chaos by self-controlling feedback. Phys. Lett. A, **170**:421–428

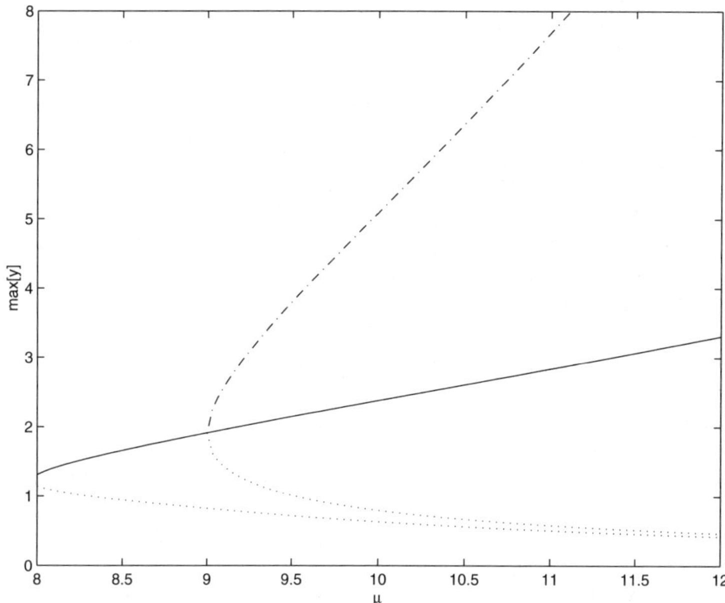

Fig.6. Uncontrolled (dash-dotted) / Controlled (solid) bifurcation diagrams.

9. Pyragas, K. (1995) Control of chaos via extended delay feedback. Phys. Lett. A, **206**:323–330
10. Bleich, M. E., Socolar, J. E. S. (1996) Stability of periodic orbits controlled by time-delay feedback. Phys. Lett. A, **210**:87–94
11. Vidyasagar, M. (1993) Nonlinear System Analysis. 2nd ed., Englewood Cliffs, NJ: Prentice-Hall
12. Khalil, H.K. (1996) Nonlinear Systems. 2nd ed., Englewood Cliffs, NJ: Prentice-Hall
13. Sastry, S. (1999) Nonlinear Systems. New York: Springer
14. Gelb, A., Vander Velde, W. E. (1968) Multiple-Input Describing Functions and Nonlinear System Design. New York: McGraw-Hill
15. Tesi, A., Abed, E. H., Genesio, R., Wang, H. O. (1996) Harmonic balance analysis of period doubling bifurcations with implications for control of nonlinear dynamics. Automatica, **32**:1255–1271
16. Siljak, D. D. (1969) Nonlinear Systems: The Parameter Analysis and Design. New York: Wiley
17. Atherton, D. P. (1975) Nonlinear Control Engineering. London: Van Nostrand Reinhold
18. Atherton, D. P. (1981) Stability of Nonlinear Systems. Chichester: Research Study Press (Wiley)
19. Mees, A. I. (1981) Dynamics of Feedback Systems. New York: Wiley
20. Moiola, J. L., Chen, G. (1993) Frequency domain approach to computation and analysis of bifurcations of limit cycles: A tutorial. Int. J. Bifur. Chaos, **3**:843–867

21. Piccardi, C. (1994) Bifurcations of limit cycles in periodically forced nonlinear systems: The harmonic balance approach. IEEE Trans. Circ. Syst.-I, **41**:315–320
22. Moiola, J. L., Chen, G. (1996) Hopf Bifurcation Analysis: A Frequency Domain Approach. Singapore: Word Scientific
23. Bonani, F., Gilli, M. (1999) Analysis and stability of bifurcations of limit cycles in Chua's circuit through the Harmonic Balance approach. IEEE Trans. Circ. Syst.-I, **46**:881–890
24. Seydel, R. (1994) Practical Bifurcation and Stability Analysis: from Equilibrium to Chaos. 2nd ed., New York: Springer-Verlag
25. Nayfeh, A. H., Balachandran, B. (1995) Applied Nonlinear Dynamics. New York: Wiley
26. Kuznetsov, Y. A. (1995) Elements of Applied Bifurcation Theory. New York: Springer-Verlag
27. Genesio, R., Tesi, A. (1992) Harmonic balance methods for the analysis of chaotic dynamics in nonlinear systems. Automatica, **28**:531–548
28. Genesio, R., Tesi, A. (1992) A harmonic balance approach for chaos prediction: Chua's circuit. Int. J. Bifur. Chaos, **2**:61–79
29. Thompson, J. M. T., Stewart, H. B. (1986) Nonlinear Dynamics and Chaos. Chichester: Wiley
30. Chen, G., Dong, X. (1998) From Chaos to Order: Methodologies, Perspectives and Applications. Singapore: World Scientific
31. Chen, G. (ed.) (1999) Controlling Chaos and Bifurcations in Engineering Systems. Boca Raton, FL: CRC Press
32. Boccaletti, S., Grebogi, C., Lay, Y. C., Mancini, H., Maza, D. (2000) The control of chaos: theory and applications. Physics Report, **329**:103–197
33. Genesio, R., Tesi, A., Wang, H. O., Abed, E. H. (1993) Control of period doubling bifurcations using harmonic balance. In Proc. 32nd IEEE Conf. Decision Control, San Antonio, TX, 492–497
34. Lee, H. C., Abed, E. H. (1991) Washout filters in the bifurcation control of high alpha flight dynamics. In Proc. Amer. Contr. Conf., Boston, 206–211
35. Basso, M., Genesio, R., Tesi, A., Abed, E. H. (1995) On controlling limit cycle bifurcations in continuous feedback systems. In Proc. 12th ECCTD, Istambul, 487–490
36. Basso, M., Genesio, R., Tesi, A. (1997) Controller design for extending periodic dynamics of a chaotic CO2 laser. Syst. Contr. Lett., **31**:287–297
37. Basso, M., Genesio, R., Stanghini, M., Tesi, A. (1997) Subharmonic control of chaos with application to a CO2 laser. Chaos, Solitons and Fractals, **8**:1449–1460
38. Di Bernardo, M., Johansson, K. H., Vasca, F. (2001) Self-oscillations and sliding in relay feedback systems: Symmetry and bifurcations. Int. J. Bifur. Chaos, **11**:1121–1140

Global Control of Complex Power Systems

David J. Hill[1], Yi Guo[2], Mats Larsson[3], and Youyi Wang[4]

[1] Department of Electronic Engineering
City University of Hong Kong, P. R. China
eedavid@cityu.edu.hk
[2] School of Electrical Engineering and Computer Science
University of Central Florida, Orlando, USA
[3] ABB Corporate Research, Baden-Dattwill, Switzerland
[4] School of Electrical and Electronics Engineering
Nanyang Technological University, Singapore

Abstract. This chapter presents an overview of recent results on an approach to total control of power systems. It is upwards compatible from any conventional or prior advanced control and provides a framework for coordinated development of control to address all major dynamical problems. The approach appears applicable to complex systems generally where behaviour is influenced by nonlinearity, large-scale, uncertainty and hybrid nature. In particular, the presence of bifurcations of various kinds is allowed for while using ideas from switching and optimal control. The approach will be illustrated by consideration of coordinated control for transient stability, voltage regulation and emergency voltage control of power systems.

1 Introduction

This chapter describes progress towards a general methodology for control of complex systems, particularly those that have large scale, are strongly not linear and have substantial uncertainty in their modelling. The work is motivated by earlier work, on transient stabilisation and network control for voltage security of power systems [1–6], and some ideas for general control. In arriving at progressively more general and flexible approaches, various techniques from advanced modern control have been absorbed such as dynamic programming, model predictive control, nonlinear control, so-called fuzzy control (as a non-linear control formula), switching and hybrid control. The overall framework to devise controllers which deal with complexity is called global control.

Models are assumed to be of a heterogeneous hybrid kind, i.e., a mixture of differential, algebraic and switching equations with different versions in different spatial and state domains. The various control elements correspond to existing physical controls, typically designed independently and assumed to be tunable, and other modules to be designed. Control is implemented in several layers of continuous and discrete actions.

Global hybrid control is presented as natural way to harness all the control elements optimally to coordinate a control response to dynamical problems

G. Chen, D.J. Hill, and X. Yu (Eds.): Bifurcation Control, LNCIS 293, pp. 155–187, 2003.

as they arise. The controlled response can be designed to achieve feasible secondary specifications such as for transient behavior, quality of supply, economy. Global control can be regarded as a concept which has evolved from techniques like gain scheduling and, more recently, multi-model control. A further idea of global control is to combine the qualitative and quantitative knowledge through some hierarchy. This connects to methods of systematic switching via membership type functions. One of the most practically successful approaches for utilizing the qualitative knowledge of a system to design a controller is fuzzy control. Results have been obtained in [9] where heterogeneous control provides combination of local control actions appropriate to different operating regions. A common feature of these methods is that they make use of mature results for the local controller design and provide necessary coordinations. A major feature added here is that the local controls are coordinated in optimal control schedules. In the development of generic techniques of control, very little recognition is made of underlying bifurcation structure, except in examples. A further feature to provide here is that the controller architecture be fully sensitive to bifurcations of various kinds.

Following development of the main ideas, two important problems in power system control will be used for illustration, namely transient stability control and network control. Power system dynamics is commonly studied in terms of bifurcations occurring as power levels are raised. Transient stability and voltage regulation are both important properties to guarantee in power system control, but they are studied via different model descriptions and related to different stages of system operation (i.e. the transient period, post-transient period, mid-term and long-term behaviour). In this chapter, we design a global controller to co-ordinate the transient stabilizer and voltage regulator in the transient period. Earlier control results deal with the two problems separately, or employ a switching strategy of two different kinds of controllers, which causes a discontinuity of system behavior. The designed controller is smooth and robust with respect to different transient faults.

To enhance the transient stability of power systems, in recent years a great deal of attention has been paid to the application of feedback linearization approaches [10], see e.g. [11–13]. Compared with use of conventional approximate linearization, which can only deal with local stability around an operating point, the controlled system can endure large disturbances and retain a steady post-fault condition. Inevitably, in order to enhance the stability, the power angle has to be one of the feedback variables whereas the generator voltage is not needed. In such transient stabilizing control, a common phenomenon is that the post-fault voltage value varies considerably from the prefault one [14]. From the practical point of view, voltage quality is a very important index of power supply in power system operation; so the post-fault value is expected to reach the normal value as closely as possible. In [14–16], voltage regulation was achieved by introducing voltage feedback. However, the voltage controllers are only effective around a working region, i.e., they work well when a small disturbance occurs and at the post-fault stage, but cannot survive a large disturbance. Different behavior of nonlinear power systems in different operating regions requires different control objectives and consequently different control actions must be employed in each operating region. How to achieve a satisfactory control performance over a wide range of anticipated operating conditions is demonstrated using global control.

The second major illustration introduces use of optimal control to achieve coordination of local physically based controllers. A method of optimal coordination is presented for load shedding, capacitor switching and tap changer operation using a dynamic system model in order to preserve or retain long-term voltage stability. Voltage stability is well-known to be related to bifurcations encountered as the loading of the system is increased. The method is based on model predictive control and a tree search. A model of the controlled system, including the network and load dynamics, is used to predict the future system behaviour based on the current state and applied control actions. The optimal control state according to these predictions is then obtained using a search method similar to those used in chess playing computers.

The structure of the chapter is as follows. Section 2 will discuss global control in general terms. Sections 3 and 4 present the specific illustrations of transient stabilisation with voltage regulation and network control with dissimilar devices respectively. Section 5 gives some conclusions. The presentation is somewhat descriptive to enable presentation of ideas which have scope for further development theoretically and in applications. Earlier versions of this chapter have been presented in conference form [7,8].

2 Global Control Ideas

2.1 Introduction

Global control is a further development of modern control design towards the capability to handle complex systems. Like power systems, many practical systems have the following four major characteristics: 1) substantial nonlinearity; 2) large-scale; 3) uncertainty in the model; and 4) hybrid or heterogeneous form. The last aspect refers to a mixture of control actions, i.e. discrete and continuous, and control requirements, possibly in different regions of operation. Adaptive, robust, intelligent and stochastic control are well-known methodologies which overcome parametric variation, unstructured uncertainty, unknown models and random disturbances respectively. Here it is of interest to note more recent developments in so-called multi-model, switching and hybrid control [20–24] as at least partially able to address the problems of complexity in a holistic way. No attempt is made to give extensive references; some representative results directly supporting this discussion only are chosen for reference. It remains to develop these methods to address systems of serious large-scale and mixed uncertainty types such as appear in power systems. In practice, a framework is needed for designing controls of a truly global kind which act in a coordinated way across the whole system geographically and for all operating situations, ie for all states and in-the-large (in the sense of stability theory) as parameters and conditions vary. This appears possible by combining ideas already developed for power systems with some of the newer techniques in control.

There are several special features of power systems which are noteworthy for the study of complex systems: 1) the system has a large-scale network structure; 2) many of the controls are embedded in the system with some having scope for tuning; further control design must allow for and enlist where possible these existing controls; 3) the overall control scheme will have a hierarchical structure; 4) the control actions available physically are already largely determined and have diverse timing, cost and priority for action; 5) the control goals are multi-objective with local and global requirements which vary with system operating state, e.g., normal and insecure states in power systems. Of course, many systems in industry, such as chemical processes, share these features. A related issue which is often mentioned in more practical discussions of general control science needs is "reconfigurability". It is recognised that the control designers have a huge toolbox of techniques for developing the models and local controllers throughout a system. There needs to be more effort on the higher levels of control to deal with failures and external events. Power systems engineers certainly need no reminding of this issue.

In general, we need a high-level version of distributed adaptive optimal control which "swarms" around the complex system attacking problems as they arise, but keeping a meta-view so that other problems are not ignored while attending to a particular one. Needless to say the implementation requires accounting for physical features like physical sparseness and time-scales as always to reduce computation demands. Thus, the approach favoured exploits model-based approaches as much as possible, only appealing to notions from intelligent control to deal with interpolation on extrapolation from known models.

2.2 Hybrid models

The use of hybrid models in power systems is already well-accepted to capture the use of mixed continuous and discrete control actions [70]. The equations often take a differential-algebraic-difference (DAD) form as follows

$$\dot{x} = f(x, w, z(k); \lambda) \tag{1}$$

$$0 = g(x, w, z(k); \lambda) \tag{2}$$

$$z(k+1) = h(x, w, z(k); \lambda) \tag{3}$$

The dynamic state variables x are variables that appear as derivatives in the differential equations and cannot change instantaneously. On the other hand, the algebraic state variables w do not appear as derivatives, and can thus change instantaneously due to changes in x or $z(k)$. The discrete state variables $z(k)$ have discrete-time dynamics and can change only at fixed time instants given by a selected sample time. The dynamic state variables relate to generator flux, continuous control and load dynamics; algebraic state variables relate to network voltages and currents, and the discrete states $z(k)$ typically arise from discrete control logic such as relay controls. The parameters are denoted λ and may include control parameters, independent parameters and action variables. We can write $\lambda = (\theta, \vartheta, u)$ where θ represents tunable parameters, ϑ represents the (structured) uncertainty and u the control variables which are yet to be designed. The basic DAD model will include control algorithms (discrete and continuous) for which no further tuning is planned.

The use of hybrid or DAD models has been discussed in more general terms for power systems in [25,26]. In particular, the discrete time actions can be generalised to allow for switching and reset events which depend on system behaviour. Recent advances in modelling software make use of hybrid models more straightforward. For instance, we can use the modelling and simulation tool Dymola, which has a power systems library [27].

As exists for differential equation models and DA models, e.g. see [28], a basic set of analysis tools for DAD systems is needed. This is currently the subject of active research. Some results have been summarised in [29,30].

2.3 Control elements

One of the problems with applying modern control methods like adaptive, fuzzy or robust control to power systems is that they are established in a generic framework which is not sensitive to the structure of particular practical systems. Hence many practitioners get the impression that these methods, while sounding appealing at first, are really impractical. Part of the problem is that they are presented as replacement rather than additional control. A more specific challenge is to blend these new ideas into the overall scheme of control which exists, i.e. excitation control, PSS, security control and network regulators such as tap-changers. Typically, much of this is already trusted and we only need some new control modules and overall co-ordination.

Co-ordination has not been a strong point of power system control in the past. Generally a sequential design approach has been taken rather than a holistic (or global) approach. It has often been the case that the solution to the last problem has caused the next problem, e.g. fast excitation systems for transient stability led to self-excited oscillations. Of course there were limitations in control technology which prevented use of a more comprehensive approach and the ingenuity of engineers to overcome these problems has been impressive. However, it might be that this approach has about met its limit with recent changes in the industry leading to much greater complexity. On the other hand, modern information technology for control, communication and computing (CCC) gives us possibilities to use intelligent distributed algorithms to complement more traditional approaches. It is worth noting that many of these issues have been dealt with in a simpler form in development of coordinated PSS schemes in the 1980's [31–33]. The PSS modules were designed largely according to local stability criteria. They could be designed sequentially or simultaneously to accommodate the effect of interactions, but this could be a large computational task either way. The simultaneous approach amounts to an optimisation of any tunable parameters within certain operating constraints to position the closed loop modes of the system with adequate damping. In Russia, such algorithms were highly developed for co-ordinating stability on large systems and robustly with respect to changes of system structure - see [34] and especially the references within. In this particular problem, it is interesting to note that system structural conditions were found, related inter alia to loading levels, under which the designs could be largely decoupled [35]. Typically, such decoupling does not apply in heavily loaded systems, but its presence in any degree maybe useful.

In looking to generalise these ideas to the whole control problem of maintaining stability while providing specified performance, we start by identifying the basic *control elements,* i.e. those existing controllers and their tunable parameters which are free to adjust for system-wide purposes.These will be based on physical devices or at least the designs for them. The controller could have the form

$$u_i = U_i(x_i, w_i, z_i(k); \theta_i, \vartheta_i) \tag{4}$$

where subscript i refers to a geographical part of the system and/or some special control task, e.g. transient stabilisation in an area of a power system. The controller is typically expressed as a function of state variables or as a dynamical system driven by certain system outputs. This controller could be anything from a simple classical PID controller to the most sophisticated nonlinear controller (including rule-based, fuzzy or model-based designs such as feedback linearisation). The controller is required to perform well in spite of the uncertainty ϑ_i. The parameters θ_i are available to implement co-ordination with other controllers.

2.4 Bifurcations and global control

The development of nonlinear control theory has generally evolved along several lines of thought. There are many books on this topic - see for instance the recent ones [10,36–38]. Emerging from mathematics of dynamical systems and the study of some specific systems, there has been an extensive study of nonlinear systems dynamics in terms of underlying bifurcations in the models [37]. This line has been extensively pursued for power systems analysis [39]. While bifurcation control has been developed for special systems of scientific interest [40], it has not been developed for power systems in any practical way. Most work on generic nonlinear control does not deal with the bifurcation structure. In fact, many assumptions used in the development of nonlinear control algorithms effectively rule out bifurcations of dynamical behaviour. When we consider the possibility of parameters crossing bifurcation boundaries, we should develop control on at least two levels, as is commonly discussed for power systems: the first level uses continuous and switching type control to establish an appropriate equilibrium operating point (or manifold); the second level deals with the system dynamics around these operating points and uses techniques from stabilisation and local bifurcation control. We could classify controllers as: 1) bifurcation avoiding; 2) bifurcation delaying; 3) bifurcation eliminating; 4) bifurcation accepting. Loosely speaking, the controller could keep the system well away from the bifurcation boundaries, as is typical for power systems, push the boundaries further away in parameter space, eliminate certain boundaries, or devise a scheme for operating across the boundaries. The study of performance of the various forms of textbook control tools in these situations is incomplete, but intuitively they will have many problems dealing with such complexity in their basic form. For example, adaptive control is clearly challenged by the unknown parameters encountering a bifurcation.

Our global control objective is to achieve good control performance over a wide range for the anticipated operating region with robustness to different faults whose sequences are not known apriori. The bifurcation boundaries define domains of operation where the dynamical behaviour is qualitatively different, see [39] for review papers. A possible control strategy, accepting the boundaries, is to design controllers for each structurally stable region and switch between them in some way. A scheme which has the abovementioned two-level structure has been presented and used for a standard example in [5]. The switching controller has a weighted Sugeno type form used in fuzzy control, but is motivated by the need to allocate control effort according to dynamical behaviour. Assuming the state-parameter space is partitioned into two domains, the control u takes the form:

$$u = u_e + \mu_1 u_1 + \mu_2 u_2 \tag{5}$$

where u_e provides the new equilibrium manifold; μ_1 and μ_2 are functions of an indicator variable expressing the closeness to a particular region or control concern; and $u_1$and u_2 are the local controllers which can have the general form (4). The overall control for a specific design is just a special form of nonlinear controller which may have tunable parameters. There may be many such controllers acting on the complex system.

More generally the domain boundaries could be defined by regions where modelling is unknown and typically behaviour changes in some major way, so switching control again becomes natural. Between these discrete models the models are incomplete and uncertainty must be allowed for. A complete model becomes an interpolation which can be expressed in various ways, e.g.

see [9,41] for use again of fuzzy-control type membership functions to switch between local controllers.

Another possible control architecture employs a neural network to choose between local controllers according to pattern recognition of dynamical behaviour. This idea has been studied for a power system application [74].

2.5 Optimal coordination and swarming

Once all the local controllers are designed and teamed up where needed, e.g. as in (5), the freedom to use them flexibly should be fully exploited. It is of no great concern that the controllers might be complex in themselves. Besides the argument that they can often be expressed in software code, at a system-wide level the controller complexity is related more to the number of parameters that need tuning. This may involve simply tuning of free parameters, as in the PSS case mentioned above, or a more complicated scheduling exercise which allows for any control redundancy to use scaling and timing according to control costs, dynamic properties and priority. The latter idea is familiar at least for static scheduling in operations research, but is quite complicated for dynamical systems as it requires the solution of the optimal control problem [42,43]. In [1,3], a technique along these lines has been developed for optimally coordinated voltage security control of power systems using differential dynamic programming (DDP) and trajectory sensitivity methods to set-up the staged use of available controls in an optimal way. However, there is a wealth of techniques which can be explored here: in particular, all ideas in modern control which boil down to use of optimal control. In Section IV, we will illustrate further work on the voltage security problem using Model Predictive Control (MPC) and search techniques as a way to set up the optimal tuning. The connection between DP and MPC can be made [44], but each has been developed with a set of techniques for implementation with various forms of complexity. There are many special ideas which can be mentioned to implement optimal coordination. One useful one, which suits the current framework, was formulated in the Russian literature and used for adaptive nonlinear control [45,46]. The control objective is expressed in memoryless or integral form. The control algorithm takes a two-level structure with a controller of the form (4) and a so-called speed-gradient parameter update.

The switching idea expressed above can be interpreted as allowing the controller to adjust the individual contributions of the local controllers according to the particular problem being faced at a given time. For instance when an instability occurs in a certain domain or in a certain variable type, e.g. voltage instability, the indicators driving the weighting functions will automatically adjust the control to emphasize the appropriate actions. When optimal coordination of many such controllers is considered, we can think of numerous controller elements in a complex system as swarming [47] to deal with problems as they arise.

It is beyond the scope of this chapter, but it is straightforward to see special forms of adaptive and learning control as specific examples of the above coordinated tuning of control elements framework. The theory of aspects of optimal control for structured nonlinear systems with variable parameters needs further development, but many ideas have already been established in the development of adaptive control- see for instance [20,48,49].

3 Global Transient Stability and Voltage Regulation

3.1 Introduction

In this section, we design global control to maintain transient stability and achieve satisfactory post-fault voltage level of a power system when subjected to severe disturbances. The control signal from the global controller has the form (5), i.e. the weighted average of the signals from the local control laws. Since the membership functions can be determined by direct measurable variables of power systems, it has several appealing properties compared with the switching control strategy proposed by [14]. For example, to achieve control action switching, the fault sequence needed to be known apriori; also as long as the switching time is fixed according to a certain fault, the system may not survive another different fault. The global control law overcomes these disadvantages by sensing the relevance of stabilisation and regulation and automatically shifting to the appropriate controller. By a smooth transition between pure control regions, our global control objective is achieved. It is important that it adopts a mature control strategy in each local region, i.e. familiar schemes can be preserved.

The results presented here are based on the paper [6].

3.2 Dynamical model of power systems

We consider the particular SMIB power system arrangement shown in Figure 1. The actual dynamic response of a synchronous generator in a practical power system when a fault occurs is very complicated including many nonlinearities such as the magnetic saturation. However, the classical third order dynamic generator model has been commonly used for designing the excitation controller. More complete models are used in the simulations to evaluate the design in the presence of other effects.

The classical third-order dynamical model of a SMIB power system (Figure 1) can be written as follows [52–54]:

Mechanical equations:

$$\dot{\delta} = \omega \tag{6}$$

$$\dot{\omega} = -\frac{D}{2H}\omega + \frac{\omega_0}{2H}(P_m - P_e) \tag{7}$$

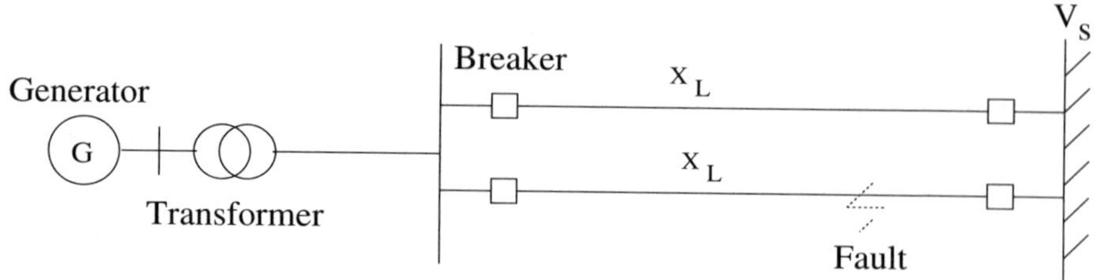

Fig.1. A single machine infinite bus power system.

Generator electrical dynamics:

$$\dot{E}_q' = \frac{1}{T_{do}'}(E_f - E_q) \tag{8}$$

Electrical equations:

$$E_q = \frac{x_{ds}}{x_{ds}'}E_q' - \frac{x_d - x_d'}{x_{ds}'}V_s \cos\delta \tag{9}$$

$$P_e = \frac{V_s E_q}{x_{ds}}\sin\delta \tag{10}$$

$$I_q = \frac{V_s}{x_{ds}}\sin\delta = \frac{P_e}{x_{ad}I_f} \tag{11}$$

$$Q_e = \frac{V_s}{x_{ds}}E_q\cos\delta - \frac{V_s^2}{x_{ds}} \tag{12}$$

$$E_q = x_{ad}I_f \tag{13}$$

$$E_f = k_c u_f \tag{14}$$

$$V_t = \frac{1}{x_{ds}}\left[x_s^2 E_q^2 + V_s^2 x_d^2 + 2x_s x_d x_{ds} P_e \cot\delta\right]^{\frac{1}{2}} \tag{15}$$

where $\delta(t)$ is the power angle of the generator (in *radian*), $\omega(t)$ is the relative speed (in *rad/s*), P_m is the mechanical input power (in *p.u.*), P_e is the active power delivered to bus (in *p.u.*), E_f is the referred field excitation voltage, E_q' is the transient EMF in the quadrature axis (in *p.u.*), and V_t is the terminal voltage of the generator (in *p.u.*). The notation for other variables and parameters are standard and readers are referred to [14,54,55].

The fault considered in this chapter is a symmetrical three phase short circuit fault which occurs on one of the transmission lines.

In the next section, we test the existing controller performances on different fault sequences, and then in Section 3.D a new global controller will be presented and tested .

3.3 Local controllers

A. Nonlinear controller for transient stability

As discussed in [10], feedback linearization is a quite appealing design method for nonlinear systems. Since it avoids the local nature of approximate linearization and transforms the system to be linear over a very wide range, it has been applied to power systems by different authors shortly after it was available – see e.g. [11–15,56–59]. In the following, we briefly describe a design based on so-called direct feedback linearization (DFL) compensators [14,13]. This more easily allows particular physical states to be preserved than the geometric algorithm version in [10].

By a DFL transformation [13], (6)-(8) becomes

$$\begin{aligned} \dot{\delta} &= \omega \\ \dot{\omega} &= -\frac{D}{2H}\omega - \frac{\omega_0}{2H} \triangle P_e \\ \triangle\dot{P}_e &= -\frac{1}{T'} \triangle P_e + \frac{1}{T'} v_f \end{aligned} \tag{16}$$

where

$$\triangle P_e = P_e - P_{m0} \tag{17}$$

$$T' = \frac{x'_{ds}}{x_{ds}} T'_{d0} \tag{18}$$

$$\begin{aligned} v_f &= \frac{V_s}{x_{ds}} \sin\delta \left[k_c u_f + T'_{d0}(x_d - x'_d)\frac{V_s}{x_{ds}} \sin\delta\ \omega \right] \\ &\quad + T' \frac{V_s}{x_{ds}} E_q \cos\delta\ \omega - P_{m0} \end{aligned} \tag{19}$$

Note that the mechanical power control is represented as a constant power P_{m0}, i.e. the governor loop is relatively slow acting.

Now, we note (16) is a linear system with the new input v_f. Robust control techniques for linear systems [60,61] can be employed. By solving an algebraic Riccati equation (ARE) – see [13,55] for details, the control law is obtained as

$$v_f = -k_\delta \delta - k_\omega \omega - k_P \triangle P_e \tag{20}$$

where k_δ, k_ω, k_P are the linear gains obtained from the solutions of ARE.

The real excitation control u_f can be obtained from (19) to give

$$\begin{aligned} u_f &= \frac{x_{ds}}{k_c V_s \sin\delta} \left[v_f - T' \frac{V_s}{x_{ds}} E_q \cos\delta\ \omega + P_{m0} \right] \\ &\quad - T'_{d0}(x_d - x'_d) \frac{V_s}{k_c x_{ds}} \sin\delta\ \omega \end{aligned} \tag{21}$$

The DFL nonlinear control (21) with (20) guarantees the transient stability of power system (6)-(8) for admissible uncertain x_L and V_s – proof is given in [13].

However, since V_t is a nonlinear function of δ, P_e and the system structure, any change in the system structure will cause the voltage to reach another post-fault equilibrium point even if δ and P_e are forced to go back to their prefault steady values. So the generator terminal voltage V_t could stay at a different post-fault state which is undesirable in practice.

From the above, we can see that although the DFL nonlinear compensator is effective for stability, it cannot guarantee voltage regulation. The simulation results shown later in this section will verify that the DFL nonlinear controller enhances the transient stability of power systems but cannot by itself achieve a satisfactory post-fault voltage level.

B. Voltage controller

Voltage regulation is an important property particularly in the post-transient period. The basic objective is to regulate the voltage to reach its nominal value. Voltage controllers have been given in [15] using LQ-optimal techniques and in [16] using a linear robust control technique. Both of them have the problem that they deteriorate transient stability over the whole operating region.

For example, as proposed in [16], differentiating equation (15) gives

$$\triangle \dot{V}_t = f_1(t)\omega + \frac{f_2(t)}{T'_{d0}} \triangle P_e + \frac{f_2(t)}{T'_{d0}} v_f \tag{22}$$

where $f_1(t)$ and $f_2(t)$ are highly nonlinear functions of δ, P_e and V_t – see [16] for details. Since $f_1(t)$ and $f_2(t)$ are dependent on the operating conditions, their bounds can be found within a certain operating region. So a new linearized system which is represented by the vector $[\triangle V_t, \omega, \triangle P_e]$ can be developed. Robust linear control techniques can be applied to obtain

$$v_f = -k_V \triangle V_t - k_\omega \omega - k_P \triangle P_e \tag{23}$$

where k_V, k_ω, k_P are linear gains dependent on the bounds of $f_1(t), f_2(t)$. The real excitation input u_f is chosen as defined in (21).

Since the voltage is introduced as a feedback variable in (23), the post-fault voltage is prevented from excessive variation. It is unnecessary to keep the power angle regulated once transient stability is assured.

However, since the design of the voltage controller involves estimating nonlinearity bounds within a certain operating region, it is only effective locally. In another words, when serious disturbances occur which cause the system to operate in a wider range outside the estimated one, the designed system may not perform well.

In conclusion, the voltage controller achieves voltage regulation, but it is only valid locally. This point will be verified in the simulations in Section 3.3.

C. Coordinated control by switching

By now it can be seen that the DFL nonlinear controller and voltage controller achieve different control objectives in different regions of the states. In [14,55], a nonlinear coordinated control scheme was proposed where a switching strategy is used between the different control actions to guarantee transient stability enhancement and voltage regulation.

The simple switching scheme is as follows:

Step 1: When the fault occurs at $t = t_0$, the DFL nonlinear controller u_f (21) with (20) is employed to maintain the transient stability of power systems;
Step 2: At $t = t_s$, the control law switches to the voltage controller u_f (21) with (23) to maintain the desired post-fault voltage level.

Since the switching time is physically fixed according to one particular fault sequence, it may not be suitable for a different fault sequence, which may then destabilize the power system. Also, the switching time fixed in the first post-transient period cannot achieve voltage regulation over the whole working region allowing for later faults. In summary, the switching strategy with the switching time fixed is not robust with respect to different faults.

How to design a universal control law which is robust with respect to uncertain faults is a challenging problem. Before stating our global controller which overcomes these obstacles, we show in the next subsection simulation results of the controllers which were discussed above.

D. Simulations of local controllers

The parameters of the SMIB power system which is shown in Figure 1 are as follows:

$$\begin{aligned}
&x_d = 1.863, \quad x_d^{'} = 0.257, \quad x_T = 0.127,\\
&T_{do}^{'} = 6.9, \quad x_L = 0.4853, \quad H = 4, \quad D = 5,\\
&K_c = 1, \quad x_{ad} = 1.712, \quad \omega_0 = 314.159.
\end{aligned} \tag{24}$$

The physical limit of the excitation voltage is taken as

$$-3 \leq k_c u_f \leq 6 \tag{25}$$

The operating point of the power system used in the simulations is:

$$\delta_0 = 72^\circ, P_{m0} = 0.9p.u., V_{t0} = 1.0p.u. \tag{26}$$

The fault sequence consists of two symmetrical three phase short circuit faults–one temporary and then one permanent (Case 3 in [6]):
Stage 1: The system is in a prefault steady state;
Stage 2: A fault occurs at $t = t_0$;

Stage 3: The fault is removed by opening the breakers of the faulted line at $t = t_1$;
Stage 4: The transmission lines are restored at $t = t_3$;
Stage 5: Another fault occurs at $t = t_4$;
Stage 6: The fault is removed by opening the breakers of the faulted line at $t = t_5$;
Stage 7: The system is in a post-fault state.

The nominal switching time for the controller has been set at $t_s = t_2$. We choose in the simulation $t_0 = 0.1s, t_1 = 0.25s, t_2 = 1s, t_3 = 1.4s, t_4 = 2.1s, t_5 = 2.25s$.

The fault location is indexed by a positive constant λ which is the fraction of the line to the left of the fault. The nominal fault location for the given fault sequence is $\lambda = 0.04$.

The controllers employed in the simulations are [16]:

$$Transient\ controller:$$
$$v_f = 22.36\ \delta + 12.81\ \omega - 82.45 \triangle P_e \tag{27}$$
$$Voltage\ controller:$$
$$v_f = -40.14 \triangle V_t + 10.11\ \omega - 30.81 \triangle P_e \tag{28}$$

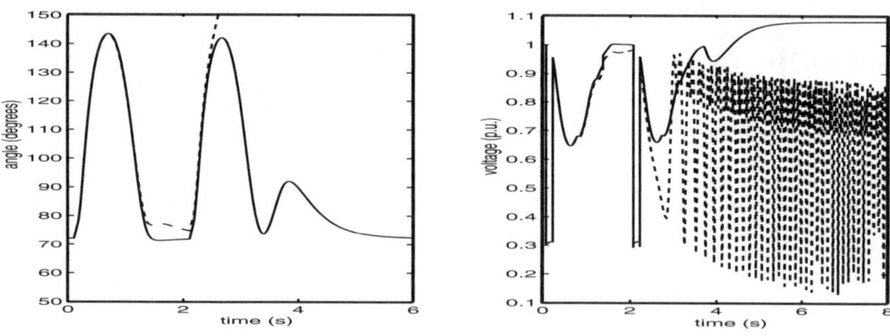

Fig.2. Power system responses: '—' DFL nonlinear controller; '- -' voltage controller.

The power system responses with the three different controllers are shown in Figures 2-3. (The highly oscillatory voltage response, caused by the unstable local controller, would be prevented by protection actions in practice.)

From the simulation results it can be observed that the transient nonlinear controller stabilizes the disturbed system but the post-fault voltage differs from the prefault value by 10%-20%, which is not acceptable in practice. The voltage controller cannot stabilize the system. For the switching controller, the switching time is seen to be important since an inappropriate one causes the loss of synchronism of generators. This point is clearly shown in Fig-

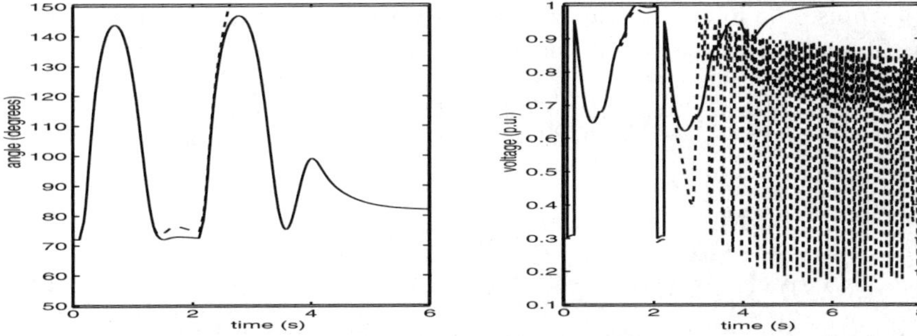

Fig.3. Power system responses for switching controller: '—' $t_s = 1.3s$; '--' $t_s = 1s$.

ure 3 where the switching controller with nominal switching time $t_s = 1s$ destabilizes the system, but $t_s = 1.3s$ stabilizes it.

The simulation results are consistent with the analysis stated early in this Section. The strategy of simply switching between different control actions is not reliable due to the non-existence of a universal switching time. In the next Section, we will propose a new global control strategy which achieves transient stability enhancement and voltage regulation simultaneously and robustly.

3.4 Global controller

Our global control objective is to achieve good control performance over a wide range for the anticipated operating region. Specifically, we have the following control task:

Global control problem: *Design a smooth nonlinear feedback control law for the excitation system (6)-(8), such that the closed-loop power system is transiently stable when subjected to a severe disturbance, and restores the steady prefault voltage value after the disturbance.*

Desired properties of the global controller include robustness with respect to different faults whose sequences are not known apriori.

A. Operating region membership function

We want to achieve both transient stability enhancement and good post-fault performance of the system. Good post-fault performance means that after the transient period we wish to control the excitation unit to regulate the generator terminal voltage V_t. According to the qualitatively distinct operating conditions and the corresponding control objectives over each region, local controllers are designed and coordinated through some strategy. We firstly need to define the operating regions and membership functions.

We use the following trapezoid-shaped like membership functions which are able to indicate different operating stages.

$$\mu_V(z) = \left(1 - \frac{1}{1+\exp(-120(z-0.08))}\right) \cdot \left(\frac{1}{1+\exp(-120(z+0.08))}\right)$$
$$\mu_\delta(z) = 1 - \mu_V \tag{29}$$

where

$$z = \sqrt{\alpha_1 \omega^2 + \alpha_2 \left(\triangle V_t\right)^2} \tag{30}$$

and α_1, α_2 are positive design constants providing appropriate scaling which can be chosen according to different sensitivity requirement of power frequency and voltage. This surprisingly simple choice of dynamical indicator function z works quite well.

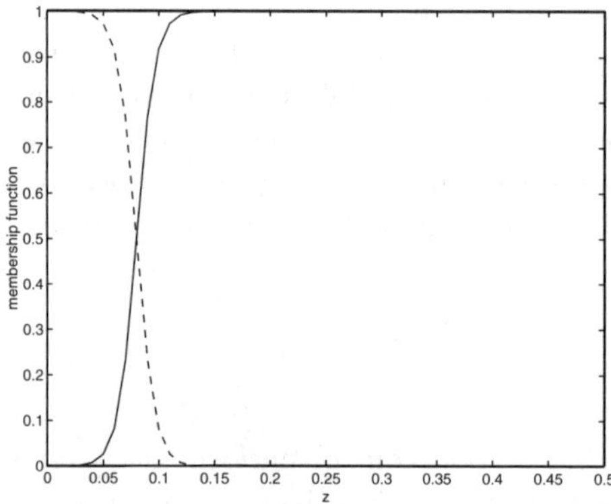

Fig.4. Membership functions: '–' μ_δ, '- -' μ_V.

Membership function (29) is plotted in Figure 4. It can be seen that $\mu_\delta(z)$ gets its dominant value when z is far away from the origin, which corresponds to the transient period; on the other hand, $\mu_V(z)$ does so when z is close to the origin, which indicates the post-transient period. Since the membership function values are determined by the directly measurable variables, ω and $\triangle V$, the fault sequence need not to be known beforehand.

Therefore, the whole operating region is partitioned into the following two subspaces by the membership functions, where S_1 indicates the transient period and S_2 indicates the post-transient period.

$$S_1 = \{(\omega, \triangle V) | \mu_V \le \mu_\delta\}$$
$$S_2 = \{(\omega, \triangle V) | \mu_V > \mu_\delta\} \tag{31}$$

The characteristic function of each subspace S_l $(l = 1, 2)$ is defined by:

$$\tau_l = \begin{cases} 1 & z \in S_l \\ 0 & \text{otherwise} \end{cases} \tag{32}$$

Note that $\tau_1 + \tau_2 = 1$.

It should be pointed out that ω and $\triangle V_t$ are chosen as the indicator variables in (30) since they sufficiently represent the operating status for the problem of transient stability and voltage regulation. If the problem under consideration is voltage stability, reactive power could be included in the indicator. Similarly, the proposed method can be extended to other power system control issues. The chosen membership functions have a trapezoid-like shape which is well known in fuzzy control to separate operating conditions. From our simulation experience, the system performance is not sensitive to different parameters α_1 and α_2.

B. Global control law

After the state space is partitioned, it is desirable that in the transient period, which corresponds to subspace S_1, the transient stability controller takes effect; while in the post-transient period, which corresponds to subspace S_2, the voltage controller dominates. The global control law is the average of the individual control laws, weighted by the operating region membership functions, i.e., the input v_f takes the form:

$$v_f = \mu_\delta v_{f1} + \mu_V v_{f2} \tag{33}$$

where v_{f1} is the nonlinear controller (20) and v_{f2} is the voltage controller (23). The real excitation control u_f can be implemented by (21).

The global control (33) has the following interpretation: in the transient period, the system is far away from the equilibrium; the primary goal is to control the states to enter a neighborhood of the equilibrium without large oscillations; then in the post-transient period around the equilibrium, the voltage needs to be tuned to reach the prefault level. The membership functions play the role of appropriate weighting and smooth interpolation of the two controllers. One of the appealing abilities of the method is that the operating status is automatically distinguished by the membership functions which are functions of directly measurable variables. The form of control law (33) is such that a smooth transfer between the local controllers is automatically achieved.

The global heterogeneous control for transient stability and voltage regulation is illustrated in Figure 5, where block #1 and #2 represents local controllers, and membership functions play the role of appropriate coordination.

The proposed global control law (33) features the following properties:

- Control action is determined by online measurement of power frequency and voltage, which makes it unnecessary to know the fault sequence beforehand;
- The controller is globally effective in the presence of different uncertain faults;

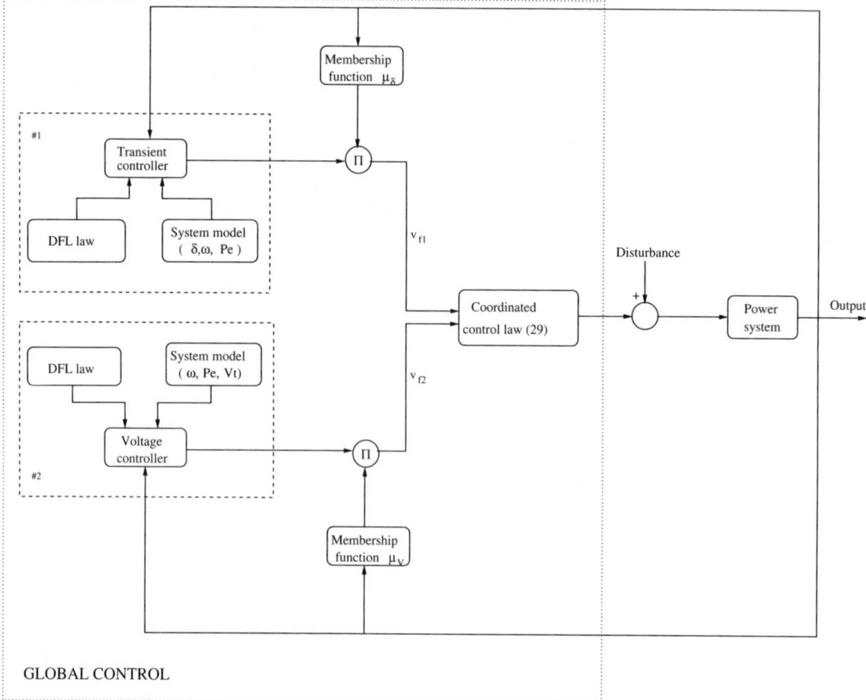

Fig.5. Global controller for transient stability and voltage regulation.

- The controller inherits the properties of local controllers, i.e., it is robust with respect to parameter uncertainties.

Simulation results follow to demonstrate the effectiveness of the heterogeneous control law.

C. Simulations

In this subsection, we exhibit the closed-loop power system performance with the global controller. The power system parameters and fault case are the same as in Section 3.3. The global control law employed is (33) with v_{f1} and v_{f2} from (27) and (28) respectively. Figure 6 shows the system performance. In contrast with Figures 2 and 3, both transient stability and normal post-fault voltage values are achieved. From the simulations, we can see the global controller achieves the proposed control task. In [6], it is also shown that the global controller is robust with respect to different fault sequences and their locations.

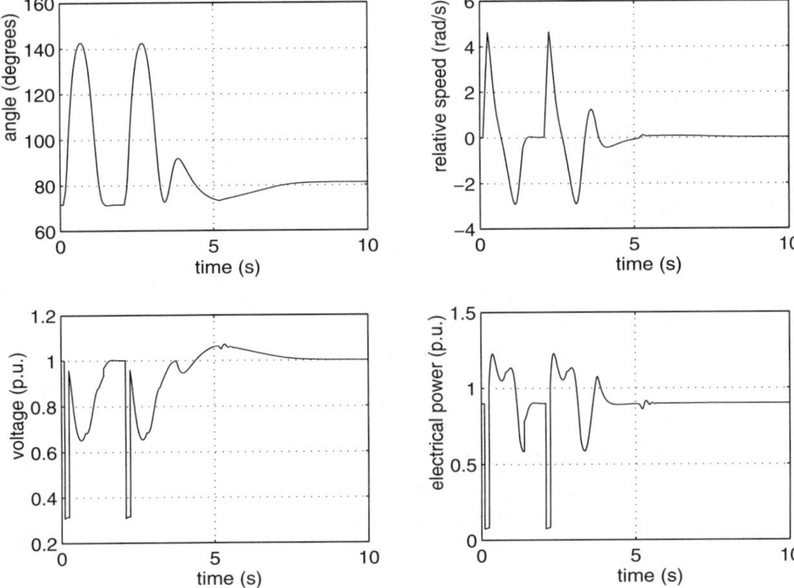

Fig.6. Power system responses: global control law.

D. Discussion

The global control design procedure involves the following three steps:

- Qualitatively distinct operating regions are identified by membership functions of directly measurable indicator variables, and the region partitionings admit overlap;
- The control law in each region is specified to be a usual type of controller developed from model-based nonlinear control techniques;
- The global control law is the weighted sum of local controllers, which achieves smooth transitions between the different operating conditions.

The full power of the control law (4), (5) has not been used here, namely the equilibrium controller u_e and the coordination/optimisation of tunable control parameters. The former is illustrated in [5]. The optimal scheduling of controller flexibility is illustrated in the next section. These ideas are clearly applicable in other application areas.

4 Emergency Voltage Control

4.1 Introduction

Electric power systems are designed, through careful planning and preventive control schemes, to survive most foreseeable disturbances [54]. Not all possible disturbances, however, can be foreseen at the planning stage and these may result in instability that will eventually lead to collapse or islanding of the system. The objective of an emergency control system is to detect such situations and carry out control actions necessary to prevent collapse of the system. Here we consider the problem of emergency voltage control in the long-term time period.

The emergency controls considered here are: switching of capacitor banks, tap changer operation and load shedding. The principles of using these as efficient emergency controls in a dynamical context have been established in [62–65]. Similar principles are used to the discussion in Section 2.4. The control includes switching to establish a viable equilibrium combined with dynamic stabilization about that equilibrium; proximity to bifurcation boundaries must be evaluated. The highly nonlinear nature of the problem calls for sophisticated methods based on a dynamic system model rather than the adhoc methods used in practice. In particular, proper coordination of all controls would minimize the amount of load shed thus minimising impact on customers and cost. Most controls are inherently discrete in their nature; for example, capacitor banks and tap changers must be switched in fixed steps.

Methods for coordination of secondary voltage control and load shedding using static system models have been described, e.g. in [66]. Coordination of dissimilar controls in the emergency state based on a dynamic system model has so far received little attention. A method based on differential dynamic programming and a security measure as the distance to the closest bifurcation has been derived in [1,3,67]. None of the above mentioned references fully take account of the discrete nature of controls or the inherent dynamics of the loads.

This section demonstrates the application of model predictive control (MPC) combined with search techniques to optimise emergency state voltage control. General model predictive control is well established for process control (see, e.g. [18,19,68]) but the application to power systems network control is novel. A model of the controlled system, including the network and load dynamics, is used to predict the future system behaviour based on the current state and the applied control actions. The optimal control actions can then be obtained from the solution of a combinatorial optimization problem by a search method. Various predictors are employed, ranging from a detailed predictor based on simulation of the system to more computationally tractable linearized versions. The results presented here are based on the paper [2].

4.2 System modelling

Again we use a single system to illustrate the ideas. Consider the power system in Figure 7. There are two loads, four capacitor banks and two on-load tap changing transformers considered as actuators. These can be operated in the discrete fixed steps given in Table 2.

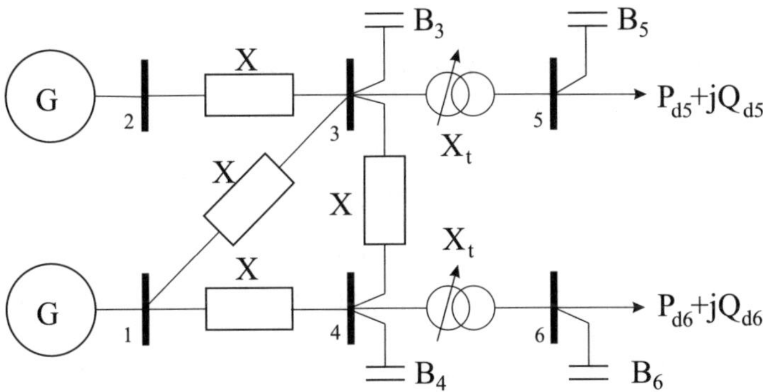

Fig.7. Example power system studied.

The generator at bus 1 is modelled as an infinite bus with fixed voltage and voltage angle while the generator at bus 2 is modelled using the standard sixth order dq-model with first order excitation system, an over-excitation limiter and a first order governor. The loads are modelled using the aggregate exponential recovery model [69]

$$T_p\dot{x}_p + x_p = P_s(V) - P_t(V) \tag{34}$$

$$P_d = k_l(x_p + P_t(V)) \tag{35}$$

where x_p is a continuous dynamic state that can be interpreted as a measure of the energy deficit in the load and $P_s(V) = P_0V^{\alpha_s}$ and $P_t(V) = P_0V^{\alpha_t}$ are the steady-state and transient voltage dependencies respectively. P_d is the actual active load power and T_p is the active power recovery time constant. For the reactive load power, a similar model is used with corresponding characteristics Q_d, x_q, $Q_s(V) = Q_0V^{\beta_s}$, $Q_t(V) = Q_0V^{\beta_t}$ and time constant T_q. The scale factor k_l models load shedding and is applied to the reactive as well as to the active load model. The transformers are modelled as pure reactances in series with an ideal turns ratio change.

For the purpose of control determination, generators and their associated control systems are modelled more simply, using their quasi-steady state approximations given in [70], i.e, the derivative terms in the mechanical and flux equations are neglected. The power system network model is conveniently expressed in the hybrid differential-algebraic-difference form given by equations (1)-(3).

4.3 Control problem formulation

Our global control objective is to coordinate all the control devices to ensure voltage regulation in the face of large disturbances. Specifically, we have:

Global Control Problem: *Design a scheduling control law for the system (expressed in the DAD form) such that the voltages are regulated to within* *a practically small region of the set-point values following a large disturbance. The voltage error will be determined by discrete controller steps.*

It is convenient to review the basic principle of Model Predictive Control (MPC) illustrated in Figure 8. A model predictive controller uses a system model to predict the future output trajectories (dotted lines) based on the current state and several different candidate input sequences. A cost function is defined based on the deviation of each predicted trajectory from a desired trajectory (dashed line). The optimal control sequence, in the sense that it minimizes the defined cost function, is then obtained by solving an optimization problem on-line each time a new instance of the control sequence is to be determined. The optimal control is then applied to the system, until the next sample time. The interval between the current time t^* and the prediction horizon $t^* + t_p$ is referred to as the prediction interval, and is chosen based on the settling time of the slowest dynamics. Usually, the prediction interval is chosen as a multiple of the sample time.

In the standard formulation, it is assumed that all control signals are continuous, i.e., can assume all values within some fixed range. In the case of a linear system this leads to an optimization problem that can be solved using quadratic programming; nonlinear programming is naturally required with a nonlinear system. The application here differs from the general nonlinear case in that the control variables are constrained to fixed discrete values which makes it necessary to resort to a combinatorial optimization method. A sequence of control inputs, for the first n samples during the prediction horizon, is obtained from the optimization. The first control step in this sequence is then applied to the plant and the rest are ignored. In order to save computation time, various devices can be used which restrict controller complexity.

A. Combinatorial optimisation

Accordingly, the problem of selecting the new control vector u^+ at the operating point $(x^*, u^-) = (x(t^*), u(t^*))$ can be formulated as the combinatorial optimization problem

$$\begin{array}{ll} \text{minimize} & J(x^*, u) \\ \text{subject to} & u \in \mathcal{S}(u^-) \end{array} \tag{36}$$

where u is the optimization variable corresponding to the new control state to be used during the interval $[t^*, t^* + t_p]$. u^- denotes the control state presently in use, i.e. before the optimization starts at time t^*. $\mathcal{S}(u^-)$ is the set of available control states at time t^* and is normally a function of u^-. The scalar function $J(x^*, u)$ is referred to as a cost function and should evaluate the candidate control state u such that a smaller result indicates a more desirable state. Typically the cost function should include a component based

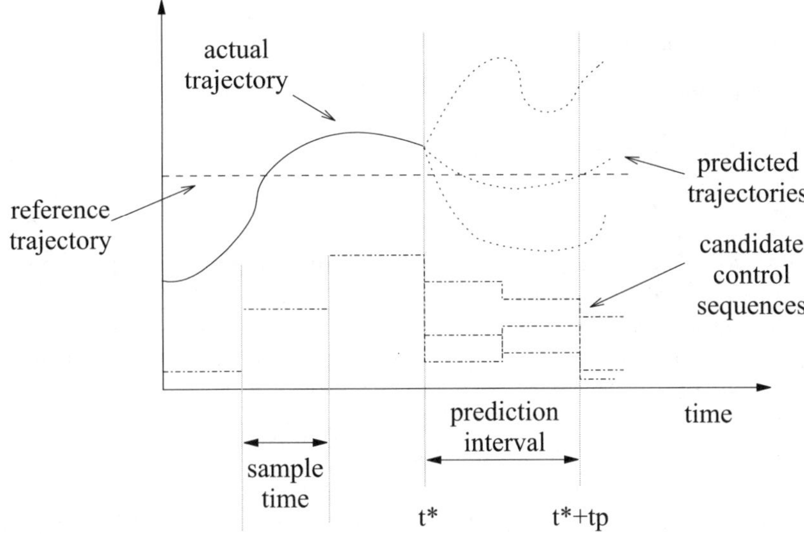

Fig.8. Principle of Model Predictive Control.

on the deviations of the chosen outputs from their reference trajectories and a component based on the control variations.

Constraints can be imposed on inputs as well as states and outputs as follows:

- Input constraints.
  Actuator limits, for example, are modelled using the set of available control states. Since the set may be a function of the present control state, actuator ramp limits can also be handled. More specific constraints can also be represented. For instance, a useful constraint is to set priorities for control actions as in [1,3,67].
- Output and state constraints.
  These could be modelled as hard constraints in the optimisation However, it is generally not desirable to impose hard constraints on the outputs or states since they may be impossible to strictly satisfy, and for large disturbances no control could be determined. Therefore output and state constraints are modelled as soft constraints by adding terms to the cost function [71]. If this cost is dominant over other terms in the cost function, the primary aim of the controller will be to minimize any constraint violation.

Control actions are divided into two classes:

- Normal control measures that will be routinely scheduled; capacitor and tap changer operations belong to this category.
- Emergency control measures that will be scheduled only when required for the satisfaction of a constraint. This control is activated only if the

constraint cannot be satisfied using any combination of the normal control measures. Load shedding belongs to this category.

The cost function is defined in the form

$$J(x^*, u^+) = \int_{t^*}^{t^*+t_p} \big((\hat{y} - y_r)^T \mathbf{Q} (\hat{y} - y_r) \\ +(u^+ - u^-)^T \mathbf{R} (u^+ - u^-) + P(t, x^*, u^+)\big)\ dt \quad (37)$$

where $\hat{y} = \hat{y}(t, x^*, u^+)$ is a prediction of the output trajectory using information up to time t^* if the new piecewise constant control u^+ is applied during the prediction interval. The control state presently in use is denoted u^-, the reference trajectory y_r, and the weighting matrices for output errors and control variations $\mathbf{Q}$ and $\mathbf{R}$ respectively. A special penalty term $P(t, u^+, x^*)$ is introduced whenever a constraint violation or a singularity-induced bifurcation is predicted to occur within the prediction horizon ($t \in [t^*, t^* + t_p]$). A suitable selection of $\mathbf{Q}$, $\mathbf{R}$ and $P(t, u^+, x^*)$ ensures that none of the emergency controls are scheduled unless necessary to remove a constraint violation during the prediction interval.

B. Predictors

The cost function is calculated from predictions of the output trajectories. Several different methods of obtaining (approximations of) these trajectories are possible [2]. The benchmark method is Nonlinear MPC (NLMPC) The system (1)–(3) is a hybrid nonlinear system and it is therefore generally impossible to find an analytical expression for the trajectories ($\hat{y}(t, u)$). The most straightforward method is to obtain them numerically. This requires simulation of the hybrid DAE model over the prediction interval for each evaluation of the cost function.

The NLMPC predictor makes use of the full simulation model and thus accounts for hybrid as well as nonlinear behaviour of the system. In [2], the performance of NLMPC is compared with various approximate prediction methods:

- Euler State Prediction (ESP);
- MPC using off-equilibrium linearisations (LMPC);
- Euler state prediction – linear output approximation (ESPLO).

The details are more to do with numerical methods than the ideas of this chapter. Nevertheless, it is an important topic for practical implementation – see [2] for further discussion.

4.4 A tree search method

The combinatorial optimization problem (36) is a search problem where the search state-space is determined by the control constraints. For illustration,

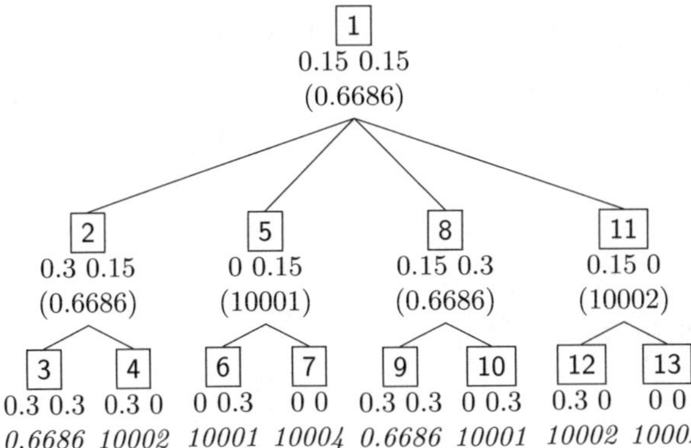

Fig.9. Search example.

the set of available control states $\mathcal{S}(u)$ is explored for a best control state using the standard recursive depth-first algorithm [72] shown in Table 1. The algorithm relies on two types of auxiliary functions:

s(u) which generates all control states , i.e., a subset of $\mathcal{S}(u)$, reachable from the current state u with a single switching of one control.

c(u) which evaluates the state u, according to the cost criteria (37).

Figure 9 shows the search tree for a simplified search problem where only the controls B_3 and B_4 in Figure 7 are enabled; the additional constraint that each control may only change once in the tree has been imposed. The number shown below each node corresponds to the result reported by the function *search*. The search starts at the initial state (1) with $B_3 = 0.15$, $B_4 = 0.15$, corresponding to the root of the tree. From here four new states can be reached (2, 5, 8, 11) by changing either B_3 or B_4 one step up or down. From each of these four states another two can be reached by changing the other control. Here the optimum is found in nodes 3 and 9 for $B_3 = 0.3$, $B_4 = 0.3$. Note that the search tree (13 nodes) is larger than the state space ($3^2 = 9$ possible combinations) since it contains not only all possible control states but also all paths to them. The example search requires 13 evaluations of the cost-function.

For this simple example, the search state space is kept small by the control constraints because of the small number of controls. However, the tree grows exponentially with the number of controls and an exhaustive search of the tree is not feasible for practical cases. With a larger number of controls, the size of the tree is limited by applying a maximum search depth which corresponds to the maximum number of controls that may change in one search.

Table 1. Pseudocode for construction and traversal of the search tree.

```
[f_out] = function search (f_in, u_in, m);
  f = call c(u_in)
  if m = 1 /* the node is a leaf node */
    best = f
  else
    best = min(f, f_in)
    for u_i = call s(u_in) do
      f_tree = call search(f_in, u_i, n − 1)
      best =min(f_tree, best)
    end
  return f_out = best
```

f_{in}	- best value found prior to search of tree
u_{in}	- initial (root) control state
m	- search depth
f_{out}	- best value found after search of tree

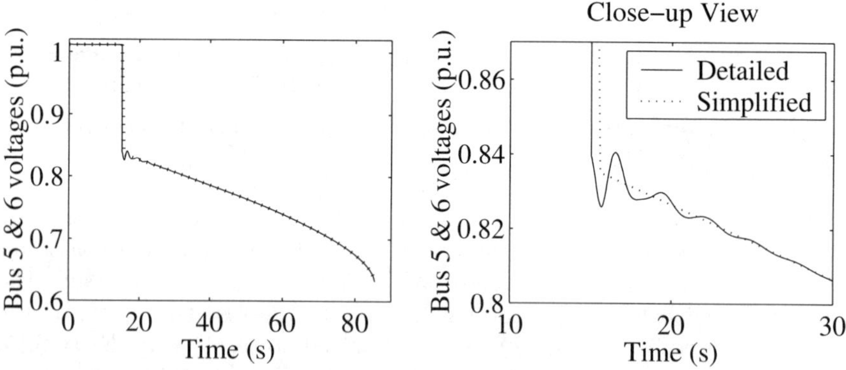

Fig.10. Voltage response without voltage control.

4.5 Simulation results

To illustrate the performance of the emergency controllers, results from simulation of the system in Figure 7 are presented in Figures 11-12. The disturbances applied are changes of the four line impedances from X=0.2 to X=1.0 p.u. When computing the new control signal, it is assumed that it will be in use during the entire prediction interval (that is, $n = 1$). If no emergency actions are taken, voltage collapse will occur at simulation time 70 s as shown in Figure 10. This figure also shows the response with a more detailed model including the fast electromechanical oscillations. The simulations have been carried out with a 30 s sampling interval of the controller,

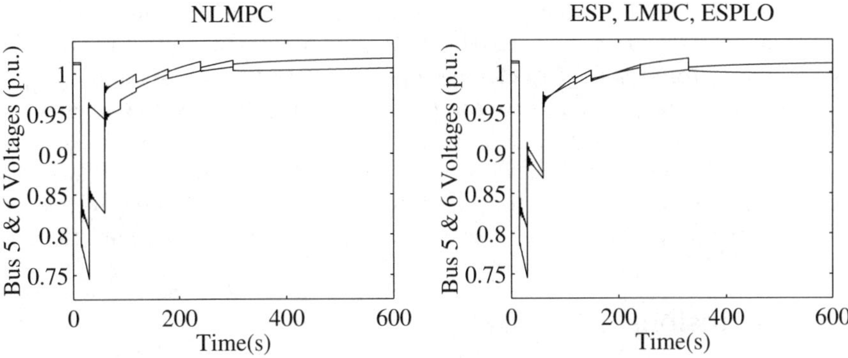

Fig.11. Voltage response at buses 5 and 6 with all controls enabled.

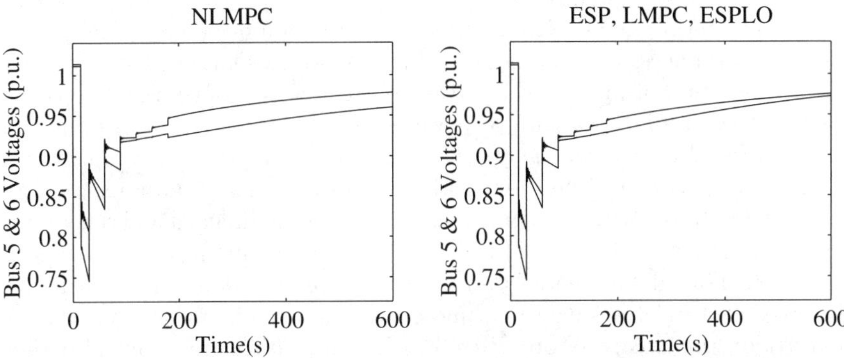

Fig.12. Voltage response at buses 5 and 6 with B_5 and B_6 disabled. The three methods in the right hand figure yield identical results.

i.e. the tree-search optimization is carried out periodically every 30 s. By the use of voltage constraints, load shedding is authorized if the voltages at bus 5 or 6 are predicted to be lower than 0.9 p.u. at the prediction horizon. The search depth is 4, the prediction horizon 120 s and the weight matrices and penalty terms were chosen according to particular tuning procedure in [2], with the output vector y containing the voltages of buses 5 and 6.

Figure 11 shows the voltage response at buses 5 and 6 with all controls enabled. In this case, it is possible to stabilize the system without resorting to load shedding. All the predictors yield near identical results. The NLMPC version stabilizes and restores the voltages close to their pre-disturbance levels using 5 capacitor bank steps and 6 tap changer operations. In [2] it is shown that more approximate predictors are capable of stabilizing the voltage with less computation, but use an extra capacitor switching for a total of 6 switchings.

Figure 12 shows the voltage response to the same disturbance but with capacitor controls B_5 and B_6 disabled. The NLMPC stabilizes the voltages using 4 capacitor steps, 5 tap changer operations and 4 load shedding steps. Here, the tap changer positions are explicitly controlled by the emergency controller. Normally, the two tap changing transformers would be under feedback control by relays such as those described by for example [73]. Thus, the emergency control applicable to tap changers would normally be tap locking.

Reference [2] discusses use of search heuristics to reduce computation.

5 Conclusions

This chapter has given a conceptual framework, called global control, for the control of complex systems. Some illustrations are given using simple power system problems. The further development for large systems is occurring [75,76]. The term global refers to scale (in dimension or geographically) and size of disturbances as is clearly needed for the world's larger power systems. The combination of ideas developed in control science, optimisation and related to the specific structure of power systems provides the possibility of more advanced control in the era of deregulation where more complexity is going to be inevitable. The framework involves hybrid system modelling, bifurcation analysis, switching control and optimal coordination and scheduling of control. Two special control tasks were investigated in more detail. Firstly, we showed how global control can even improve the coordination of multi-objective control of a single machine on an infinite bus power system. We defined our global control objective as achieving satisfactory control performance over a wide range of anticipated operating conditions; specifically, transiently stabilizing the power system when subjected to a severe disturbance and retaining good voltage level after the disturbance. The controller was of the switching kind using membership functions. The applicability of the ideas to more complex systems should be clear even if there are nontrivial issues to resolve in any design.

The second illustration demonstrates use of coordinated control by the application of Model Predictive Control combined with search techniques to emergency voltage control. The controller optimally coordinates capacitor switching, tap changer operation, and load shedding in different geographic locations, ensuring minimum usage of emergency controls such as load shedding. A search method is employed to minimize a cost function based on predictions of the system's future behaviour. The results presented here consider emergency state operation. The same methodology can be used in an implementation that provides control coordination aimed at loss minimization in the normal state and arrest of voltage collapse in the emergency state. The cost function would be based on system active power losses and two sets of voltage constraints would be used, for the normal and emergency states respectively. In principle the methodology described here could be applied to any kind of stability problem in power systems and similar problems that arise in other applications.

Acknowledgements.
The work for this chapter was initiated when all the authors were at the University of Sydney and continued at City University of Hong Kong. The work was supported by the Australian Research Council and City University.

Appendix: System data

Table 2. Control Variables (All data given in p.u.)

symbol	control	initial value	lower limit	upper limit	control step
B_3	cap. 3	0.00	0.00	0.30	0.15
B_4	cap. 4	0.00	0.00	0.30	0.15
B_5	cap. 5	0.00	0.00	0.30	0.15
B_6	cap. 6	0.00	0.00	0.30	0.15
k_{t35}	tap 3-5	1.04	0.80	1.20	0.02
k_{t46}	tap 4-6	1.06	0.80	1.20	0.02
k_{l5}	load 5	1	0.85	1	0.05
k_{l6}	load 6	1	0.85	1	0.05

Table 3. Network, Generator and Load Parameters

Load	**Generator (at bus 2)**
$P_{05} = P_{06} = 0.6$ p.u.	$V_r = 1.05$ p.u.
$Q_{05} = Q_{06} = 0.3$ p.u.	$P_m = 0.3$ p.u.
$T_{p5} = T_{p6} = 60$ s	$H = 3.54$ MVA/MWs, $D = 0.0$ p.u.
$T_{q6} = T_{q6} = 60$ s	$r_a = 0.00327$ p.u.
$\alpha_{s5} = \alpha_{s6} = 0$	$x_d = 1.76$ p.u., $x_q = 1.58$ p.u.,
$\alpha_{t5} = \alpha_{t6} = 2$	$x'_d = 0.42$ p.u., $x'_q = 0.30$ p.u.
$\beta_{s5} = \beta_{s6} = 0$	$x''_d = 0.10$ p.u., $x''_q = 0.20$ p.u.
$\beta_{t5} = \beta_{t6} = 2$	$T'_{d0} = 6.66$ s, $T'_{q0} = 0.44$ s
Network	$T''_{d0} = 0.02$ s, $T''_{q0} = 0.03$ s
$n_{max} = 1.2$ p.u.	
$n_{min} = 0.8$ p.u.	**Exciter & AVR (at bus 2)**
$X_t = 0.1$ p.u.	$T_{exc} = 2.5$ s
$T_t = 30$ s	$K_{avr} = 100$
$X = \begin{cases} 0.2, & time < 15 \text{ s} \\ 1.0, & time > 15 \text{ s} \end{cases}$	$E_{fmax} = 3.5$ p.u.
$V_1 = 1.05\angle 0$ p.u.	$E_{fmin} = -3.5$ p.u.

References

1. Popović, D. H., Hill, D. J., Wu, Q. (2003) Optimal voltage security control of power systems. Int. J. Electr. Power Energy Syst., 24(2):121–130
2. Larsson, M., Hill, D. J., Olsson, G. (2003) Emergency voltage control using searching and predictive control. Int. J. Electr. Power Energy Syst., 24(4):305–320
3. Popović, D. H., Hill, D. J., Wu, Q. (1998) Coordinated static and dynamic voltage control in large power systems. In Proc. Bulk Power System Voltage Phenomena -IV: Restructuring. Fink, L. H., Vournas, C. D. (eds.), Tech. Univ. Athens, Greece, 391–405
4. Makarov, Y. V., Popović, D. H., Hill, D. J. (1998) Stabilization of transient processes in power systems by an eigenvalue shift approach. IEEE Trans. Power Syst., **13**(2):382–388
5. Shahrestani, S., Hill, D. J. (2000) Global control with application to bifurcating power systems. Syst. Contr. Lett., **41**(3):145–155
6. Guo, Y., Hill, D. J., Wang, Y. (2001) Global transient stability and voltage regulation for power systems. IEEE Trans. Power Syst., **16**(4):678–688
7. Hill, D. J., Guo, Y., Larsson, M., Wang, Y. (2001) Global hybrid control of power systems. In Proc. Bulk Power System Dynamics and Control -V, August, Japan, 374–394
8. Hill, D. J., Guo, Y., Larsson, M., Wang, Y. (2002) Global hybrid control for large power systems. In Proc. 4th Asian Contr. Conf., Singapore, 851–856
9. Kuipers, B., Aström, K. (1994) The composition and validation of heterogeneous control laws. Automatica, **30**:233–249
10. Isidori, A. (1995) Nonlinear Control Systems: An Introduction to Communications and Control Engineering, 3rd ed., New York: Springer-Verlag
11. Chapman, J. W., Ilic, M. D., King, C. A., Eng, L., Kaufman, H. (1993) Stabilizing a multimachine power system via decentralized feedback linearizing excitation control. IEEE Trans. Power Syst., **8**:830–839
12. Lu, Q., Sun, Y. (1989) Nonlinear stabilizing control of multimachine systems. IEEE Trans. Power Syst., **4**:236–241
13. Wang, Y., Xie, L., Hill, D. J., Middleton, R. H. (1992) Robust nonlinear controller design for transient stability enhancement of power system. In Proc. 31st IEEE Conf. Decision Control, Tuscon, Arizona, 1117–1122
14. Wang, Y., Hill, D. J., Middleton, R. H., Gao, L. (1993) Transient stability enhancement and voltage regulation of power systems. IEEE Trans. Power Syst., **8**:620–627
15. Gao, L., Chen, L., Fan, Y., Ma, H. (1992) A nonlinear control design for power systems. Automatica, **28**:975–979
16. Zhu, C., Zhou, R., Wang, Y. (1997) A new nonlinear voltage controller for power systems. Int. J. Electr. Power Energy Syst., **19**:19–27
17. Hauer, J. F. (1993) Robustness issues in stability control of large electric power systems. In Proc. 32nd IEEE Conf. Decision Control, San Antonio, TX, 2329–2334
18. Clarke, D. W. (1994) Advances in Model-Based Predictive Control. Oxford: Oxford Univ. Press
19. Maciejowski, J. M. (2002) Predictive Control with Constraints. London: Prentice-Hall

20. Narendra, K. S., Balakrishnan, J. (1997) Adaptive control using multiple models. IEEE Trans. Auto. Contr., **42**(2):171–187
21. Bemporad, A., Morari, M. (1999) Control of systems integrating logic,dynamics and constraints. Automatica, Special Issue on Hybrid Systems, **35**:407–427
22. Tomlin, C. J., Lygeros, J., Sastry, S. S. (2000) A game theoretic approach to controller design for hybrid systems. Proceedings of the IEEE, Special Issue on Hybrid Systems: Theory and Applications, **88**(7):949–970
23. Branicky, M. S., Borkar, V. S., Mitter, S. K. (1998) A unified framework for hybrid control: Model and optimal control theory. IEEE Trans. Auto. Contr., **43**(1):31–45
24. Gokbayrak, K., Cassandros C. G. (2000) A hierarchical decomposition method for optimal control of hybrid systems. In Proc. 39th IEEE Conf. Decision Control
25. Hiskens, I. A., Pai, M. A. (2000) Hybrid systems view of power systems modelling. In Proc. IEEE Int. Symp. Circ. Syst., Geneva, **II**:228–231
26. Chen, L., *et.al.* (2000) Stability analysis for digital controls of power systems. Electric Power Systems Research, **55**:79–86
27. Larsson, M. (2000) A Modelica library for power system stability studies. Modelica Workshop 2000. Lund University, Lund, Sweden. http://www.modelica.org/workshop2000/proceedings/Larsson.pdf
28. Hill, D. J., Mareels, I. M. Y. (1990) Stability theory for differential/algebraic systems with application to power systems. IEEE Trans. on Circ. Syst.-I, **37**(11):1416–1423
29. DeCarlo, R., Branicky, M., Pettersson, S., Lennartson, B. (2000) Perspectives and results on the stability and stabilizability of hybrid systems. Proceedings of the IEEE, Special Issue on Hybrid Systems: Theory and Applications, **88**(7):1069–1082
30. Michel, A. N., Hu, B. (1999) Towards a stability theory of general hybrid dynamical systems. Automatica, Special Issue on Hybrid Systems, **35**:371–384
31. Abdalla, O. H., Hassan, S. A., Tweig, N. T. (1984) Coordinated stabilization of a multimachine power system. IEEE Trans. Power Apparatus Syst., **103**:483–494
32. Fleming, R. J., Mohan, M. A., Parvatism, K. (1981) Selection of parameters of stabilizers in multimachine power systems. IEEE Trans. Power Apparatus Syst., **100**:2329–2333
33. Lim, C. M., Elangovan, S. (1985) A new stabilizer design technique for multimachine power systems. IEEE Trans. Power Apparatus Syst., **104**:2393–2400
34. Makarov, Y. V., Maslennikov, V. A., Hill, D. J. (1996) Revealing loads having the biggest influence on power system small disturbance stability. IEEE Trans. Power Syst., **11**(4):2018-2023
35. Lam, D. M., Yee, H. (1998) A study of frequency responses of generator electrical torques for power system stabilizer design. IEEE Trans. Power Syst., **13**(3):1136–1142
36. Sepulchre, R., Jankovic, M., Kokotovic, P. (1997) Constructive Nonlinear Control. London: Springer-Verlag
37. Sastry, S. (1999) Nonlinear Systems: Analysis, Stability and Control. New York: Springer-Verlag

38. Astrom, K. J., *et al.* (eds.) (2001) Control of Complex Systems. London: Springer-Verlag
39. Hill, D. J. (ed.) (1995) Special Issue on Nonlinear Phenomena in Power Systems: Theory and Practical Implications, Proceedings of the IEEE, **83**(11):1439–1587
40. Chen, G., Moiola, J. L., Wang, H. O. (2000) Bifurcation control: Theories, Methods and Applications. Int. J. Bifur. Chaos, **10**(3):511–548
41. Cao, S. G., Rees, N. W., Feng, G. (1997) Analysis and design for a class of complex control systems – Part II: Fuzzy controller design. Automatica, **33**:1029–1039
42. Bryson, A. E. (1999) Dynamic Optimization. Lonoman, CA: Addison-Wesley
43. Teo, K. L., Goh, C. J., Wong, K. H. (1991) A Unified Computational Approach to Optimal Control Problems. England: Longman Scientific and Technical
44. Meadows, E. S. (1997) Dynamic programming and model predictive control, In Proc. Amer. Contr. Conf., Albuquerque, NM, 1635–1639
45. Andrievsky, B., Stotsky, A., Fradkov, A. L. (1988) Velocity-gradient algorithms in control and adaptation problems. Auto. Remote Contr., **49**:1533–1564
46. Seron, M. M., Hill, D. J., Fradkov A. L. (1995) Nonlinear adaptive control of feedback passive systems. Automatica, **31**(7):1053–1060
47. Bonabeau, E., Dorigo, M., Theraulaz, G. (1999) Swarm Intelligence: From Natural to Artificial Systems. Oxford: Oxford Univ. Press
48. Middleton, R. H., Goodwin, G. C., Hill, D. J., Mayne, D. Q. (1988) Design issues in adaptive control. EEE Trans. Auto. Contr., **33**(1):50–58
49. Johansson, R. (1990) Quadratic optimization of motion coordination and control. IEEE Trans. Auto. Contr., **35**(11):1197–1208
50. Ilic, M., Galiana, F., Fink, L. (1998) Power System Restructuring: Engineering and Economics. Boston: Kluwer
51. Vournas, C. (ed.) (1998) Proceedings International Symposium on Bulk Power Systems Dynamics and Control-IV Restructuring. Santorini, Aug. 23-28, 391–405
52. Anderson, P. M., Fouad, A. A. (1994) Power System Control and Stability. New York: IEEE Press
53. Bergen, A. R. (1986) Power Systems Analysis. Englewood Cliffs, NJ: Prentice-Hall
54. Kundur, P. (1994) Power System Stability and Control. New York: McGraw-Hill
55. Wang, Y., Hill, D. J. (1996) Robust nonlinear coordinated control of power systems. Automatica, **32**:611–618
56. Illic, M. D., Mak, F. K. (1989) A new class of fast nonlinear voltage controllers and their impact on improved transmission capacity. In Proc. Amer. Contr. Conf., Pittsburgh, PA, 1246–1251
57. Mak, F. K. 1992) Design of nonlinear generator excitors using differential geometric control theories. In Proc. 31st IEEE Conf. Decision Control, Tuscon, AZ, 149–1153
58. Marino, R. (1984) A example of a nonlinear regulator. IEEE Trans. Auto. Contr., **29**:276–279
59. Mielczarski, W., Zajaczkowski, A. (1989) Nonlinear controller for synchronous generator. In Proc. IFAC Nonl. Contr. Syst. Design Symp., Capri

60. Khargonekar, P. P., Petersen, I. P., Zhou, K. (1990) Robust stabilization of uncertain linear systems: Quadratic stabilizability and H_∞ control theory. IEEE Trans. Auto. Contr., **35**:356–361
61. Xie, L., Fu, M., de Souza, C. E. (1992) H_∞ control and quadratic stabilization of systems with parameter uncertainty via output feedback. IEEE Trans. Auto. Contr., **37**:1253–1256
62. Vu, K. T., Liu, C.-C. (1992) Shrinking stability regions and voltage collapse in power systems. IEEE Trans. Circ. Syst.-I, **39**(4):271–89
63. Xu, W., Mansour, Y. (1994) Voltage stability analysis using generic dynamic load models. IEEE Trans. Power Syst., **9**(1):479–93
64. Popović, D. H., Hiskens, I. A., Hill, D. J. (1996) Investigations of load-tap changer interaction. Int. J. Electr. Power Energy Syst., **18**(2):81–97
65. Arnborg, S., Andersson, G., Hill, D. J., Hiskens, I. A. (1998) On influence of load modelling for undervoltage load shedding studies. IEEE Trans. Power Syst., **13**(2):395–400
66. Popović, D. S., Levi, V. A., Gorečan, Z. A. (1997) Co-ordination of emergency secondary-voltage control and load shedding to prevent voltage instability. IEE Proceedings. Generation, Transmission & Distribution, **144**(3):293–300
67. Wu, Q, Popović, D. H., Hill, D. J., Parker, C. J. (1999) System security enhancement against voltage collapse via coordinated control. Power Eng. Soc. 1999 Winter Meeting, IEEE, **1**:755–60
68. Garcia, C. E., Prett, D. M., Morari, M. (1989) Model predictive control: theory and practice – a survey. Automatica, **25**(3):335–48
69. Karlsson, D., Hill, D. J. (1994) Modelling and identification of nonlinear dynamic loads in power systems. IEEE Trans. Power Syst., **9**(1):157–163
70. Van Cutsem, T., Vournas, C. (1998) Voltage Stability of Electric Power Systems. Boston: Kluwer
71. Mayne, D. Q., Rawlings, J. B., Rao, C. V., Scocaert, P. O. M. (2000) Constrained model predictive control: Stability and optimality. Automatica, **36**(6):789–814
72. Russell, S. J., Norvig, P. (1995) Artificial Intelligence - A Modern Approach. Englewood Cliffs, NJ: Prentice-Hall
73. Sauer, P. W., Pai, M. A. (1994) A comparison of discrete vs. continuous dynamic models of tap-changing-under-load transformers. In Proc. Bulk power System Voltage Phenomena - III: Voltage Stability, Security and Control, Davos, Switzerland
74. Zakrzewski, R. R., Mohlor, R. R., Kolodziej, W. J. (1994) Hierarchical intelligent control with flexible AC transmission system application. Contr. Eng. Practice, **2** (6):979–987
75. Wu, Q., Popović, D. H., Hill D. J., Parker, C. (2001) Voltage security enhancement via coordinated control. IEEE Trans. Power Syst., **16**(1):127–135
76. Leung, J., Hill, D. J., Ni, Y., Hui, R. (2002) Global power system control using a unified power flow controller. In Proc. 14th Power Syst. Comput. Conf., Sevilla, Spain

Preserving Transients on Unstable Chaotic Attractors

Tomasz Kapitaniak and Krzysztof Czolczynski

Division of Dynamics, Technical University of Lodz
Stefanowskiego 1/15, 90-924 Lodz, Poland
tomaszka@ck-sg.p.lodz.pl

Abstract. In this chapter we present a controlling method which allows the preservation of transient chaotic evolution in the desired region of the phase space. The concept of practical stability for the perturbed chaotic attractors and the connection between practical and asymptotic stability is described. Our controlling procedure allows asymptotically unstable chaotic attractors to become practically stable in such a way that transients on the unstable chaotic attractors or in their neighborhoods do not decay. Illustrative applications are presented.

1 Introduction

The main idea of chaos control is to replace the evolution of strange chaotic attractor by desired periodic behaviour [1–6]. For some particular applications, like for example communication [8], it is important to preserve chaos or keep the trajectory in the desired region of the phase space when the dynamical system is under the external perturbation.

One of the fundamental problems in practical applications of chaotic dynamics is the problem of stability of the chaotic attractor A. The basin of attraction $\beta(A)$ is the set of points whose ω -limit set is contained in A. In Milnor's definition of an attractor [7] the basin of attraction need not include the whole neighborhood of the attractor, i.e., we say that A is a Milnor attractor if $\beta(A)$ has positive Lebesgue measure. For example, riddled basins which have recently been found in practical physical systems [9,10,12,13], have a positive Lebesgue measure but do not contain any neighborhood of the attractor. If the basin of attraction contains the neighborhood of A, then the attractor is asymptotically stable.

Most of the chaotic attractors which can be met in the practical engineering systems are quasiattractors, i.e., the limiting sets enclosing the periodic orbits of different topological types, structurally unstable homoclinic trajectories, etc. Practical systems are mainly quasihyperbolic [15], i.e., many different types of attractors co-exist in the phase space.

Existing definitions of the stability of chaotic attractors cannot always give sufficient practical information about the behaviour of the real engineering system which is under the influence of both permanently acting and short time impulse-like perturbations. The main problems in stability analysis of chaotic engineering systems are as follow:

G. Chen, D.J. Hill, and X. Yu (Eds.): Bifurcation Control, LNCIS 293, pp. 189–203, 2003.

- Basin of attraction of the asymptotically stable chaotic attractor can be so small that perturbations can take trajectory out of it to the basin of another attractor.
- The system operates in finite time T during which the system cannot reach attractor, but for $t < T$ system evolves in the limited part of the phase space which does not necessarily include the final attractor of the system [14].

In this chapter we described the concept of the practical stability and practical stability in finite time for the chaotic attractors [16,17]. Assuming that appropriate control applied to the dynamical systems evolving on the unstable attractor (repeller) can be considered as a perturbation we contracted the controlling procedure which allows asymptotically unstable chaotic attractors to become practically stable. Our approach is different from other controlling procedures [1–6] as our goal is not to replace chaotic evolution by the desired periodic behaviour but to keep the dynamical system to evolve permanently in the neighborhood of the chaotic repeller. The applied controlling procedure creates attractors which do not exist in the uncontrolled systems.

This chapter is organized as follows. In Section 2 we recall definitions of practical stability of chaotic attractors and discuss similarities and differences between asymptotic and practical stability of chaotic attractors. In Section 3 we describe the controlling procedure which allows some asymptotically unstable attractors to be practically stable. Examples of the applications of controlling are given in Section 4. Finally, we summarize our results in Section 5.

2 Practical Stability

2.1 Definitions

Consider a dynamical system given by

$$\frac{dx}{dt} = f(x,t)\,, \tag{1}$$

which for initial conditions $x(t_0) = x_0 \in \omega$, where ω is an open set, has asymptotically stable attractor $A \in \mathcal{R}^n$.

Definition 1

1. Let the system (1) be under the influence of permanently acting perturbations $p(x,t)$ so the perturbed system is in the form

$$\frac{dx}{dt} = f(x,t) + p(x,t)\,. \tag{2}$$

2. Let perturbation function $p(x,t)$ fulfill condition

$$||p(x,t)|| \leq \delta \,,$$

where $\delta > 0$.

3. Let Ω be a closed, bounded set such that $A \in \Omega$ and $\omega \in \Omega$.

If for all initial conditions $x(t_0) = x_0 \in \omega$, all functions $p(x,t)$ and all $t \geq t_0$, $x(t) \in \Omega$ then attractor A is practically stable (in relation to sets ω, Ω and perturbations $p(x,t)$). ♢

In this definition function $p(x,t)$ describes all continuously acting perturbations. Set ω defines limits of both uncertainties in initial conditions and short time perturbations. Perturbed trajectories $x(t)$ of the system (2) evolve in the region of the phase space given by the set Ω which is usually larger than attractor A of unperturbed system (1).

If attractor A is a fixed point [18] and in the absence of permanently acting perturbations ($p(x,t) = 0$) definition 1 is equivalent to the definition of the stability in the sense of Lagrange in relation to the set Ω. In every case the practical stability is independent of the stability in the sense of Lyapunov.

Many engineering systems operate in finite time and for stability investigations of such systems we can introduce a weaker definition.

Definition 2

Let conditions (1)-(3) of Definition 1 be fulfilled.

If for all initial conditions $x(t_0) = x_0 \in \omega$, all functions $p(x,t)$ and all $t_0 \leq t < T$, $x(t) \in \Omega$ then attractor A is practically stable in finite time T (in relation to sets ω, Ω and perturbations $p(x,t)$). ♢

It is easy to understand the definition, looking in Fig. 1, which shows a Poincaré map of the system with two co-existing stable attractors A and B (shown in black) and the boundary between their basins of attraction (respectively $\beta(A)$ and $\beta(B)$. Each trajectory (not perturbed) which starts from the basin of attraction, tends to the appropriate attractor. The grey area around each attractor is the set ω, from which the perturbed trajectories may start. How big the set ω is, depends on impulse perturbations which we have to predict (expect), taking into considerations the conditions in which our system operates. Let us assume that the sets ω are inside the basins of attraction, so they do not cross the boundary between the basins. The white areas around the sets ω are the sets Ω. The perturbed trajectory which starts from ω stays inside the set Ω and tends to the attractor, when the set Ω is located inside one of the basins of attraction. The attractor A is practically stable in relation to the sets Ω and ω. When the set Ω crosses the boundary between the basins of attraction, the perturbed trajectory which starts in the set ω of attractor B may fall into the part of ω lying in the basin of attraction of attractor A and tend to it, instead of attractor B. This means that in relation to sets ω and Ω, attractor B is practically unstable. In such a case, to retain stability of the attractor, a controlling procedure must be employed.

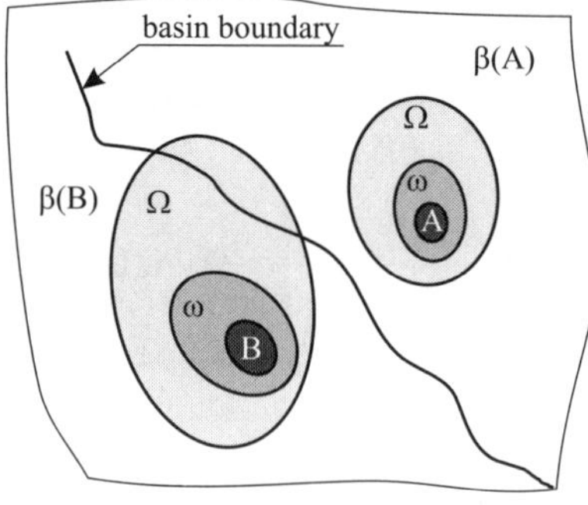

Fig.1. Practically stable (A) and unstable (B) attractors

3 Controlling Procedure

In this section, we present the controlling procedure which can allow us to make asymptotically unstable attractor practically stable, i.e. to keep the system trajectory in the desired region of the phase space. To achieve this goal we assume that one of the system parameters, say d can be adjusted finely around a nominal value d_0, i.e., $d \in [d_0 + \Delta p, p_0 - \Delta p]$, where $\Delta p/p_0 \ll 1$.

Observation of the system behavior allows the determination of the phase space region which a trajectory $\Gamma(t)$ must enter shortly before going to the undesired region of the phase space. This region will be called a dangerous zone and will be indicated by $\mathcal{D}$. In the neighborhood of $\mathcal{D}$ one can identify a number, K say, of trajectories $\gamma_k(t)$ going out of this neighborhood and evolving further in the appropriate region of the phase space. We can further identify a set $\mathcal{S}$ of points γ_k^* on these trajectories which we called a safe set. A schematic representation of zone $\mathcal{D}$ and set $\mathcal{S}$ is shown in Figure 2.

$\mathcal{D}$ can be estimated also when the equations of motion are unknown and our knowledge of the system is based on a scalar time series. In this case, one can construct a return map $x_{n+1} = f(x_n)$ and identify preimages of points going straight to the fixed points. There, preimages define the dangerous zone (see, e.g. [21]).

Our controlling method consists of three steps:

- estimation of the dangerous zone $\mathcal{D}$;
- identification of trajectories $\gamma_k(t)$ in the neighborhood of $\mathcal{D}$ which allow further evolution in the desired region of the phase space and creation of the safe set $\mathcal{S}$ of points γ_k^* in the ϵ -neighborhood of $\mathcal{D}$;

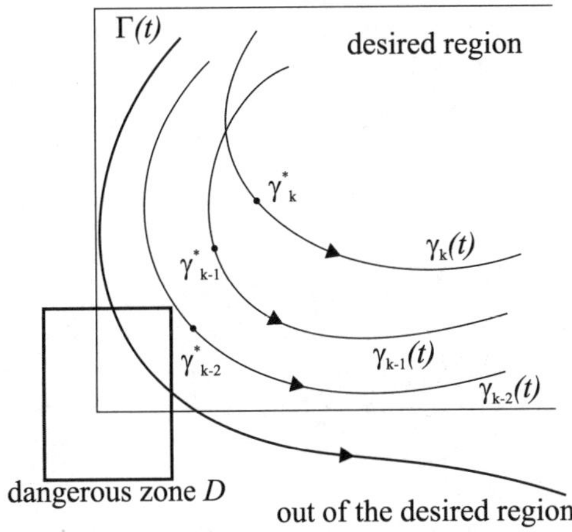

Fig.2. Idea of controlling procedure

- introduction of a small temporal change of one of the system parameters to allow the system to switch from actual trajectory $\gamma(t) \in \mathcal{D}$ to one of the safe $\gamma_k(t)$ trajectories.

To control a trajectory $\Gamma(t)$ which enters the dangerous zone $\mathcal{D}$ we use a simple feedback procedure. Suppose that the trajectory of the N-dimensional map $x_{n+1} = M(x_n, p)$ (the numerical solution of Eq. (4) can be considered as an example of such a map), entering the set $\mathcal{D}$ falls into a ϵ-neighborhood of a point γ_k^* in the safe set $\mathcal{S}$, i.e., $|x_n - y_n| \leq \epsilon$, where y_n, representing the point γ_k^*, and its future iterates $y_{n+1}, y_{n+2}...$ constitute a safe trajectory going away from the dangerous set $\mathcal{D}$. In the neighborhood of x_n, we can consider the following linearized dynamics

$$\Delta x_{x+1} = DM(x_n, d)\Delta x_n + \frac{\partial M}{\partial d}\Delta d_n, \tag{3}$$

where $\Delta x_n = x_n - y_n$, $\Delta d_n = d_n - d_0$, and the Jacobian $DM(x_n, d)$ and the vector $\partial M/\partial d$ are calculated at $x_n = y_n$ and $d_n = d_0$. If we choose a unit vector u in the phase space and let $u\Delta x_{n+1} = 0$, we obtain the d- parameter perturbation necessary to achieve control

$$\Delta d_n = -\frac{uDM(x_n, d)\Delta x_n}{u\frac{\partial M}{\partial d}}. \tag{4}$$

The unit vector can be chosen arbitrarily provided that (i) it is not orthogonal to x_{n+1}, and (ii) the denominator in (4) is not zero.

4 Applications

4.1 Controlling spiral attractor of Chua's circuit

As an example let us consider the dynamics of Chua's circuit [19,20]. Chua's circuit is RLC circuit with four linear elements (two capacitors, one resistor, and one inductor) and a nonlinear diode, which can be modelled by a system of three differential equations. The equations for this circuit are

$$\begin{aligned}\frac{dx}{dt} &= \alpha(y - x - g(x)),\\ \frac{dy}{dt} &= x - y + x,\\ \frac{dz}{dt} &= \beta y,\end{aligned} \tag{5}$$

where the piecewise linear function

$$g(x) = bx + 0.5(a - b)(|x + 1| - |x - 1|),$$

describes three different voltage-current regimes of nonlinear diode. Parameters α, β, a and b are constant.

Two of the best-known attractors of system (5) the spiral and double-scroll attractors are shown in Fig. 3(a), (b). They can be observed for $\beta = 14.87$, $a = -1.42$, $b = -0.68$, $\alpha = 7.7$ (spiral) and $\alpha = 10$ (double-scroll). In the case of spiral attractor there are two symmetrical co-existing attractors A_1 and A_2. Attractors A_1, A_2 and B are asymptotically stable.

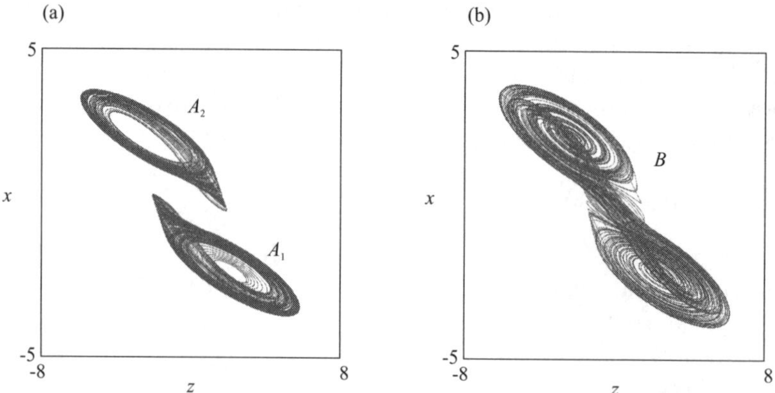

Fig.3. Attractors of Chua's circuit: $\beta = 14.87$, $a = -1.42$, $b = -0.68$; (a) $\alpha = 7.7$ (spiral attractor), (b) $\alpha = 10$ (double-scroll attractor). The two-dimensional $x - z$ cross-sections of the three dimensional basins of attraction defined by $y = 0$ are shown. Attractors A^+, A^- and B are projected into these cross-sections

If we define set ω and Ω in the way as in Fig. 4(a) and allow perturbations $p(x, t)$ to evolve only in Ω attractor A_1 is practically stable in relation to sets ω, Ω and perturbations $p(x, t)$. If these sets ω and Ω are too small from practical point of view and we have to consider larger sets like these in Fig. 4(b) attractor is no more practically stable.

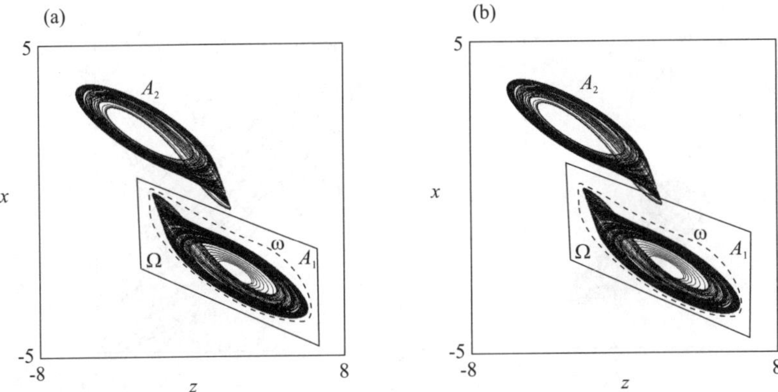

Fig.4. Examples of practically stable (a) and practically unstable (b) attractors of Chua's circuit

In the case of $\alpha = 10$, double-scroll attractor is the only chaotic attractor of the system (3) and both spiral attractors are unstable (they do not exist). Application of the controlling procedure described in the previous section allows the preservation of the system evolution in the neighborhood of, say, A_1 (attractor A_1 can be made practically stable). Let us consider the evolution on double scroll attractor shown in Fig. 5(a) and (b). We can define the dangerous zone $\mathcal{D} \in B$ and $\mathcal{D} \in \Omega$ which the system visits before executing the undesirable evolution on A_2 as shown in Fig. 5(a) and (b). Our goal is to diverge trajectory entering $\mathcal{D}$-region out of it to the safe set S ($\gamma_k^* \in S$) – Fig. 5. To achieve this goal we assume that one of the system parameters, let us say α can be adjusted finely around a nominal value α_0, i.e., $\alpha \in [\alpha_0 + \Delta\alpha, \alpha_0 - \Delta\alpha]$, where $\Delta\alpha/\alpha_0 << 1$.

The α parameter change given by Eq.(4) is an occasional feedback control which applied to the system of Fig. 2(b) allows us to make originally unstable spiral attractors A_1 or A_2 practically stable. In this example our method stabilizes a smaller unstable chaotic attractor embedded into larger stable one.

In the numerical experiment we have been able to control spiral attractor like one in Fig. 3(a). The changes of the controlling parameters $\partial\alpha$ were smaller than 4% of its nominal value $\alpha = 10$. The controlled attractor shown in Fig. 5 is not identical as spiral attractor A_1 of Fig. 3(a) but with the definition 1 the evolution shown in this figure can be considered as the evolution on practically stable attractor A_1.

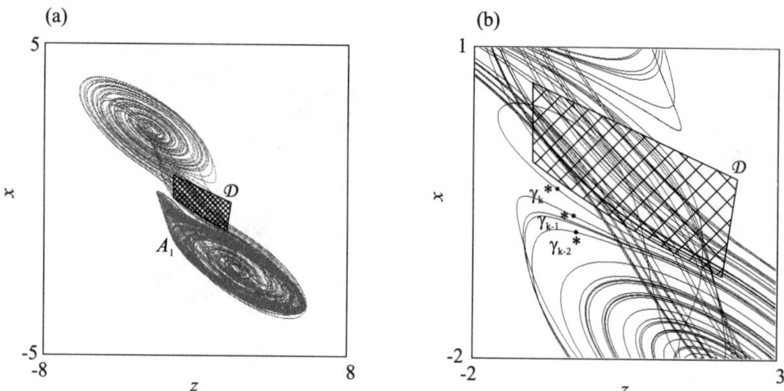

Fig.5. Control of asymptotically unstable spiral attractor A_1

4.2 Preserving transient chaos of the Lorenz system

As the second example, we consider the dynamics of the Lorenz system

$$\begin{aligned} \frac{dx}{dt} &= -\sigma(x-y)\,, \\ \frac{dy}{dt} &= -xz + rx - y\,, \\ \frac{dz}{dt} &= xy - bz\,, \end{aligned} \tag{6}$$

where $\sigma, b, r \in \mathcal{R}$ are constant. It is well-known that the system (5) exhibits transient chaotic behaviour for $r \in (13.96, 24.74)$. In this case a system trajectory evolves for a significantly long period of time on a strange chaotic repeller (with a shape similar to the well-known Lorenz attractor) before converging to one of the fixed points $C_{1,2} = (\pm[b(r-1)]^{1/2}, \pm[b(r-1)]^{1/2}, r-1)$. Such a trajectory is shown in Figure 6, where the trajectory ultimately converges to $C_2 = (-[b(r-1)]^{1/2}, -[b(r-1)]^{1/2}, r-1)$. Our controlling goal in this example is to preserve transient chaos, i.e., do not allow system trajectory to converge to the regular attractor [22].

In the Lorenz system the identification of the dangerous zone $\mathcal{D}$ is straightforward as it is well-known [23] that the occurrence of the transient chaos is connected with the breakdown of the homoclinic orbit of the unstable point $(0,0,0)$, and zone $\mathcal{D}$ therefore lies in the neighborhood of the unstable manifold of the fixed point $(0,0,0)$. Alternatively $\mathcal{D}$ can be identified in the neighborhoods of stable fixed points $C_{1,2}$.

We now use the approach described in Section 3 to control the system (6). The dangerous set $\mathcal{D} = \{(x,y,z); -0.02 < x,y < 0.02, 3.5 < z < 4.5\}$ was taken as in Fig. 7 (without control, after the transient chaotic evolution of an average lifetime, here taken to be $\tau = 690$, trajectories converge to one of the fixed points). We assume that the accessible parameter r can be slightly

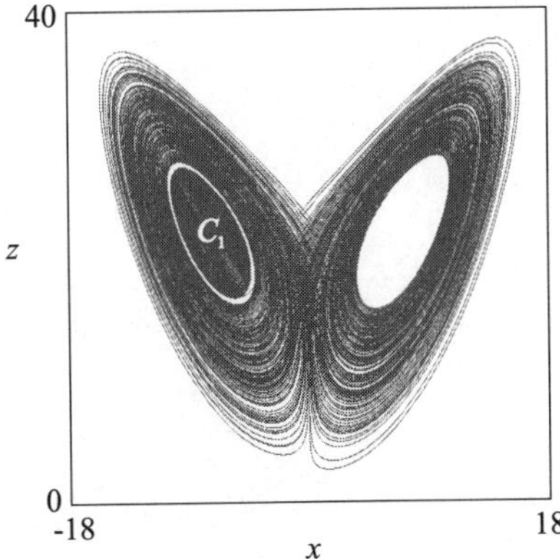

Fig.6. Transient chaotic evolution of system (5): $\sigma = 10, b = 8/3, r = 23.2, x(0) = 1, y(0) = 0, z(0) = 45$

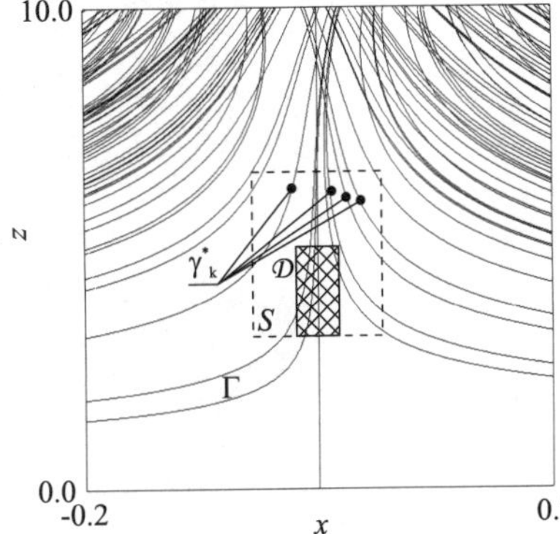

Fig.7. The dangerous zone $\mathcal{D}$ and safe set $\mathcal{S}$ for the system (5)

perturbed around its nominal value $r_0 = 23.2$, and we take the maximum allowed parameter perturbation Δr_{max} to be 10^{-1}. We then create a safe set $\mathcal{S}$ which consists of 200 points $\gamma_k^* \in \{(x, y, z) : -0.04 < x, y < 0.04, 3.49 < z < 4.51\} - \mathcal{D}$, and in the controlling procedure take $\epsilon = 10^{-2}$. With such a control we insure the preservation of transient chaos. An example of controlled evolution is shown in Fig. 8, where shown attractor is not asymptotically stable but it is practically stable according to Definition 1 with perturbations given by the controlling procedure.

4.3 Controlling impact oscillator

Consider the linear oscillator shown in Fig. 9, with the mass m, the stiffness coefficient of the spring k, and the coefficient of viscous damping c [24]. The symbol δ denotes the gap between the bottom surface of the mass and the unmovable basement. The vibrations of the system are forced by the external harmonic force $F sin(\omega t)$. During the vibrations the mass of the oscillator may (but does not have to) impact on the basement.

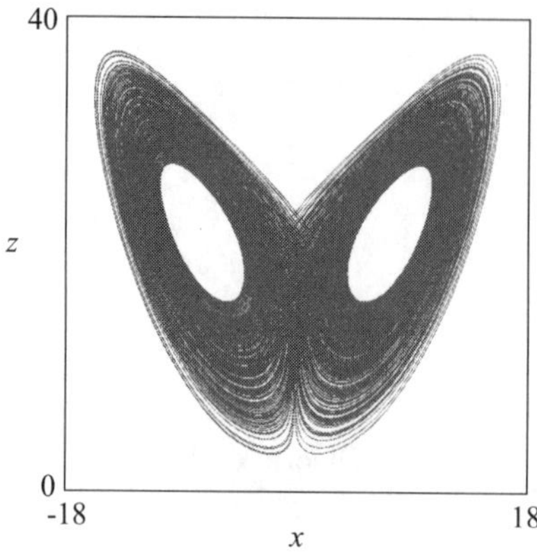

Fig.8. Controlled trajectory of the system (5); $x(0) = 1, y(0) = 0, z(0) = 45$

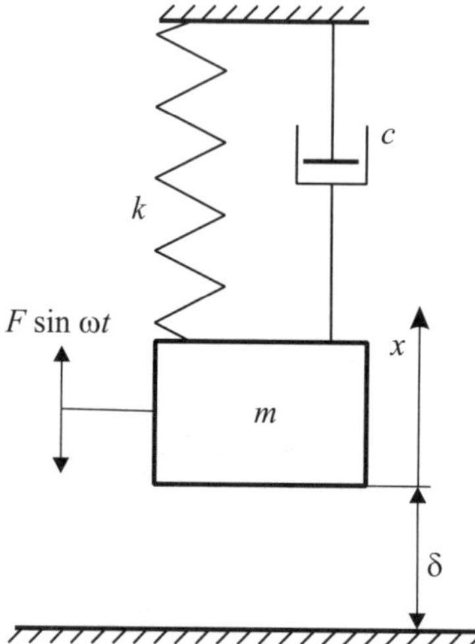

Fig.9. Impact oscillator

The mathematical model of the system consists of the differential equation of motion

$$m\frac{d^2x}{dt^2} + c\frac{dx}{dt} + kx = F\sin(\omega t) \tag{7}$$

which may be also written in the dimensionless form

$$\ddot{x} + 2\gamma\dot{x} + x = sin\eta\tau\,, \tag{8}$$

where

$$\begin{aligned} \tau = \alpha t, \qquad \alpha^2 = \frac{k}{m}\,, \\ \gamma = \frac{c}{2\sqrt{km}}, \qquad \tau = \frac{\omega}{\alpha}\,. \end{aligned} \tag{9}$$

The equations of impacts which are based on the well known Newton's law with the coefficient of restitution k_r:

$$
\begin{aligned}
m\left(\frac{dx}{dt}\right)' - m\frac{dx}{dt} &= S_d\,,\\
\left(\frac{dx}{dt}\right)' &= -k_r\frac{dx}{dt}\,,
\end{aligned}
\tag{10}
$$

where S_d denotes the impulse of the force and $(')$ denotes the velocity just before the impact. Equation (10) may be written in the dimensionless form:

$$
\begin{aligned}
\dot{x}_1' - \dot{x}_1 &= S_d\,,\\
\dot{x}_1' &= k_r\dot{x}_1'\,.
\end{aligned}
\tag{11}
$$

The basic dimensionless data of the system considered in our numerical studies are as follows: the damping coefficient γ_1 is equal to 0.1 of the critical one, the gap δ is equal to 0.2 of the static deflection of the spring under the weight of the mass m_1, the coefficient of restitution $k_r = 0.9$.

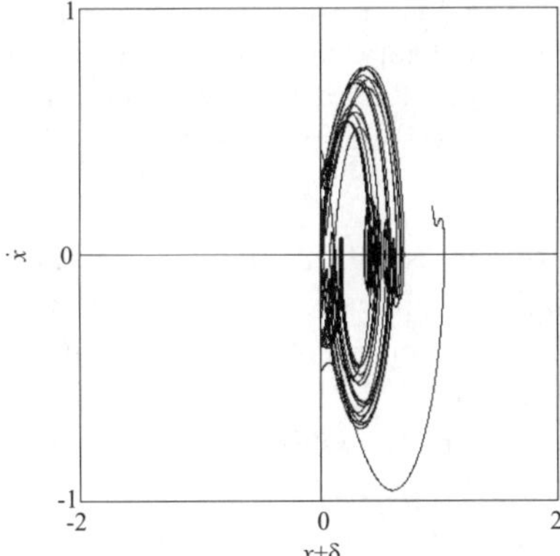

Fig.10. Chaotic motion of the oscillator (6), $x_0 = 0.75, \dot{x}_0 = 0.2$

In the neighborhood of the frequency of the exciting force $\eta_1 = 4$ one observes depending on the initial conditions of the motion two different types of the behavior of the considered system. Figure 10 shows the phase plane of the system oscillations for the initial conditions $x_0 = 0.75$ and $\dot{x}_0 = 0.2$. On this plane we observe the motion with permanent impacts; the shape of the attractor proves that the motion has a chaotic character. The second possibility is a regular, harmonic motion without impacts. For this case we have calculated the basins of attraction of both attractors i.e., the map of the initial conditions which lead to a regular or chaotic motion of the system shown in Figure 11. If the motion starts from the pair $(x_0, \dot{x}_0)$ belonging to the white area (at the zero phase of the exciting force), the undisturbed trajectory of the motion will always cross this map in the white area and tend to the chaotic attractor. The black points mark the points in which the trajectory starting from the initial conditions $x_0 = 0.75, \dot{x}_0 = 0.2$ crosses the map after each period of the external force. Contrary to this case, if the motion starts from the grey area, the trajectory will cross the map also in grey areas, tending to the second, coexisting attractor; i.e., impacts-less periodic solution represented by the point $(-0.075, -0.2)$.

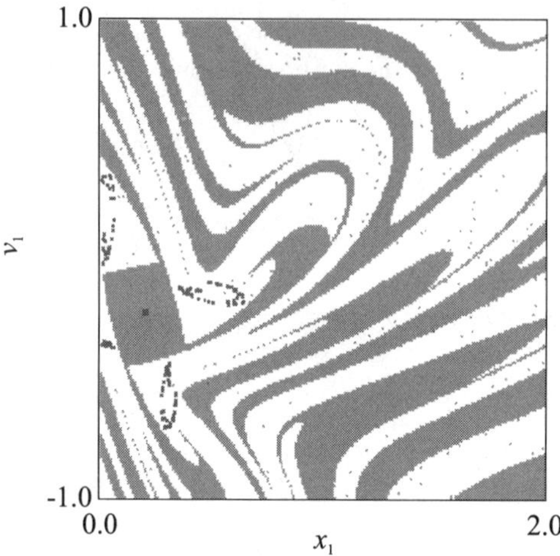

Fig.11. Basin of attraction of the chaotic attractor (white) and basin of attraction of periodic impactless motion (grey)

Let us assume that the chaotic motion of the oscillator at the frequency of the forcing $\eta_1 = 4$ is our desired dynamical behavior. Assume that the system evolves to the chaotic attractor. Let us consider the perturbation of the motion caused by a temporary lack of power in the exciter. Due to this perturbation the trajectory crosses the map in the white and grey areas tending to the point $(0, 0)$. After switching on the power, the trajectory can go to the one of the coexisting attractors depending on its position at the moment when the power is switched on. This implies that desired chaotic attractor is not practically stable in the relation to the perturbation caused by the temporal lack of power. For example, in the case shown in Fig. 12 after the lack of power for the time equal to 5 periods of the external force, the trajectory tends to the periodic attractor. In the case of the longer break (55 periods) the trajectory goes to the chaotic attractor (Fig. 13).

There is a possibility to control the system, in which the lack of energy is the reason of disturbances: after switching off the exciter the controlling system prevents switching it on, until the trajectory leaves the grey zone, tending to the point $(0, 0)$ in the white zone. This type of control requires the constant monitoring of the trajectory position and can be practically difficult. In the second approach we proposed to add to the system an impulse perturbation moving the trajectory to the appropriate point in the white area. The coordinates of such system can be stored in the controlling computer

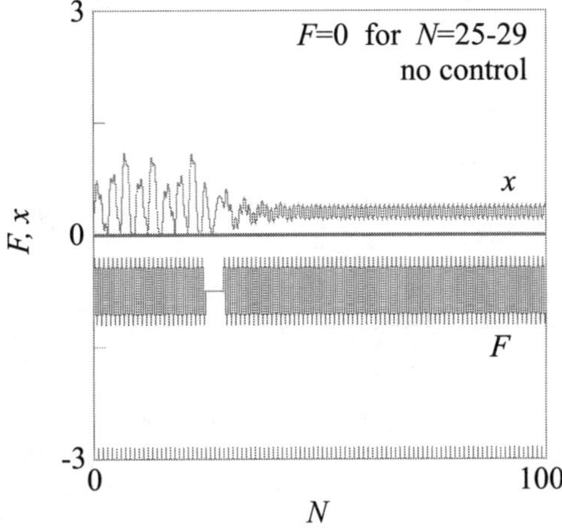

Fig.12. Time series of $x_1(t)$ and external force, 5 periods of external force lack of power

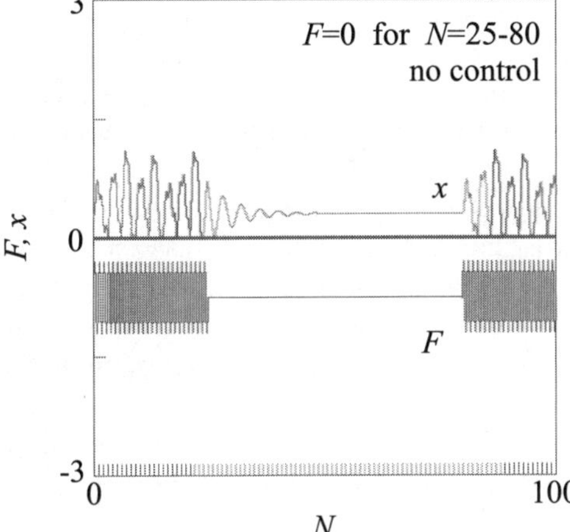

Fig.13. Time series of $x_1(t)$ and external force, 55 periods of external force lack of power

and the impulse perturbation is introduced at the moment when the power is switched on. Evolution of the system trajectory with such a control is shown in Fig. 14.

Designing oscillatory systems it is necessary to identify all possible attractors and their basins of attraction. In the case of coexistence of two or more attractors, their practical stability should be checked in relation to the predicted sets Ω and ω describing possible perturbations. When the attractor representing the working conditions is practically unstable, a controlling procedure should be employed. The method of controlling depends on the particular properties of the system.

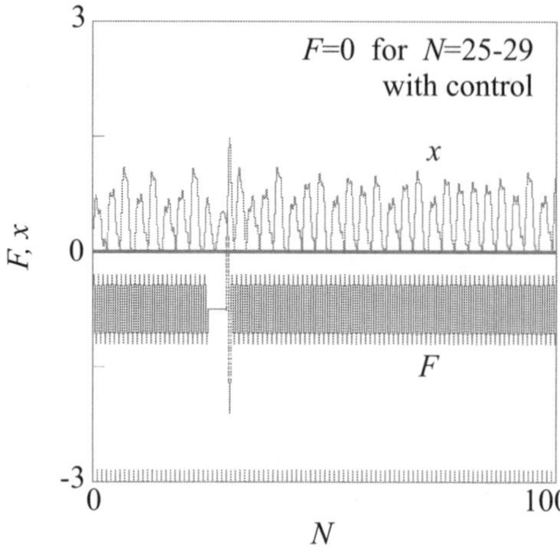

Fig.14. Time series of $x_1(t)$ and external force, 5 periods of external force lack of power with applied control

5 Conclusions

We hope that the concept of practical stability can be very useful in the study of chaotic systems. Particularly, it can be useful when the appropriate control of asymptotically unstable chaotic attractors can make them practically stable.

The controlling procedure which allows to make the asymptotically unstable chaotic attractors practically stable has been described. This method allows the system evolution in the desired region of the phase space in the neighborhood of a repeller. Alternatively it can be applied to control transients in the dynamical systems with co-existing attractors and irregular boundaries between basins of attraction.

To achieve our goals we have to have knowledge of the system dynamics, but not necessarily the equation of motion, which is usually the case in practical systems, and must be able to apply a small temporal change in one of the system parameters.

References

1. Ott, E., Grebogi, C., Yorke, J. A. (1990) Controlling chaos. Phys. Rev., Lett., **64**:1196–1199
2. Kapitaniak, T. (1996) Controlling Chaos. London: Academic Press
3. Lakshmanan, M., Murali, K. (1996) Chaos in Nonlinear Oscillations: Controlling and Synchronization. Singapore: World Scientific
4. Chen, G., Dong, X. (1998) From Chaos to Order: Methodologies, Perspectives and Applications. Singapore: World Scientific
5. Kapitaniak, T., (2000) Chaos for Engineers: Theory, Applications and Control. Berlin: Springer-Verlag
6. Chen, G. (ed.) (2000) Controlling Chaos and Bifurcation in Engineering Systems. Boca Raton, FL: CRC Press

7. Milnor, J. (1985) On the concept of attractor. Commun. Math. Phys., **99**:177–195
8. Carroll, T. L., Heagy, J. F., Pecora, L. M. (1996) Transforming signals with chaotic synchronization. Phys. Rev., **E54**:4676–4680
9. Alexander, J. C., Kan, I., Yorke, J. A., You, Z. (1992) Riddled basins. Int. J. Bifur. Chaos, **2**:795–812
10. Ott, E., Sommerer, J. C., Alexander, U. I., Kan, I., Yorke, J. A. (1994) The transition to chaotic attractors with riddled basins. Physica D, **76**:384–410
11. Heagy, J. F., Carroll, T., Pecora, L. (1996) Experimental and numerical evidence for riddled basins in coupled chaotic systems. Phys. Rev. Lett., **73**:3528–3531
12. Kapitaniak, T. (1995) Experimental observation of riddled behaviour. J. Phys. A, **28**:L63–66
13. Kapitaniak, T., Chua, L. O. (1996) Locally-intermingled basins of attraction in coupled Chua's circuits. Int. J. Bifur. Chaos, **6**:357–366
14. Czolczynski, K., Kapitaniak, T. (1996) Controlling Hopf bifurcation in transients. Machine Vibrations, **5**:89–91
15. Anishchenko, V. S., Kapitaniak, T., Safonova, M., Sosnovtseva, O. (1994) The birth of double- double-scroll attractor. Phys. Lett. A, **181**:207–214
16. Kapitaniak, T., Brindley, J. (1998) Practical stability of chaotic attractors. Chaos, Solitons and Fractals, **9**:43–50
17. Kapitaniak, T., Czolczynski, K., Brindley, J. (1999) Practical stability of synchronized chaotic attractors and its control. In Proc. 7th IEEE Mediterranien Conf. Contr. Auto., Haifa, 1224–1230
18. Bogusz, W. (1996) Technical Stability. Warsaw: PWN
19. Madan, R. (ed.) (1994) Chua's Circuit. Singapore: World Scientific
20. Kennedy, M. P. (1992) Robust op amp realization of Chua's circuit. Frequenz, **46**:66–80
21. Parmanada, P., Eiswirth, M. (1996) Suppressing large excursions to a chaotic attractor using occasional feedback control. Phys. Rev. E, **54**:1036–1039
22. Kapitaniak, T., Brindley, J. (1998) Preserving transient chaos. Phys. Lett. A, **241**, 41–45
23. Ott, E. (1993) Chaos in Dynamical Systems. Cambridge: Cambridge Univ. Press
24. Blazejczyk-Okolewska, B., Czolczynski, K., Kapitaniak, T. (2000) Controlling attractors in the systems with impacts. In Proc. Int. Conf. Advanced Problems in Vibration Theory and Applications, Zhang, J. H., Zhang, X. N. (eds), Beijing: Science Press, 548–551

Bifurcation Control in Feedback Systems

Diego M. Alonso[1,3], Daniel W. Berns[2], Eduardo E. Paolini[1], and Jorge L. Moiola[1,3]

[1] Dpto. de Ing. Eléctrica y de Computadoras
Univ. Nacional del Sur, Av. Alem 1253
B8000CPB Bahía Blanca, Argentina
[2] Dpto. de Ing. Electrónica
Univ. Nacional de la Patagonia S. J. Bosco
Ciudad Universitaria, Ruta Prov. Nro. 1, Km. 4
9005 Comodoro Rivadavia, Argentina
[3] CONICET, Argentina jmoiola@criba.edu.ar

Abstract. In this chapter the mathematical tools from bifurcation theory are used within the framework of feedback control systems. The first part deals with a simple example where the amplitude of limit cycles and the appearance of period-doubling bifurcations are controlled using a method derived from the frequency domain approach. In the second part, bifurcation theory is used to analyze the dynamical behavior of an inverted pendulum with saturated control. The main objective is to find appropriate values of the controller parameters to achieve the stabilization of the pendulum at the inverted position and, at the same time, to obtain the largest basin of attraction.

1 Introduction

Bifurcation control is a new research field in which the qualitative changes of the behavior of dynamical systems after varying a distinguished parameter are modified conveniently by a certain though specific control. In this regard, the mathematical tools of bifurcation theory are readapted to provide a suitable framework aiming to smooth the qualitative changes of the associated local bifurcation diagrams, change the type of bifurcation, create a limit cycle in a certain region of the state space, modify its amplitude or frequency, delay the appearance of the bifurcation or its effects, enlarge the basin of attraction of the operating points or attracting sets, etc. In other words, bifurcation control uses the knowledge of the qualitative features of different bifurcations to help to design suitable controllers for "nonconventional" control in nonlinear systems. Some tutorial papers and books are [1,18–20]; the references therein provide a kind of historical account about the development of different methodologies and applications.

The research in this field have been triggered after the local feedback stabilization for elementary bifurcations proposed by [2,3]. In those papers the stabilization of equilibrium points or periodic orbits is addressed using the tools for classifying the dynamics in the simplest critical cases, i.e. when one

G. Chen, D.J. Hill, and X. Yu (Eds.): Bifurcation Control, LNCIS 293, pp. 205–228, 2003.

or two eigenvalues of the linearized system cross the imaginary axis and provoke the change of the stability of the equilibrium point. Since then, several papers appeared covering distinct control objectives for continuous-time and discrete-time nonlinear systems. Regarding the control objectives, it is worth to mention the stabilization of the period-doubling bifurcation for discrete-time systems [4] as well as the delay of the appearance of period-doubling bifurcation for continuous-time nonlinear systems [44]. Besides the stabilization of equilibrium points and periodic orbits, the amplitude control of oscillations has been another key issue (see [16,32]). In those works the authors have taken advantage of the normal form of Hopf bifurcations and proposed a feedback law to achieve the desired amplitude control. In this case, the computation of the stability index of the emerging periodic solutions is crucial (see also [14] for a closely related implementation). On the other hand, other approaches deal with the control of the oscillation amplitude, as for example [45], but using optimization techniques.

There is a wide range of promising applications of bifurcation control, notably in the improvement on reliability and efficiency of aeroengines [25,46,47], avoiding the conditions for voltage collapse in power systems [1,33], improving the performance of the oscillatory behavior of tethered satellites [28], to name just a few. Moreover, some important developments on the theoretical side have emerged in order to propose a unified approach of the so-called normal forms for continuous-time nonlinear control systems [23,24] and for discrete-time nonlinear control systems [12,27]. Finally, the creation of a certain type of bifurcation with desired properties and at a preferred location has been addressed in [7,17] for the case of Hopf bifurcation.

In practical applications, however, physical limitations of actuators impose an important constraint to the controller design [8,10,35,46,47]. A control law derived neglecting these limitations will surely result in a global behavior completely different to that expected. For example, actuator's saturation may cause the appearance of new equilibria, periodic solutions, homoclinic connections and even chaos. In this chapter an oscillatory mechanical system with a restricted control action is addressed using analytic and numerical tools from nonlinear dynamics and bifurcation theory. The system is an underactuated pendulum with a rotating disk mounted in its free extreme. The "almost" global stabilization problem at the inverted position with continuous state feedback as solved in [34,36,38] requires a control action capable of dominating the gravity torque. This hard constraint is currently bypassed using a switched control strategy: a nonlinear controller for swinging-up the pendulum, and a linear one to stabilize it at the inverted position. This approach has been successfully applied to many different classes of inverted pendula (some of these can be found, for example, in [11,21,29,40,41,48]). For the particular case of the pendulum treated in this chapter, the swinging-up approach has been performed using methods from bifurcation theory and energy-type considerations [6].

Although the switched strategy is effective when dealing with actuators with insufficient torque, the stabilization with a bounded continuous state feedback is still a challenging problem. Towards this end, a saturated, continuous control law is proposed here, and conditions over controllers gains to ensure the stability of the inverted position are derived using classical tools. The dynamic behavior of the closed loop system is analyzed as a function of a control (bifurcation) parameter. The closed loop system exhibits a rich dynamics, including Hopf and cyclic fold bifurcations, as well as heteroclinic connections of saddle points and homoclinic orbits for certain values of the control parameter. These results complete a previous work done by the authors [8] to find a set of parameter values where the pendulum is swung-up from the downward position to the inverted position.

In this chapter a simple example is shown for controlling the amplitude of limit cycles and for delaying the bifurcations of the cycles using the so-called frequency domain approach. This method is also used in the second part of the chapter when analyzing the dynamics of the inverted pendulum via a saturated control law. In the following section the frequency domain method is briefly described with applications in amplitude control and delay of limit cycle bifurcations. Then, the stabilization of the inverted position of the pendulum with a saturated control law is analyzed using bifurcation theory.

2 Hopf and Period-Doubling Bifurcation via a Frequency Domain Method

2.1 Preliminaries

Consider the following parameterized nonlinear system

$$\begin{aligned} \dot{x}(t) &= f(x;\mu) = A(\mu)x(t) + B(\mu)g(C(\mu)x(t);\mu), \\ y(t) &= C(\mu)x(t), \end{aligned} \tag{1}$$

where A, B and C are $n \times n$, $n \times r$ and $m \times n$ matrices, respectively, $\mu \in R$ is the main bifurcation control parameter, $x \in R^n$ is the state vector, $y \in R^m$ is the system output, $g : R^m \to C^{2q+1}(R^r)$ is the system feedback (a smooth nonlinear function), $f : R^n \to C^{2q+1}(R^n)$ is the system vector field (a smooth nonlinear function), and n, m, q and r are positive integers. Define

$$G(s;\mu) = C(\mu)\left[sI - A(\mu)\right]^{-1} B(\mu),$$

and let $z = -y$ with $g(y;\mu) := h(z;\mu)$. Then, the equilibrium solution of (1) can be obtained by solving the following equation

$$G(0;\mu)\, h(\hat{z};\mu) + \hat{z} = 0. \tag{2}$$

Also, define

$$J_{\hat{z}} = \left. \frac{\partial h(z)}{\partial z} \right|_{z=\hat{z}} .$$

A Hopf bifurcation occurs when one of the eigenvalues of the linearized system transfer matrix $G(i\omega;\mu)\, J_{\hat{z}}$, denoted $\hat{\lambda}$, satisfies

$$\hat{\lambda}(i\omega_0;\mu_0) = -1 + i0\,, \qquad i = \sqrt{-1}\,, \tag{3}$$

for some values of ω_0 and μ_0. Then, a periodic branch arises from criticality and continues to develop as μ is varied, in accordance with the result of the Hopf bifurcation theorem.

The Graphical Hopf Theorem [30] provides the following qth-order prediction of the limit cycle

$$z(t) \approx z_q(t) = \hat{z} + \mathbf{Re}\left\{ \sum_{k=0}^{2q} Z_q^k\, e^{ik\omega_q t} \right\}, \tag{4}$$

where ω_q and Z_q^k are the frequency and complex amplitude of the k-harmonic predicted by a q-order harmonic balance approximation [31]. The method requires to solve the following set of equations

$$Z_q^k = -G(ik\omega_q,\mu)\, H_q^k, \qquad k \in [0,1,...,2q]. \tag{5}$$

These equations are solved in terms of $Z_q^1 = Z_q^1(v,\theta_q)$, where v is the right eigenvector of $G(i\omega,\mu)J_{\hat{z}}$ associated with the eigenvalue $\hat{\lambda}$, and θ_q is a measure of the amplitude of the periodic solution. In simple terms, H_q^k is a function of Z_q^k which involves partial derivatives of the generic function $h(z;\mu)$ with respect to z. For a multivariable nonlinear system the explicit formulas up to the eighth-order (involving ninth-order partial derivatives) were given in [31].

The amplitude and frequency of the periodic solution are evaluated by computing the complex functions $\xi_q(v,\omega)$, $q = 1,2,3,4$ and solving the following equation in a way reminiscent to the describing function method

$$\hat{\lambda}(i\omega;\mu) = -1 + \sum_{k=1}^{q} \xi_k(v,\omega)\, \theta^{2k} .$$

Finally, the predicted periodic solution can be obtained by placing the solution pair $(\hat{\omega}_q,\hat{\theta}_q)$, known as the L_q-approximation solution, in Eq. (4).

2.2 Stability analysis of the limit cycles

Suppose that a limit cycle, $x_H(t)$, has been generated by the Hopf bifurcation mechanism. Define a perturbed trajectory, $x_p(t)$, by

$$x_p = x_H + x_D,$$

where x_D is a perturbation of x_H. Taking the time derivative gives

$$\begin{aligned}\dot{x}_p &= \dot{x}_H + \dot{x}_D \\ &= A\left(x_H + x_D\right) + Bg\left(C\left(x_H + x_D\right)\right),\end{aligned}$$

where the dependence of A, B and g on μ is removed from the notation for the sake of simplicity. Then,

$$\dot{x}_D = A\left(x_H + x_D\right) + Bg\left(C\left(x_H + x_D\right)\right) - \dot{x}_H,$$

and after some algebraic manipulations, we arrive at

$$\dot{x}_D = Ax_D + B\left[g\left(C\left(x_H + x_D\right)\right) - g\left(Cx_H\right)\right].$$

It is possible to demonstrate the growth or decay of x_D and, thereafter, the instability or stability of x_H by using an appropriate Poincaré section and looking for a fixed point solution of the so-called return map [39]. The stability of this fixed point depends on the position, relative to the unit circle, of the eigenvalues of the *monodromy matrix* M, defined as

$$\begin{cases}\dot{X}\left(t\right) = J_D(t)X\left(t\right), \\ X\left(0\right) = I, \\ M = X\left(\frac{2\pi}{\omega_H}\right),\end{cases} \tag{6}$$

where $J_D(t)$ is a periodic matrix defined by

$$J_D(t) = \left.\frac{\partial \dot{x}_D\left(t\right)}{\partial x_D}\right|_{x_D=0} = A + BJ_{x_H}\left(t\right),$$

with

$$J_{x_H}\left(t\right) = \left.\frac{\partial g\left(Cx\right)}{\partial x}\right|_{x=x_H(t)},$$

and ω_H is the frequency of the limit cycle.

The eigenvalues of M are called the characteristic multipliers or Floquet multipliers, and they are denoted as λ_i, $i = 1, ..., n$. One of them, λ_1, is always equal to $+1$. In planar systems ($n = 2$), there is only one possibility in which the remaining eigenvalue can cross the unit circle, through the point $1 + i0$, leading for example to a cyclic-fold, transcritical or pitchfork bifurcation of cycles. When $n = 3$, there are two additional possibilities: a Neimark-Sacker bifurcation (two complex conjugate characteristic multipliers crossing the unit circle simultaneously), or a period-doubling bifurcation (one real characteristic multiplier crossing the unit circle at the point $-1 + i0$). With respect to the stability, a periodic orbit becomes stable when all the characteristic multipliers (except the one at $+1$) stay within the unit circle, and it becomes unstable when they move out.

One problem in the application of matrix M is the need to integrate two dynamical systems simultaneously: the original nonlinear system and the variational equation (6). Moreover, transients of the nonlinear system may be very large, leading to long simulation times before the actual calculation of M can be completed. In order to drastically reduce the computational burden, it is possible to calculate instead of the original matrix M approximate matrices M_q (where $2q$, $q = 1, 2$ and 3, are the numbers of harmonics of the periodic solution), so that it is only needed to deal with the integration of the (approximate) variational equation

$$\begin{cases} \dot{Y}(t) = J_{D_q}(t) Y(t) \,, \\ Y(0) = I \,, \\ M_q = Y\left(\frac{2\pi}{\omega_q}\right) \,, \end{cases}$$

where $J_{D_q}(t)$ is a periodic matrix (obtained by using the information of L_q-approximation for the limit cycle), defined by

$$J_{Dq}(t) = A + BJ_{x_q}(t) \,,$$

with

$$J_{x_q}(t) = \left. \frac{\partial g(Cx)}{\partial x} \right|_{x = x_q(t)} .$$

2.3 The scheme for detecting Hopf and period-doubling bifurcations

The following procedure uses the Hopf bifurcation theorem in the frequency domain and provides limit cycle prediction with higher-order approximations near the Hopf bifurcations points. The detection of period-doubling bifurcation is obtained using an approximation of the monodromy matrix. The required steps are:

1. Find all the equilibria of the given system and the conditions for Hopf bifurcations.
2. Choose an equilibrium, and find the corresponding L_1 prediction and approximate monodromy matrix M_1 .
3. If a value of μ can be found such that a characteristic exponent $\hat{\lambda}$ crosses the unit circle, then evaluate L_2 and L_3 predictions, as well as the approximations of the monodromy matrices M_2 and M_3, in a neighborhood of μ, to improve the accuracy of the detection (see [15] for more details about the approximation errors).

In the following section, the preceding algorithm is used to provide an accurate detection of the onset of a period-doubling bifurcation in a nonlinear feedback control system. The procedure also helps to derive conditions for delaying the appearance of the period-doubling bifurcation [44] and finally, to modify the feedback law for controlling the amplitude of oscillation.

Table 1. Characteristic multipliers of M_1, M_2 and M_3 for detecting a period-doubling bifurcation of system (7)

	μ	λ_1	λ_2	λ_3
	−0.53	1.00558	−0.999267	−0.046272
M_1	−0.52	1.00601	−1.11676	−0.043808
	−0.51	1.00646	−1.23802	−0.041813
	−0.51	1.00065	−0.87955	−0.05950
M_2	−0.50	1.00072	−0.98074	−0.05649
	−0.49	1.00078	−1.08462	−0.05408
	−0.4900	1.00022	−0.98558	−0.05962
M_3	−0.4885	1.00023	−1.00027	−0.05926
	−0.4875	1.00023	−1.01009	−0.05902

2.4 Application example: A nonlinear feedback control system

Let us consider the dynamical system

$$\begin{aligned} \dot{x} &= Ax + B[g(Cx) + u], \\ y &= Cx, \end{aligned} \tag{7}$$

where

$$A = \begin{bmatrix} 0 & 1 & 0 \\ 0 & 0 & 1 \\ -1-\alpha & -1.2 & \mu \end{bmatrix}, \quad B = \begin{bmatrix} 0 \\ 0 \\ 1 \end{bmatrix}, \quad C = \begin{bmatrix} 1\,0\,0 \end{bmatrix},$$

and

$$g(Cx) = x_1^2 + \alpha x_1,$$

where $u = 0$, $\mu \in [-5/6, 0)$ is the bifurcation parameter and α is chosen for a convenient frequency domain realization, and without loss of generality can be set equal to 1. The equilibria of this system are $x_a = (0,0,0)$ and $x_b = (1,0,0)$.

A branch of periodic solutions emerges from x_a when a supercritical Hopf bifurcation takes place at the value $\mu_{HB} = -5/6$. The continuation of limit cycles has been performed using LOCBIF [26] from μ_{HB} until the first period-doubling bifurcation takes place at $\mu_{PD} \approx -0.481045$. Evaluating the approximations of the monodromy matrices M_1, M_2 and M_3 by using the periodic predictions L_1, L_2, L_3 (involving two, four and six harmonics, respectively), the characteristic multipliers of Table 1 are obtained. Notice that, as a result of the technical requirement for the construction of the Poincaré map, one of the eigenvalues must be equal to 1.000; deviation from this value is related to the approximation error of the method. According to the values of the characteristic multiplier λ_1 in Table 1, it is clear that higher order (L_3) solution provides smaller approximation errors. This table also reveals that

Table 2. Characteristic multipliers of M_1, M_2 and M_3 for system (9) when $\mu = -0.37$

	λ_1	$\hat{\lambda}_2$	λ_3
M_1	1.00353	−1.00018	−0.13269
M_2	1.00069	−0.85263	−0.15366
M_3	1.00021	−0.801848	−0.162747

the approximate monodromy matrix M_3 has one multiplier $\hat{\lambda}_2 \approx -1 + i0$ –the defining condition for the period-doubling bifurcation– when μ reaches the critical value $\mu_{PDM_3} \approx -0.4885$ which is in close agreement to the value obtained using LOCBIF ($\mu_{PD} \approx -0.481045$).

To delay the appearance of the period-doubling bifurcation an outer feedback

$$u = \varepsilon_1 x_2^3, \tag{8}$$

where ε_1 is a control parameter, is applied to system (7). Notice that this external loop does not change the equilibrium solution, but just delays the period-doubling bifurcation, as shown in [44]. The closed-loop system may be written as

$$\begin{aligned} \dot{x} &= A_1 x + B g_1 (C_1 x) \ , \\ y &= C_1 x \, , \end{aligned} \tag{9}$$

with

$$A_1 = \begin{bmatrix} 0 & 1 & 0 \\ 0 & 0 & 1 \\ -1-\alpha & -1.2-\alpha & \mu \end{bmatrix}, \quad B = \begin{bmatrix} 0 \\ 0 \\ 1 \end{bmatrix}, \quad C_1 = \begin{bmatrix} 1\,0\,0 \\ 0\,1\,0 \end{bmatrix},$$

and

$$g_1 (C_1 x) = \alpha (x_1 + x_2) + x_1^2 + \varepsilon_1 x_2^3 = g (x_1) + \alpha x_2 + \varepsilon_1 x_2^3 \, .$$

Once again μ is the main bifurcation parameter and $\alpha = 1$.

Let us consider $\varepsilon_1 = -0.3$. Computing the approximations of the monodromy matrices M_1, M_2 and M_3 at $\mu = -0.37$, the characteristic multipliers shown in Table 2 are obtained. For accurate prediction of the first period-doubling bifurcation, monodromy matrices M_2 and M_3 are computed for $\mu = -0.35$, -0.33 and -0.31, and the characteristic multipliers shown in Table 3 are found. Note that the eigenvalue $\hat{\lambda}_2$ of M_1 (and of M_2 and M_3) crosses the unit circle at $-1+i0$, predicting a period-doubling bifurcation for a critical value of μ_{PD}. The most likely value is $\mu_{PDM_3} \approx -0.33$ (see Table 3); simulations performed with LOCBIF detect a period-doubling bifurcation occurring at $\mu_{PD} \approx -0.3228$.

Table 3. Characteristic multipliers of M_2 and M_3 of system (9) for a range of the main bifurcation parameter μ

	M_2			M_3		
μ	λ_1	$\hat{\lambda}_2$	λ_3	λ_1	$\hat{\lambda}_2$	λ_3
-0.35	1.00077	-0.96338	-0.15234	1.00024	-0.91184	-0.16028
-0.33	1.00085	-1.07449	-0.13293	1.00028	-1.02231	-0.16004
-0.31	1.00093	-1.18641	-0.15501	1.00032	-1.13393	-0.16146

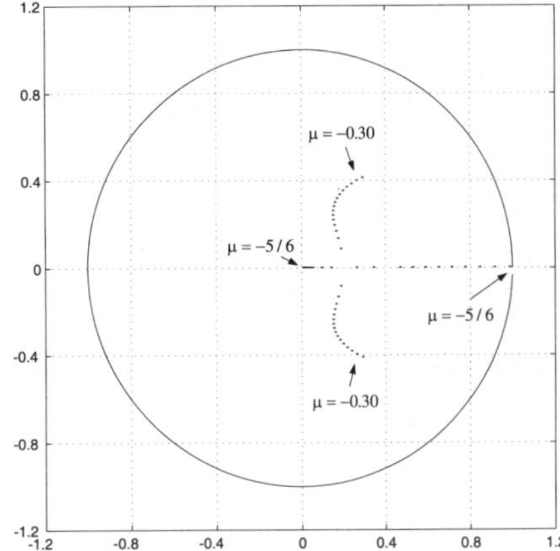

Fig.1. Geometric loci of the characteristic multipliers λ_2 and λ_3 of M_3, $\varepsilon_1 = -1$ (the remaining multiplier λ_1 stays very close to 1, indicating an accurate prediction)

Now, set $\varepsilon_1 = -1$. The loci of the non trivial characteristic multipliers of M_3 for μ varying from $-5/6$ to -0.3 is shown in Fig. 1. The first multiplier λ_1 stays very close to 1 for all values of μ, (the worst case is $\lambda_1 = 0.99673$ for $\mu = -0.31$). Numerical continuation of the amplitude of the limit cycle predicts a crossing of the complex multipliers as in a Neimark-Sacker scenario, but for larger values of μ ($\mu = 0$), satisfying the control objective of delaying the instability of the limit cycle.

Control law (8) depends on the internal state x_2 which is the derivative of the output $y = x_1$ of the system. A feedback law depending only on the output could be obtained replacing x_2 by the "dirty derivative" of y, computed with a washout (highpass) filter. Now the external (dynamic) feedback loop is given by

$$\dot{w} = \varepsilon_2 x_1 - \varepsilon_2 w := z\,,$$
$$u = \varepsilon_1 z^3\,.$$

Under this feedback the closed-loop system results

$$\dot{x} = A_2 x + B_2 g_2 (C_2 x)\,, \tag{10}$$
$$y = C_2 x,$$

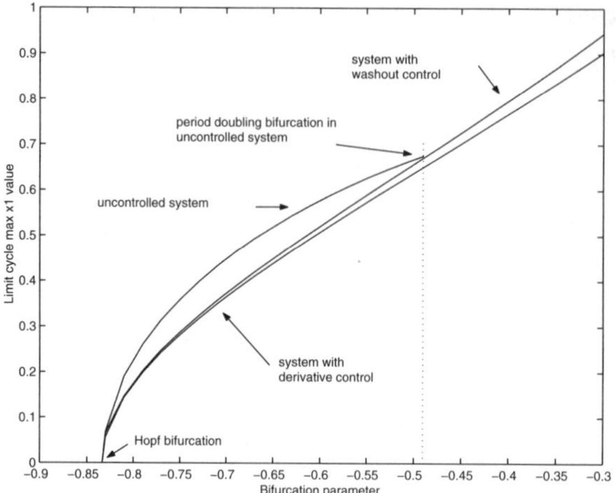

Fig.2. Comparison of the maximum amplitudes of the limit cycles obtained by numerical simulation. The uncontrolled system (7) is simulated for values of μ between the Hopf bifurcation and the period doubling bifurcation (indicated by the dotted line)

where

$$A_2 = \begin{bmatrix} 0 & 1 & 0 & 0 \\ 0 & 0 & 1 & 0 \\ -1-\alpha & -1.2 & \mu & 0 \\ \varepsilon_2 & 0 & 0 & -\varepsilon_2 \end{bmatrix}, \quad B_2 = \begin{bmatrix} 0 \\ 0 \\ 1 \\ 0 \end{bmatrix}, \quad C_2 = \begin{bmatrix} 1\,0\,0\,0 \end{bmatrix},$$

and

$$g_2\left(C_2 x\right) = \alpha x_1 + x_1^2 + \varepsilon_1(\varepsilon_2(x_1 - x_4))^3 .$$

With the parameter values $\varepsilon_1 = -1$ and $\varepsilon_2 = 20$ the system is very stiff, because the magnitude of the pole ε_2 of the washout filter is much greater than the real part of the pair of complex conjugate eigenvalues that originates the Hopf bifurcation. The values for ε_1 and ε_2 should be chosen carefully to obtain reasonable simulation times and accurate application of the Hopf bifurcation theorem in the frequency domain. Decreasing ε_1 to $\varepsilon_1 = -10$, numerical simulation crashes and the frequency domain method looses precision due to numeric instabilities.

Figure 2 displays a comparison of the maximum amplitudes of limit cycles obtained by numerical simulation for the uncontrolled system (7), the system with derivative control (9), and the system with a washout filter in the controller (10). Notice that both controlled systems effectively delay the appearance of the limit cycle bifurcations, and hence the beginning of complex dynamics. The derivative control provides the best performance for a given value of parameter ε_1, and tolerates larger ε_1 values. However, large

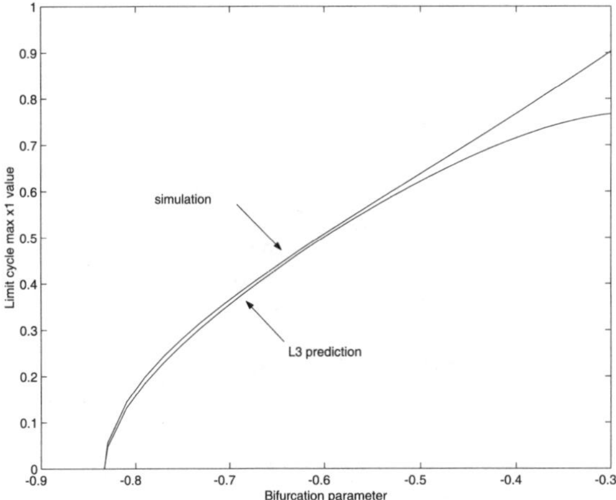

Fig.3. Comparison of the maximum amplitude of oscillation obtained by numerical simulation and predicted by the frequency domain approach for the system (9)

values of ε_1 and/or ε_2 can lead to a potential dangerous situation, because they enlarge the amplitudes of the impulse response of the washout filter at $t = 0$ (peaking phenomenon).

Finally, a comparison of the maximum amplitudes of the limit cycles for the best prediction of the frequency domain method and simulation results for the system with derivative control are shown in Fig. 3.

3 Bifurcation Theory in the Stabilization Problem

The global stabilization of an operation point which is unstable at open-loop has been subject of active research in control theory. An ubiquitous example used to illustrate the problem is the inverted pendulum, in its many different versions. To achieve global stabilization of the inverted position a switched control strategy is commonly used: a nonlinear controller for the swing-up stage, and a linear one for the local stabilization at the inverted position. Applications of this strategy can be found, for example, in [11,29,37,40–43,48]. This technique is most useful when the control action is of restricted amplitude.

These switched control laws do effectively overcome the stabilization problem. Nevertheless, the stabilization with a continuous control law is still a challenging problem. In [9] this problem is treated on the well known pole-cart system, where a control law based on the energy shaping and a smooth switching between two dissipation functions of opposite signs is proposed. The stabilization achieved is almost global, in the sense that some isolated points (unstable equilibria) must be excluded from the initial conditions. For

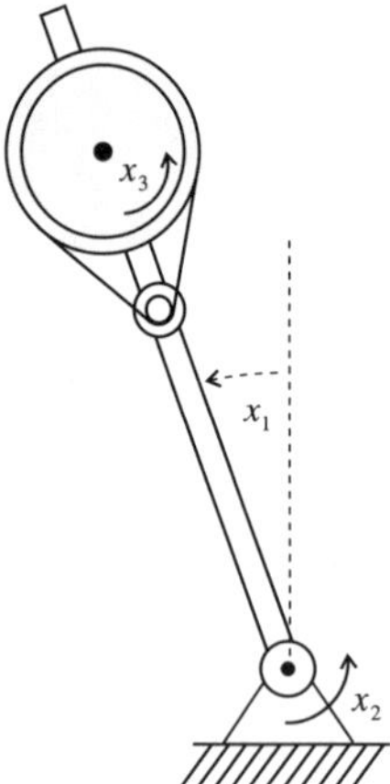

Fig.4. Inertia wheel pendulum

the inertia wheel pendulum described in the following subsection, the stabilization may be achieved with a control law based on the IDA-PBC method [36], or by one derived from a key change of coordinates [34].

The problem of stabilization with a continuous control law when the amplitude of the control action is bounded is cumbersome. In [38] this problem was treated on the inertia wheel pendulum, and the proposed controller is unable to ensure global stabilization of the inverted position when the torque delivered by the motor is lower than the gravity torque. This limitation also occurs with the controllers proposed in [34,36]. In this topic, bifurcation theory may be useful to provide alternative tools to overcome the problem. Bifurcation theory plays an important rôle in analyzing the effects caused by the saturation of the actuator in control systems (see for example [10,35]). In addition, it is useful to analyze attraction basins in systems with saturation [13].

In [8] the problem of stabilizing the inertia wheel pendulum at the inverted position with a bounded control action was introduced. The proposed control law is simple and it is bounded by a smooth function. Using bifurcation theory the domain of attraction of the inverted position is analyzed qualitatively in terms of the values of control gains. It is shown that for certain values of these gains it is possible to drive the pendulum from an initial condition near the rest position to the inverted one.

3.1 Review of previous results

Let us consider the inertia wheel pendulum (Fig. 4). This underactuated mechanical system consists of a simple pendulum with a rotating disk in the free extreme. The pendulum can rotate freely around the pivot point and the disk is actuated by means of a DC motor. The model of the system is given by

$$\begin{aligned}\dot{x}_1 &= x_2,\\ \dot{x}_2 &= q_1 \sin x_1 + q_2 x_3 - q_3 u,\\ \dot{x}_3 &= -q_1 \sin x_1 - q_2 (1+\rho) x_3 + q_3 (1+\rho) u,\end{aligned} \tag{11}$$

where x_1 and x_2 are, respectively, the angular position and angular velocity of the pendulum, x_3 is the angular velocity of the disk relative to the pendulum, q_1, q_2, q_3, ρ are positive parameters, and u is the control input. Note that $x_1 = 0$ at the inverted position. Let us consider the bounded control

$$u = u_{\max} \tanh (k_1 \sin x_1 + k_2 x_2 + k_3 x_3), \tag{12}$$

where $u_{\max}$ is related to the maximum voltage applied to the motor. The equilibrium points of the closed-loop system (11)-(12) are $x_{\mathrm{e}} = (x_{1\mathrm{e}}, 0, x_{3\mathrm{e}})$, where $x_{1\mathrm{e}}$ can be $x_{10} = 2k\pi$ (inverted position) or $x_{1\pi} = (2k+1)\pi$ (hanging position), and $x_{3\mathrm{e}}$ is obtained by solving

$$q_2 x_{3\mathrm{e}} - q_3 u_{\max} \tanh (k_3 x_{3\mathrm{e}}) = 0. \tag{13}$$

The equation (13) has a trivial solution $x_{3\mathrm{e}} = 0$, and provided that $k_3 > k_3^* := q_2 / (q_3 u_{\max})$ it has an additional pair of antisymmetric solutions noted as $\pm \hat{x}_{3\mathrm{e}}$. Therefore, the trivial solution of (13) corresponds to the equilibrium points

$$x_{00} := (x_{10}, 0, 0), \quad x_{\pi 0} := (x_{1\pi}, 0, 0).$$

In addition, x_{00} and $x_{\pi 0}$ develop a pitchfork bifurcation when $k_3 = k_3^*$, and thus for $k_3 > k_3^*$ the following equilibrium points do arise

$$x_{0+} := (x_{10}, 0, \hat{x}_{3\mathrm{e}}), \quad x_{0-} := (x_{10}, 0, -\hat{x}_{3\mathrm{e}}),$$

$$x_{\pi+} := (x_{1\pi}, 0, \hat{x}_{3\mathrm{e}}), \quad x_{\pi-} := (x_{1\pi}, 0, -\hat{x}_{3\mathrm{e}}).$$

The pitchfork bifurcation is common in physical systems with symmetry, and the boundedness of the control action often causes this kind of phenomena (see [10]). The additional equilibria do correspond to the physical situations in which the pendulum's arm is at the inverted position x_{0+} (x_{0-}), or at the hanging position $x_{\pi+}$ ($x_{\pi-}$) but with the disk spinning at constant velocity $\hat{x}_{3\mathrm{e}}$ ($-\hat{x}_{3\mathrm{e}}$). With the above notation, the control objective can be stated as follows: to stabilize x_{00} obtaining, at the same time, the largest domain of attraction. To this end conditions over control gains k_1, k_2 and k_3 ensuring local stability of x_{00} are first derived. These conditions are

$$\begin{aligned}&k_3 > k_3^*,\\ &k_2 > (1+\rho)(k_3 - k_3^*),\\ &k_1 > k_1^* \frac{k_2 - (k_3 - k_3^*)}{k_2 - (1+\rho)(k_3 - k_3^*)},\end{aligned} \tag{14}$$

Table 4. Stability of the equilibrium points as a function of k_1

Equilibrium	Type of Stability	k_1
x_{00}	Unstable Stable	$k_1 < k_{1\text{HB1}}$ $k_1 > k_{1\text{HB1}}$
x_{0+}, x_{0-}	Unstable	$\forall\, k_1$
$x_{\pi 0}$	Unstable	$\forall\, k_1$
$x_{\pi+}$, $x_{\pi-}$	Stable Unstable	$k_1 < k_{1\text{HB2}}$ $k_1 > k_{1\text{HB2}}$

Table 5. Critical values of the main bifurcation parameter k_1

$k_{1\text{HB1}}$	$k_{1\text{HET1}}$	$k_{1\text{FB2}}$	$k_{1\text{HOM1}}$	$k_{1\text{HET2}}$	$k_{1\text{FB1}}$	$k_{1\text{HOM2}}$	$k_{1\text{HB2}}$
23.084	57.88	94.99	107	187	233.3	300.5	$2.9\ 10^6$

where $k_1^* := q_1/\left(q_3 u_{\max}\right)$. Bifurcation theory is used to characterize the dynamical behavior of the feedback system and to analyze the domain of attraction of x_{00} as a function of the control parameters. For simplicity, k_2 and k_3 are fixed so that they satisfy their respective inequalities in (14), and the behavior of the system is studied considering variations of k_1. The stability of the equilibrium points as a function of k_1 is pointed out in Table 4. Note that for $k_1 = k_{1\text{HB1}}$, where

$$k_{1\text{HB1}} := k_1^* \frac{k_2 - (k_3 - k_3^*)}{k_2 - (1+\rho)\,(k_3 - k_3^*)},$$

the stability of x_{00} changes since it undergoes a Hopf bifurcation. The stability of the emerging limit cycle depends on the values adopted for k_2 and k_3. Meanwhile, the change in the stability of $x_{\pi+}$ and $x_{\pi-}$ is also owed to a Hopf bifurcation, experimented by these equilibria at

$$k_{1\text{HB2}} := \frac{k_1^*}{\alpha} \frac{k_2 - (k_3 - k_3^*/\alpha)}{k_2 - (k_3 - k_3^*/\alpha)\,(1+\rho)}.$$

The global analysis of the dynamical behavior is carried out numerically and the results are resumed in the qualitative bifurcation diagram of Fig. 5, in which the period of the limit cycles are plotted against the main bifurcation parameter k_1. In this picture Hopf bifurcations HB1 and HB2 experimented by x_{00}, and by $x_{\pi+}$ and $x_{\pi-}$ respectively, are shown. Also, saddle-node bifurcations of periodic orbits FB1 and FB2, homoclinic bifurcations HOM1 and HOM2, and heteroclinic bifurcations HET1 and HET2 are depicted. For the parameter values $q_1 = 30$, $q_2 = 0.0245$, $q_3 = 0.0393$, $\rho = 250$ and $u_{\max} = 60$, and fixing $k_2 = 50$ and $k_3 = 0.1$, the corresponding values of k_1 are shown in Table 5. More details on this numerical analysis can be found in [5,8].

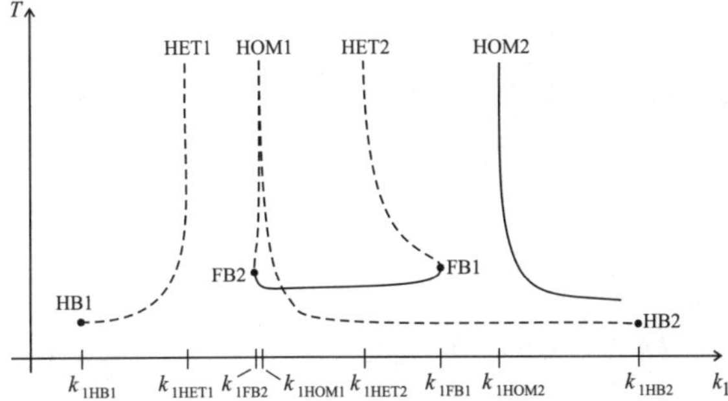

Fig.5. Qualitative bifurcation diagram. (– –) Unstable cycles; (—) stable cycles

3.2 Domain of attraction of x_{00}

The main objective of the analysis is to find values of the controller parameters such that the domain of attraction of the operating point can be enlarged. Based on the qualitative diagram of Fig. 5 it is possible to determine two intervals for k_1 which, at first sight, appear to be appropriate.

The first interval is $(k_{1\text{HET1}}, k_{1\text{FB2}})$, where there are no cycles and the desired behavior may be achieved. In this range, the equilibrium x_{00} (inverted position) is stable, but the equilibria $x_{\pi+}$ and $x_{\pi-}$ (the pendulum in the hanging position and the disk spinning at constant velocity in one or other direction) are also stable. Thus there exists a strong competence between the different basins of attraction. By means of numerical simulations it is possible to determine that the basin of attraction of x_{00} is very small for k_1 in this interval, and thus the system exhibits a behavior different from that expected. The second interval is $(k_{1\text{FB1}}, k_{1\text{HOM2}})$. For k_1 within this range, the equilibria $x_{\pi+}$ and $x_{\pi-}$ are also stable but they are surrounded by unstable (antisymmetric) limit cycles. Both cycles are of saddle type and, as it will be shown in the following, they restrict the basin of attraction of the associated equilibria. Performing numerical simulations with k_1 in the considered interval ($k_1 = 250$) the approximate shape of the domains of attraction of the stable equilibria x_{00}, $x_{\pi+}$ and $x_{\pi-}$ can be obtained. These simulation results do suggest the existence of an attractive manifold, such that trajectories started outside the manifold are attracted to it. This fact is supported by Fig. 6, in which trajectories for four different initial conditions (after a brief transient) seem to lie in the attractive manifold.

Simulations with initial conditions on the planes $x_3 = \pm\hat{x}_{3\text{e}}$ have been performed to characterize the basins of attraction of the equilibrium points $x_{\pi+}$ and $x_{\pi-}$. In the space region delimited by these planes, the domains of attraction have a cylindrical shape as shown in Fig. 7. This figure comprises different trajectories corresponding to initial conditions on a circle of radius slightly larger than the radius of the cylinder delimiting the domain of attraction of $x_{\pi+}$ and $x_{\pi-}$. Trajectories go toward the attractive manifold where

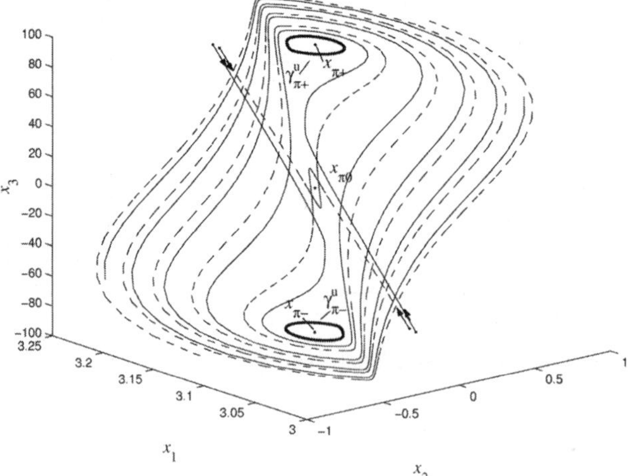

Fig.6. Trajectories with initial conditions outside the attractive manifold

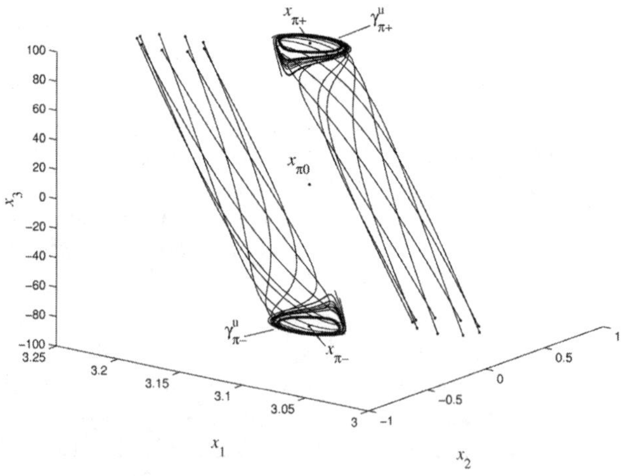

Fig.7. Bounds for the domains of attraction of the equilibrium points $x_{\pi+}$ and $x_{\pi-}$. Trajectories starting inside the cylinders end up in $x_{\pi+}$ or $x_{\pi-}$, respectively

they are rejected by the unstable limit cycles $\gamma^{\mathrm{u}}_{\pi+}$ and $\gamma^{\mathrm{u}}_{\pi-}$ and evolve over the manifold to other regions. On the other hand, a trajectory with initial condition inside the cylinders will go towards the manifold to end up in the equilibrium $x_{\pi+}$ or $x_{\pi-}$, as corresponds. Note that cycles $\gamma^{\mathrm{u}}_{\pi+}$ and $\gamma^{\mathrm{u}}_{\pi-}$ lie at the intersections of the cylinders with the invariant manifold.

Trajectories with initial conditions outside the cylinders end up in the point x_{00}, satisfying the control objective. A simulation performed with $k_1 = 250$ ($k_2=50$, $k_3=0.1$) and initial condition $(\pi, 0, 20)$, is shown in Fig. 8. Note the oscillation of increasing amplitude until the pendulum is stabilized at the inverted position with zero velocity of the disk.

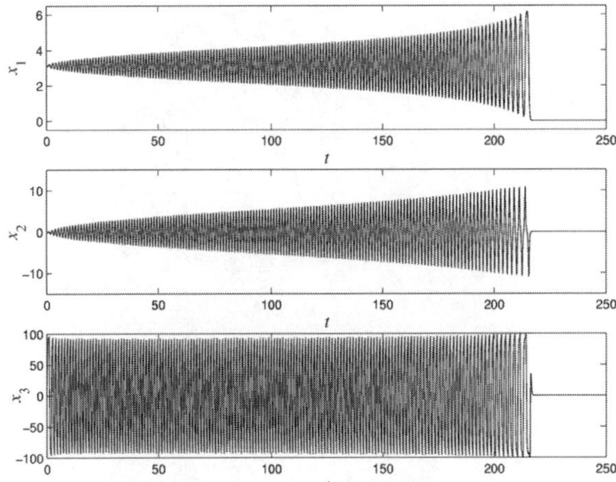

Fig.8. Stabilization at the inverted position with gains values $k_1 = 250$, $k_2 = 50$ and $k_3 = 0.1$. Initial condition $(\pi, 0, 20)$

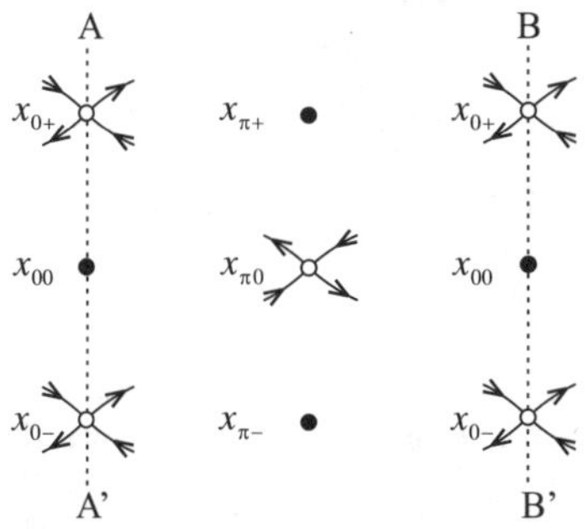

Fig.9. Equilibrium points on the attractive manifold Λ

3.3 Qualitative analysis of the dynamical behavior on a two-dimensional manifold

Based on extensive simulation results, the following conjecture is made:

Conjecture 1. There exists a two-dimensional attractive manifold Λ containing all the equilibrium points of the system, as well as orbits such as limit cycles and homoclinic and heteroclinic trajectories experimented by the system when the control parameter k_1 is varied. In addition, the invariant manifolds of the equilibrium points are such that two of them are contained in Λ, while the one which is not embedded in Λ is stable for all the equilibria and it drives all the trajectories of the system through Λ.

Considering trajectories evolving on Λ and their invariant manifolds it is possible to identify those local and global bifurcations detected analytically

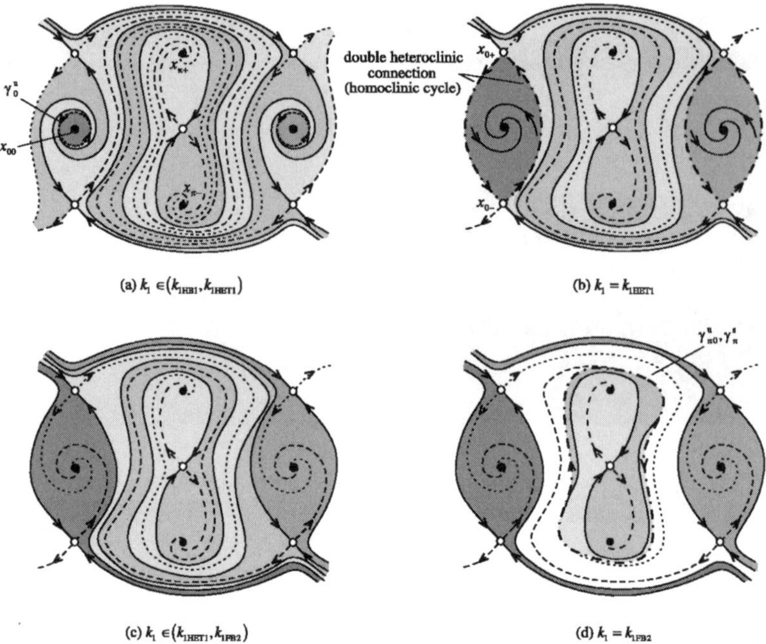

Fig.10. Qualitative diagram of the trajectories and domains of attraction of the equilibria of the system on the manifold Λ for different values of parameter k_1

and numerically in previous sections. These bifurcations and the corresponding dynamic scenarios arising when the diagram of Fig. 5 is run from left to right (increasing values of k_1) are described by means of qualitative phase diagrams on the manifold Λ. In these graphics two sets of points x_{00}, x_{0+} and x_{0-} are considered, as shown in Fig. 9. One set is associated to $x_1 = 0$ (segment AA') and the other to $x_1 = 2\pi$ (segment BB'). Although physically they are the same set of points (considering $x_1 \in S^1$, the segment AA' of Fig. 9 is homologous to segment BB') this representation is chosen for simplicity.

The dynamical scenario after Hopf bifurcation HB1, corresponding to $k_1 \in (k_{1\mathrm{HB1}}, k_{1\mathrm{HET1}})$, is shown in Fig. 10(a). The unstable cycle γ_0^{u} surrounding x_{00} restricts its basin of attraction, which is identified with different shades, since one corresponds to trajectories with initial conditions in the neighborhood of $x_1 = 0$ and the other to initial conditions in the neighborhood of $x_1 = 2\pi$. In addition, the attraction basins of $x_{\pi+}$ and $x_{\pi-}$ are shaded. Note that both attraction basins are delimited by the stable manifolds of the saddle points.

By increasing k_1, the cycle γ_0^{u} grows until it collides with the saddle points x_{0+} and x_{0-} for $k_1 = k_{1\mathrm{HET1}}$. This situation is sketched in Fig. 10(b), where the stable manifold of x_{0+} meets the unstable manifold of x_{0-}, and vice versa, forming a double heteroclinic connection or homoclinic cycle. The unstable

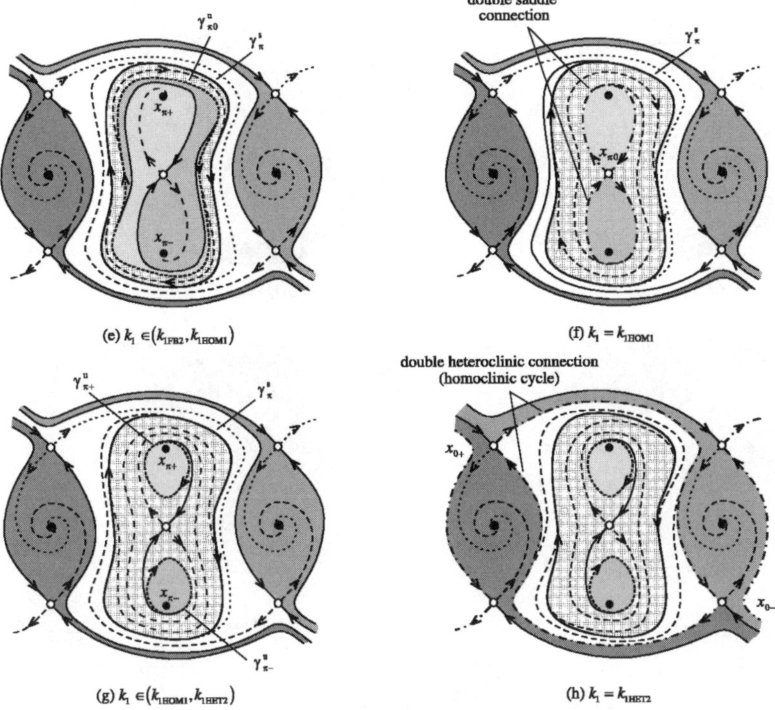

Fig. 11. Qualitative diagram of the trajectories and domains of attraction of the equilibria of the system on the manifold Λ for different values of parameter k_1

manifold of x_{0-} which was connected to $x_{\pi+}$ for $k_1 < k_{1\text{HET}1}$, after this global bifurcation is connected to x_{00}, as it is shown in Fig. 10(c).

For increasing values of k_1, the next observed phenomenon is a saddle-node bifurcation of periodic orbits, occurring at $k_1 = k_{1\text{FB}2}$. In Fig. 10(d) this fact is represented as a semi-stable cycle. Because of this bifurcation two limit cycles arise for $k_1 > k_{1\text{FB}2}$: an unstable cycle ($\gamma_{\pi 0}^{\text{u}}$), confining the equilibria $x_{\pi 0}$, $x_{\pi+}$ and $x_{\pi-}$, and a stable one (γ_{π}^{s}) confining the unstable cycle and the equilibria named before. This situation is depicted in Fig. 11(e). Notice that the basin of attraction of $x_{\pi+}$ and $x_{\pi-}$ is restricted, when compared to the case with $k_1 \in (k_{1\text{HET}1}, k_{1\text{FB}2})$ of Fig. 10 (c).

Increasing k_1 the amplitude of the stable cycle γ_{π}^{s} has no significative changes (k_1 close to $k_{1\text{FB}2}$). Nevertheless, the cycle $\gamma_{\pi 0}^{\text{u}}$ shrinks, and thus the stable and unstable manifolds of $x_{\pi 0}$ are contracted, until both manifolds join each other. In this way, a double saddle connection or homoclinic bifurcation takes place, as is shown in Fig. 11(f).

For $k_1 > k_{1\text{HOM}1}$, the unstable cycle $\gamma_{\pi 0}^{\text{u}}$ which has collapsed in the homoclinic bifurcation is split in two unstable cycles ($\gamma_{\pi+}^{\text{u}}$ and $\gamma_{\pi-}^{\text{u}}$) surrounding

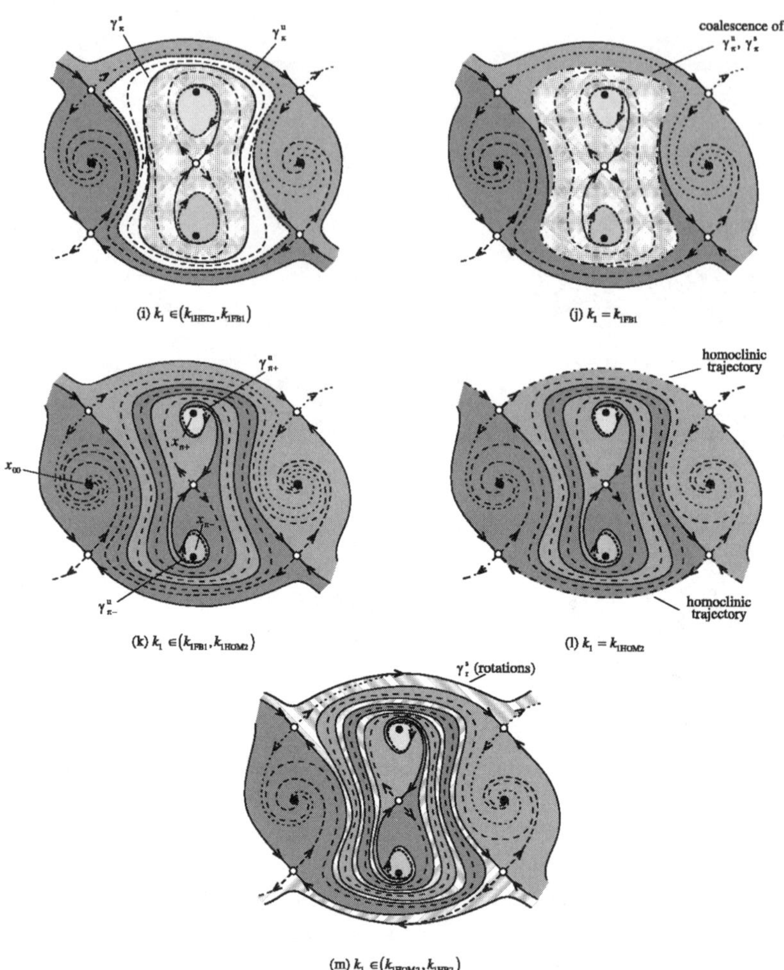

Fig.12. Qualitative diagram of the trajectories and domains of attraction of the equilibria of the system on the manifold Λ for different values of parameter k_1

$x_{\pi+}$ and $x_{\pi-}$, respectively. Thus, the basins of attraction of these equilibria are shrunk as depicted in Fig. 11(g).

When k_1 reaches the value $k_1 = k_{1\mathrm{HET2}}$ an heteroclinic bifurcation HET2 takes place: the unstable manifold of x_{0+} joins the stable manifold of x_{0-}, and vice versa, forming a double heteroclinic connection or homoclinic cycle as shown in Fig. 11(h). Therefore, the unstable manifold of x_{0+} ending in the cycle γ_π^{s} for $k_1 < k_{1\mathrm{HET2}}$ (Fig. 11(g)), ends in x_{00} for $k_1 > k_{1\mathrm{HET2}}$. This situation is depicted in Fig. 12(i), where an unstable cycle (γ_π^{u}) can also be distinguished, as well as the enlargement of the domain of attraction of x_{00}. The unstable cycle γ_π^{u} shrinks until it collides with the stable cycle γ_π^{s} for $k_1 = k_{1\mathrm{FB1}}$ (see Fig. 12(j)). Thus, for $k_1 > k_{1\mathrm{FB1}}$ both cycles disappear and the larger domain of attraction for x_{00} is obtained (for the adopted values of k_2 and k_3) as shown in Fig. 12(k). Note that the stable equilibria $x_{\pi+}$ and $x_{\pi-}$ are confined by the cycles $\gamma^{\mathrm{u}}_{\pi+}$ and $\gamma^{\mathrm{u}}_{\pi-}$.

This situation changes for $k_1 = k_{1\mathrm{HOM2}}$, since an homoclinic trajectory joining the stable and unstable manifolds of x_{0+} and x_{0-} takes place, producing the scenario given in Fig. 12(l). As a consequence, for $k_1 > k_{1\mathrm{HOM2}}$ two stable limit cycles do emerge, corresponding to rotations of the pendulum as shown in Fig. 12(m). Note that the basins of attraction of these cycles are interleaved with that of x_{00}.

Hopf bifurcation HB2, which produces the coalescence of $\gamma^{\mathrm{u}}_{\pi+}$ and $\gamma^{\mathrm{u}}_{\pi-}$ with the equilibria $x_{\pi+}$ and $x_{\pi-}$, is not shown in this series of diagrams. For the adopted numerical values, this Hopf point results in a very large value of $k_{1\mathrm{HB2}}$ ($k_{1\mathrm{HB2}} \simeq 2.9 \times 10^6$) and the behavior of the system for k_1 in the range $(k_{1\mathrm{HOM2}}, k_{1\mathrm{HB2}})$ has not been explored numerically.

3.4 Codimension two bifurcations

Some of the scenarios described on the two-dimensional manifold Λ allow us to identify some typical phenomena associated with a Bogdanov-Takens bifurcation. For example, focusing the attention on the equilibria $x_{\pi 0}$, $x_{\pi+}$ and $x_{\pi-}$ and their associated trajectories in Fig. 10(c)-(d)and Fig.11(e)-(g), one may recognize phase diagrams corresponding to the unfolding of a Bogdanov-Takens bifurcation undertaken by $x_{\pi 0}$. This unfolding corresponds to systems which preserves rotational symmetries (see [22]). On the other hand, by considering the equilibria x_{00}, x_{0+} and x_{0-} in the diagrams of Fig. 10(a)-(c), another Bogdanov-Takens bifurcation at the equilibrium x_{00} is recognized. A more complete analysis of these bifurcations [5], which involves the variation of a second parameter, will be reported elsewhere.

4 Conclusions

The frequency domain approach of the Hopf bifurcation theorem has been used as a tool for controlling the amplitude of limit cycles and delaying the appearance of a period-doubling bifurcation in a simple example. Bifurcation theory has also been applied for characterizing the basins of attraction for the stabilization of a pendulum in its inverted position. In this case, the dynamical behavior of the controlled system has been interpreted based on a planar qualitative analysis which agrees with previous analytical and numerical results. Both examples indicate the potential application of bifurcation theory as a tool to modify –via simple feedback laws– the behavior of nonlinear systems to achieve demanding control objectives.

References

1. Abed, E. H. (1995) Bifurcation-theoretic issues in the control of voltage collapse. In Proc. IMA Workshop on Systems and Control Theory for Power Systems, Chow, J. H., Kokotovic, P. V., Thomas, R. J. (eds.), New York: Springer, 1–21
2. Abed, E. H., Fu, J. H. (1986) Local feedback stabilization and bifurcation control – I. Hopf bifurcation. Syst. Contr. Lett., **7**:11–17
3. Abed, E. H., Fu, J. H. (1987) Local feedback stabilization and bifurcation control – II. Stationary bifurcation. Syst. Contr. Lett., **8**:467–473
4. Abed, E. H., Wang, H. O., Chen, R. C. (1994) Stabilization of period doubling bifurcation and implications for control of chaos. Physica D, **70**:154–164
5. Alonso, D. M. (2002) Análisis y Control de Sistemas No Lineales: Aplicación a un Péndulo Subactuado. PhD Thesis, Universidad Nacional del Sur, Argentina
6. Alonso, D. M., Paolini, E. E., Moiola, J. L. (2000) On anticontrol of Hopf bifurcations. In Proc. 39th IEEE Conf. Decision Control, Sydney, Australia, 2072–2077
7. Alonso, D. M., Paolini, E. E., Moiola, J. L. (2001) An experimental application of the anticontrol of Hopf bifurcations. Int. J. Bifur. Chaos, **11**:1977–1987
8. Alonso, D. M., Paolini, E. E., Moiola, J. L. (2002) Controlling an inverted pendulum with bounded controls. Dynamics, Bifurcations and Control, Colonius, F., Gruene, L. (eds.), Berlin: Springer, 3–16
9. Angeli, D. (2001) Almost global stabilization of the inverted pendulum via continuous state feedback. Automatica, **37**:1103–1108
10. Aracil, J., Ponce, E., Álamo, T. (2000) A frequency-domain approach to bifurcations in control systems with saturations. Int. J. Syst. Sci., **31**:1261–1271
11. Aström, K. J., Furuta, K. (1996) Swinging up a pendulum by energy control. In Proc. 13th IFAC World Congress, San Francisco, CA, 37–42
12. Barbot, J.-P., Monaco, S., Normand-Cyrot, D. (1997) Quadratic forms and feedback linearization in discrete time. Int. J. Contr., **67**:567–586
13. Barreiro, A., Aracil, J., Pagano, D. (2002) Detection of attraction domains of non-linear systems using bifurcation analysis and Lyapunov functions. Int. J. Contr., **75**:314–327
14. Berns, D. W, Moiola, J. L., Chen, G. (2000) Controlling oscillation amplitudes via feedback. Int. J. Bifur. Chaos, **10**:2815–2822
15. Berns, D. W., Moiola, J. L., Chen, G. (2001) Detecting period-doubling bifurcation: an approximate monodromy matrix approach. Automatica, **37**:1787–1795
16. Chen, D., Wang, H. O., Howle, L. E., Gustafson, M. R., Meressi, T. (1998) Amplitude control of bifurcations and application to Rayleigh-Bénard convection. In Proc. 37th IEEE Conf. Decision Control, Tampa, FL, 1951–1956
17. Chen, D. S., Wang, H. O., Chen, G. (2001) Anti-control of Hopf bifurcations. IEEE Trans. Circ. Syst.-I, **48**:661–672
18. Chen, G. (ed.) (1999) Controlling Chaos and Bifurcations in Engineering Systems. Boca Raton, FL: CRC Press
19. Chen, G., Dong, X. (1998) From Chaos to Order: Methodologies, Perspectives and Applications. Singapore: World Scientific
20. Chen, G., Moiola, J. L., Wang, H. O. (2000) Bifurcation control: Theories, methods and applications. Int. J. Bifur. Chaos, **10**:511–548

21. Furuta, K., Yamakita, M., Kobayashi, S. (1992) Swing-up control of inverted pendulum using pseudo-state feedback. Proc. Instn. Mech. Engrs. I, **206**:263–269
22. Guckenheimer, J., Holmes, P. (1993) Nonlinear Oscillations, Dynamical Systems, and Bifurcations of Vector Fields. New York: Springer
23. Kang, W., Krener, A. J. (1992) Extended quadratic controller normal form and dynamic state feedback linearization of nonlinear systems. SIAM J. Contr. Optim., **30**:1319–1337
24. Kang, W. (2000) Bifurcation control via state feedback for systems with a single uncontrollable mode. SIAM J. Contr. Optim., **38**:1428–1452
25. Kang, W., Gu, G., Sparks, A., Banda, S. (1999) Bifurcation test functions and surge control for axial flow compressors. Automatica, **35**:229–239
26. Khibnik, A. I., Kuznetzov, Y. A., Levitin, V. V., Nikolaev, E. V. (1993) Continuation techniques and interactive software for bifurcation analysis of ODE's and iterated maps. Physica D, **62**:360–371
27. Krener, A. J., Li, L. (2002) Normal forms and bifurcations of discrete time nonlinear control systems. SIAM J. Contr. Optim., **40**:1697–1723
28. Liaw, D. C., Abed, E. H. (1990) Stabilization of tethered satellites during station keeping. IEEE Trans. Auto. Contr., **35**:1186–1196
29. Lozano, R., Fantoni, I., Block, D. (2000) Stabilization of the inverted pendulum around its homoclinic orbit. Syst. Contr. Lett., **40**:197–204
30. Mees, A. I. (1981) Dynamics of Feedback Systems. Chichester, UK: Wiley
31. Moiola, J. L., Chen, G. (1996) Hopf Bifurcation Analysis. A Frequency Domain Approach. Singapore: World Scientific
32. Moiola, J. L., Berns, D. W., Chen, G. (1997) Feedback control of limit cycle amplitudes. In Proc. 36th IEEE Conf. Decision Control, San Diego, CA, 1479–1485
33. Nayfeh, A. H., Harb, A. M., Chin, C. M. (1996) Bifurcation in power system model. Int. J. Bifur. Chaos, **6**:497–512
34. Olfati-Saber, R. (2001) Global stabilization of a flat underactuated system: The inertia wheel pendulum. In Proc. 40th IEEE Conf. Decision Control, Orlando, FL, 3764–3765
35. Ortega, M. G., Aracil, J., Gordillo, F., Rubio, F. (2000) Bifurcation analysis of a feedback system with dead zone and saturation. IEEE Contr. Syst. Magazine, **20**(4):91–101
36. Ortega, R., Spong, M. W., Gómez-Stern, F., Blankenstein, G. (2002) Stabilization of a class of underactuated mechanical systems via interconnection and damping assignment. IEEE Trans. Auto. Contr., **47**:1218–1233
37. Pagano, D. J. (1999) Bifurcaciones en Sistemas de Control No Lineales. PhD Thesis, Escuela Superior de Ingenieros, Universidad de Sevilla, Spain
38. Praly, L., Ortega, R., Kaliora, G. (2001) Stabilization of nonlinear systems via forwarding $mod\{L_gV\}$. IEEE Trans. Auto. Contr., **46**:1461–1466
39. Seydel, R. (1994) Practical Bifurcation and Stability Analysis - From Equilibrium to Chaos. Berlin: Springer
40. Shiriaev, A., Pogromsky, A., Ludvigsen, H., Egeland, O. (2000) On global properties of passivity-based control of an inverted pendulum. Int. J. Robust Nonl. Contr., **10**:283–300
41. Spong, M. W. (1995) The swing up control problem for the acrobot. IEEE Contr. Syst. Magazine, **15**(1):49–55

42. Spong, M. W., Corke, P., Lozano, R. (2001) Nonlinear control of the reaction wheel pendulum. Automatica, **37**:1845–1851
43. Spong, M. W., Praly, L. (1996) Control of underactuated mechanical systems using switching and saturation. In Control Using Logic-Based Switching, Morse, A. S. (ed.), New York: Springer
44. Tesi, A., Abed, E. H., Genesio, R., Wang, H. O. (1996) Harmonic balance analysis of period-doubling bifurcations with implications for control of nonlinear dynamics. Automatica, **32**:1255–1271
45. Wang, H.-H., Krstić, M. (1998) Extremum seeking for limit cycle minimization. In Proc. 37th IEEE Conf. Decision Control, Tampa, FL, 2438–2442
46. Wang, Y., Murray, R. M. (2002) Bifurcation control of rotating stall with actuator magnitude and rate limits: Part I - model reduction and qualitative dynamics. Automatica, **38**:597–610
47. Wang, Y., Yeung, S., Murray, R. M. (2002) Bifurcation control of rotating stall with actuator magnitude and rate limits: Part II - control synthesis and comparison with experiments. Automatica, **38**:611–625
48. Wei, Q., Dayawansa, W. P., Levine, W. S. (1995) Nonlinear controller for an inverted pendulum having restricted travel. Automatica, **31**:841–850

Emerging Directions in Bifurcation Control

Hua O. Wang[1] and Dong S. Chen[1,2]

[1] Department of Aerospace and Mechanical Engineering
Boston University, Boston, MA 02215, USA
wangh@bu.edu

[2] Department of Electrical and Computer Engineering
Duke University, Durham, NC 27708, USA

Abstract. In this chapter, we present some emerging directions in bifurcation control in continuous-time dynamical systems. Specifically we discuss two unconventional bifurcation control problems: anti-control of bifurcations and nonsmooth bifurcation control. Anti-control of bifurcation means introducing a new bifurcation at desired location with preferred properties by appropriate control. Nonsmooth bifurcation control via the so-called fractional power control technique, on the other hand, introduces new nonsmooth bifurcation phenomena. In particular, a new border-collision bifurcation, namely a trumpet bifurcation is induced via a fractional power control law for continuous dynamical systems. These emerging directions offer new insights, capabilities and flexibilities in the analysis and control of bifurcations in nonlinear dynamical systems.

1 Introduction

The important role played by concepts from bifurcation theory in the sciences, engineering and the social sciences is well-established. Nonlinear phenomena such as bifurcations to limit cycles and new steady states, and transitions to chaotic behaviors have been observed and studied for a great variety of systems. Only recently have issues of the control of bifurcations and chaos been given serious considerations. All these have led to a remarkable change in the engineers and scientists interacting with bifurcations and chaos. It is now recognized that the ability to manipulate nonlinear dynamical behaviors such as bifurcations and chaos can result in significant practical benefits.

Bifurcation is defined as a change in the number of candidate operating conditions of a nonlinear system that occurs as one or more parameters are quasistatically varied [3]. The candidate operating condition is either an equilibrium point, a periodic solution, or other invariant subset of its limit set, without regard to its stability properties. The parameter being varied is referred to as the *bifurcation parameter*. A nonlinear dynamical system can exhibit many different kinds of bifurcations as one or more parameters are varied. The complex behaviors associated with bifurcations can be understood through bifurcation analysis.

In general, bifurcation control deals with the modification of the bifurcation characteristics of a parameterized nonlinear system by a control input.

G. Chen, D.J. Hill, and X. Yu (Eds.): Bifurcation Control, LNCIS 293, pp. 229–248, 2003.

The control can be a static or dynamic feedback, or an open-loop control law. The objective of control can be stabilization and/or delay of a given bifurcation, reduction of the amplitude of bifurcated solutions, optimization of a performance index near bifurcation, re-shaping of a bifurcation diagram, or a combination of these. The work has been applied to control problems in high incidence flight, stall of compression system in jet engines, voltage collapse in power systems, oscillatory behavior of tethered satellites, magnetic bearing systems, rotating chains, thermal convection loop, and cardiac alternans in heart rhythms (see [3,6] and references therein).

Our main motivation for the study of control of bifurcation and chaos relates to a performance vs. stability trade-off that appears in a variety of forms in various applications. It is often the case that significant improvement in performance is achieved by operation near the stability boundary. From the remarks above, such operation may well lead to bifurcation phenomena in the presence of small disturbances. Achieving increases in performance while maintaining an acceptable safety margin is an important current engineering challenge. An essential aspect of this challenge is the design of controllers which facilitate operation of systems in nonlinear regimes with a negligible margin of stability. It is important to note that linearized models are not adequate for prediction or control of a system's response near the stability boundary [3].

Consider a general finite-dimensional continuous-time system with one parameter:

$$\dot{x}(t) = F\big(x(t); \mu\big). \tag{1}$$

Here $x(t) \in \mathbb{R}^n$ is the state vector, $\mu \in \mathbb{R}$ is the bifurcation parameter. The vector field F is smooth in x and μ. $F(0; \mu) = 0$, i.e., $x = 0$ is always a fixed point of the system.

Suppose that the Jacobian $L_0 = \frac{\partial F}{\partial x}\Big|_{x=0,\mu=0}$ is singular at $\mu = 0$, and the fixed point $x = 0$ loses stability as μ changes its sign, i.e., a bifurcation occurs at $\mu = 0$. As the fixed point loses stability, the system either gives rise to new branches of fixed points or converge to limit cycles nearby. Then we call the system is bifurcated at critical value $\mu = 0$. If any bifurcated orbit is stable and on the other side of critical value, or *supercritical*, the system can evolve smoothly as the bifurcation parameter evolves slowly through critical value. If all the bifurcated orbit is on the same side of critical value, or *subcritical*, none of them can be stable. In this case, as the parameter μ cross the critical value, the system cannot evolve smoothly anymore and have to jump to other orbits far away. Obviously, subcritical bifurcation is catastrophical and destructive while supercritical bifurcation is evolutionary and preferable. In bifurcation control, we usually want to delay the subcritical bifurcation or convert it to a supercritical one. Other control objectives may include modifying shape and amplitude of the bifurcated solution.

If the critical eigenvalue of the Jacobian is real and changes sign in the bifurcation point, this bifurcation is called *static bifurcation*. Fixed points exchange stability in static bifurcation without creating any limit cycle. Depending on the shape and nature of bifurcation branches, static bifurcation can be classified as transcritical, saddle-node, cusp, and pitchfork, etc (see [15]).

Hopf bifurcation occurs when the system Jacobian has a pair of conjugated eigenvalues moving from the left-half plane to the right, crossing the imaginary axis, while all the other eigenvalues remain stable. At the moment of crossing, the existing equilibrium changes from being stable to unstable and a periodic orbit is born. The asymptotic stability of the bifurcated limit cycle p_ε is governed by a characteristic exponent given by a real smooth even function

$$\beta(\varepsilon) = \beta_2 \varepsilon^2 + \beta_4 \varepsilon^4 + \dots, \tag{2}$$

where ε is a measure of distance to the bifurcation point.

It is known that the limit cycle p_ε is orbitally asymptotically stable if $\beta(\varepsilon) < 0$ but is unstable if $\beta(\varepsilon) > 0$. The local stability of the bifurcated periodic solution p_ε, or the *stability of the bifurcation*, is determined by the sign of leading coefficient, which is called the *bifurcation stability coefficient*. The coefficient is usually β_2 if $\beta_2 \neq 0$. The computation of β_2 is relatively complicated, but a scheme is available for its evaluation (see [1] for more details).

Control of bifurcations by state feedback was first investigated by Abed and Fu [1,2]. Wang and Abed applied washout-filter-aided dynamic feedback controller to the problem of bifurcation control [20]. In [19], the washout-filter-aided controller was further extended to discrete-time systems. Wang *et al.* [22] considered the problem of dynamic feedback control of bifurcations by replacing washout filter with arbitrary high-pass filter. The control effects to Hopf bifurcation are obtained analytically. Kang investigated the problem of bifurcation control with normal form and invariants[16]. The effects of nonlinear control on stationary bifurcations from the differential geometry point of view. Hamzi and Kang [11] later extended the method to the control of transcritical and saddle-node bifurcation without transforming into normal forms. Gu *et al.* [9] extended the state feedback controller to output feedback case and showed some interesting results on the equivalence of linear and nonlinear output feedback control of bifurcations. Recently Chen *et al.* [7] also studied the local robustness of Hopf Bifurcation control and the characterizing conditions.

In this chapter, two emerging bifurcation control problems are presented. First, we consider the problem of anti-control of bifurcations, analogous to the anti-control problem for chaos. Anti-control of chaos means that chaos is created or enhanced when it is healthy and beneficial. Similarly, anti-control of bifurcations refers to the situation where a certain type of bifurcation is created at a preferred location with certain desired properties by appropriate controls.

Our work on anti-control of bifurcations is motivated by observations that in some applications, it may be advantageous to introduce new bifurcations to the nominal branch of system output. These new bifurcated solutions may serve as new and more desirable operating conditions; they may also serve as warning signals of impending collapse or catastrophe. Alternatively, they may be judiciously combined with existing dynamical features of the system to extend the operating region via an enlargement of the system parameter ranges which otherwise cannot be accomplished by conventional control methods. More precisely, we present some results on the introduction of new Hopf bifurcations into a given system via feedback control. Hopf bifurcations are common in many nonlinear oscillations and limit cycle behaviors, in biological, ecological, social, economic, and engineering systems. Within this context, anti-control of Hopf bifurcations can be viewed as one approach to designing limit cycles with specified oscillatory behaviors into a system by appropriate feedback controls. In the controller design, we employ dynamic feedback control laws incorporating washout filters. The use of washout filters ensures that lower-frequency orbits of the system are retained in the closed-loop system, while the transient dynamics and higher-frequency orbits are modified. Washout-filter-aided design techniques have been used to control the location and stability of Hopf bifurcations [20,21,23], which are employed in this article to derive a systematic procedure for the design of some effective control laws to create Hopf bifurcations in a given system.

Nonsmooth analysis and control of bifurcations have received less attention in the literature. Some results do exist, though a number of interesting phenomena, such as bonder-collision bifurcations, have been found, mostly for discrete-time systems only. According to [17,18], a border-collision bifurcation occurs when the fixed point of a piecewise smooth map crosses the the "border" which separates the two regions where the map is smooth, causing at least one of the eigenvalues of the fixed point to cross the unit circle. Unlike in the familiar bifurcations such as the saddle node, the period doubling (halving), and the Hopf bifurcations, the critical eigenvalue crosses the unit circle discontinuously in a border-collision bifurcation. This endows border-collision bifurcations with distinctive rich dynamics that has been observed in many experimental and numerical studies (e.g. [8]): bifurcation from period-1 attractor to period-n attractor, from period-1 attractor to chaotic attractor, and from periodic attractor to n-piece chaotic attractor. Moreover, it was shown in [5] that the cardiac system also undergoes a border-collision bifurcation at this bifurcation point, where the control objective is to suppress the cardiac alternans utilizing a form of dynamic bifurcation control directed at the border-collision bifurcation.

In the second part of this chapter, we focus on a new type of border-collision bifurcation, which is introduced by a class of nonsmooth feedback laws consisting of terms with fractional powers, referred to as (nonsmooth) fractional power control (FPC), for one-parameter nonlinear systems. These

FPC controllers are effective in changing the shape of bifurcations and improving the system behavior. The work begins with the recognition of the effectiveness of nonsmooth technique applied to local bifurcations in order to control post-instability behavior. The difficulty associated with nonsmooth designs has much to do with the lack of tools for nonsmooth analysis. The FPC technique has previously been employed in finite-time control of systems [13].

2 Anti-Control of Hopf Bifurcation

In this section, our objective is to design a controller, u, that is capable of creating a Hopf bifurcation at a desired point $(\boldsymbol{x}, \mu)$ and moreover endowing the Hopf bifurcation certain preferable characteristics.

To begin with we note that the use of a static state feedback law $u = u(x)$ has potential disadvantages in control and anti-control of bifurcations. To explain this, consider the case of an equilibrium $\boldsymbol{x}^0(\mu)$ as the nominal operating condition. In general, a static state feedback $u = u(\boldsymbol{x} - \boldsymbol{x}^0(\mu))$ designed with reference to the nominal equilibrium path $\boldsymbol{x}^0(\mu)$ of the original system will affect not only the stability of this equilibrium but also the location and stability of other equilibriums. Since any model is usually only an approximation to the physical system of interest, the nominal equilibrium branch will also be altered by the feedback. A main disadvantage of such an effect is the wasted control energy that is associated with the forced alteration of the system equilibrium structure. Other disadvantages are that system performance is often degraded by operating at an equilibrium which differs from the one at which the system is designed to operate.

For these reasons, to achieve our control (or rather anti-control) objective, we employ a form of dynamic feedback that exactly preserves all systems equilibria, regardless of any model uncertainty that does not violate the finite dimensionality of the system. This is achieved by incorporating filters called "washout filters" into the controller architecture. A washout-filter-aided control law preserves all system equilibria, and does so without the need for an accurate system model.

In the following, we start from the simple cases of one-dimensional and two-dimensional systems, and then generalize the results to the general n-dimensional case.

2.1 Bifurcation creation in 1-D systems

Consider a one-dimensional system described by a parameterized ordinary differential equation,

$$\dot{x} = f(x; \mu) + u, \tag{3}$$

where μ is a parameter and u is the control input. To achieve this control objective, we design the controller with the aid of a washout filter that preserves the equilibrium structure of the original system. Moreover, since a Hopf bifurcation can only take place in a system with dimension great or equal to two, a dynamic feedback such as the one introduced below is also necessary for the creation of a Hopf bifurcation in a one-dimensional system.

The washout-filter-aided controller assumes the following structure:

$$\begin{aligned} \dot{w} &= x - dw \triangleq y, \\ u &= g(y; \mathbf{K}), \end{aligned} \tag{4}$$

where $\mathbf{K}$ is the control gain vector, considered as a parameter vector of the nonlinear control function g. d is the washout filter time constant. The following constraints should be fulfilled:

- $d > 0$, which guarantees the stability of the washout filter.
- $g(0, \mathbf{K}) = 0$, which preserves the original equilibrium points.

Due to the nature of washout filter, $u = 0$ when $\dot{w} = y = 0$, so all the equilibria remain unchanged when control actions are applied. Denote the equilibrium point as $(x^0, y^0, w^0)_\mu$, in which the index μ indicates the parametrical dependence of these values. From the system equations, we know that $y^0 = 0$ and $w^0 = x^0/d$.

Next, we use linearization to determine the stability of the equilibria of this autonomous controlled system. The Jacobian of the controlled system (3)–(4) is

$$J = \begin{pmatrix} f_1(\mu) + g_1 & -dg_1 \\ 1 & -d \end{pmatrix}, \tag{5}$$

where $f_1(\mu) = \frac{\partial f}{\partial x}(x^0; \mu)$, $g_1 = \frac{\partial g}{\partial y}(0)$.

Using the Hopf bifurcation theorem [10], Hopf bifurcation takes place when the following conditions are satisfied:

1. (Crossing eigenvalues condition)

$$g_1 = d - f_1(\mu^0) \quad \text{and} \quad f_1(\mu^0) < 0. \tag{6}$$

2. (Transversality condition)

$$\frac{\partial \alpha}{\partial \mu}(\mu^0) = \frac{\partial f_1}{\partial \mu}(\mu^0) \neq 0. \tag{7}$$

Note that in (6) we obtain a constraint, $f_1(\mu^0) < 0$, which simply means that the bifurcation point μ^0 should be stable in the original system. Hence, it is impossible to create a Hopf bifurcation off an unstable fixed point by using this controller alone. However, there are a couple of ways to deal with bifurcation creation off an unstable equilibrium point. A straightforward approach is to employ a feedback controller that stabilizes this equilibrium point before applying the washout-filter-aided controller. Another approach is to utilize unstable washout filters in the development of the control laws. Usually, the created bifurcation is only useful if the bifurcated solution is stable. This can be achieved through nonlinear items in control function g. The details can be found in [4].

The control laws described above lead to the creation of a Hopf bifurcation and associated limit cycles in the extended state space (x, w). Since the washout-filter-aided dynamic feedback control laws preserve the equilibrium structure of the original system, the resulting limit cycles, at least locally, live in the neighborhood of the equilibrium point of the original system.

2.2 Bifurcation creation in 2-D systems

The problem of creating a Hopf bifurcation in a two-dimensional system is a little more complicated than the one-dimensional setting. For simplicity, we only consider the case with a pair of conjugated complex eigenvalues. In Subsection 2.3, we will see that the case with two real eigenvalues can be decomposed into two one-dimensional subsystems and then solved by the procedure proposed in Subsection 2.1.

Suppose that the controlled system equation has the form

$$\dot{x}_1 = f_1(x_1, x_2; \mu) + u, \tag{8}$$
$$\dot{x}_2 = f_2(x_1, x_2; \mu). \tag{9}$$

As in the 1-D case, a washout filter in x_2 is used,

$$\dot{w} = x_2 - dw \triangleq y, \tag{10}$$
$$u = g(y; \mathbf{K}), \tag{11}$$

where $\mathbf{K}$ is the gain vector of this controller and d is the time constant of the washout filter.

The linearized system matrix at the equilibrium (x_1^0, x_2^0) is

$$A(\mu) = \begin{pmatrix} f_{11}(\mu) & f_{12}(\mu) + g_1 & -dg_1 \\ f_{21}(\mu) & f_{22}(\mu) & 0 \\ 0 & 1 & -d \end{pmatrix}, \tag{12}$$

where $f_{mn}(\mu) = \frac{\partial f_m}{\partial x_n}(x_1^0, x_2^0; \mu)$, $(m, n = \{1, 2\})$ and $g_1 = \frac{\partial g}{\partial y}(0)$.

The transversality condition turns out to be more complicated for this 2-D problem, but it is more closely related to the validation rather than the design problem for bifurcation control. It would be better to check it numerically rather than analytically. Only the crossing eigenvalues condition is the key here to the creation of the intended Hopf bifurcation.

The value of g_1 can be obtained by determining the eigenvalues of $A(\mu)$. The result is

$$g_1 = -\frac{(f_{11}+f_{22})(d^2 - d(f_{11}+f_{22}) + f_{11}f_{22} - f_{12}f_{21})}{(d - f_{11} - f_{22})f_{21}} \tag{13}$$

The dumping coefficient has to satisfy two inequalities: $d > 0$ and $d - f_{11} - f_{22} > 0$. If the designated bifurcation point in the original system is stable, then $f_{11} + f_{22} < 0$. Thus, the condition is automatically satisfied. If the bifurcation point is unstable, then we have to choose d large enough ($d > f_{11} + f_{22} = 2\mathrm{Re}[\lambda_u]$) to overcome this instability, where λ_u is the unstable eigenvalue of the open loop system (8)–(9).

In this problem, it is also necessary to control the stability of the bifurcated solution. To do so, since only the linear, quadratic and cubic terms occurring in a nonlinear system undergoing a Hopf bifurcation influence the value of β_2, one can further specify the controller (11) to take a linear-plus-nonlinear form of where the gains can be obtained through a combination of the bifurcation creation conditions outlined above and the bifurcation stability coefficient β_2.

2.3 Bifurcation creation in n-D systems

With the preliminary results in one- and two- dimensional cases, we can now proceed to deal with n-dimensional systems. From the results in Sections 2.1 and 2.2, we assume that the designated bifurcation point $(\boldsymbol{x}^0, \mu^0)$ is stable, or the system Jacobian has only two unstable conjugated complex eigenvalues with multiplicity one (all the others are stable). In order to analyze this n-dimensional system, we employ the real canonical Jordan form to decompose the whole system into interconnected subsystems [14].

First, rewrite the system equation in linearized form

$$\dot{\boldsymbol{x}} = A(\mu)\boldsymbol{x} + B(\mu)u + \mathbf{F}(\boldsymbol{x}, u; \mu), \tag{14}$$

where $\mathbf{F}(\boldsymbol{x}, u; \mu)$ denotes the second and higher-order terms in $\boldsymbol{x}$. A is the Jacobian of the system. Note that A, B, and $\mathbf{F}$ all depend on μ. In the following discussion, we only use their values at $\mu = \mu^0$ and drop all the parameter dependency for notational simplicity.

According to the matrix theory [14], we know that for a real system matrix A having eigenvalues $\lambda_1, \ldots, \lambda_p$, $\alpha_{p+1} \pm i\omega_{p+1}, \ldots, \alpha_r \pm i\omega_r$ with multiplicities $n_1, \ldots, n_r$, where $\lambda_k \in \mathbb{R}$, $\alpha_k \in \mathbb{R}$, $\omega_k \in \mathbb{R}$, and $n_k \in \mathbb{N}$, $1 \le k \le r$, then a real matrix P can be found to transform A into the real Jordan canonical form:

$$\hat{A} = P^{-1}AP = \begin{pmatrix} J_{n_1}(\lambda_1) & & & & & \\ & \ddots & & & 0 & \\ & & J_{n_p}(\lambda_p) & & & \\ & & & C_{n_{p+1}}(\alpha_{p+1}, \omega_{p+1}) & & \\ & 0 & & & \ddots & \\ & & & & & C_{n_r}(\alpha_r, \omega_r) \end{pmatrix}, \tag{15}$$

where each $J_{n_k}(\lambda_k)$ is a real Jordan block and each $C_{n_k}(\alpha_k, \omega_k)$ is a complex Jordan block.

Using the transformation

$$\boldsymbol{x} = P\boldsymbol{z}, \tag{16}$$

the system (14) can be transformed into its Jordan canonical form,

$$\begin{aligned} \dot{\boldsymbol{z}} &= P^{-1}AP\boldsymbol{z} + P^{-1}bu + P^{-1}\mathbf{F}(P\boldsymbol{z}, u) \\ &= \hat{A}\boldsymbol{z} + \hat{b}u + \hat{\mathbf{F}}(\boldsymbol{z}, u), \end{aligned} \tag{17}$$

where $\hat{A} = P^{-1}AP$, $\hat{b} = P^{-1}b$, and $\hat{\mathbf{F}}(\boldsymbol{z}, u) = P^{-1}\mathbf{F}(P\boldsymbol{z}, u)$. Also breakdown $\boldsymbol{z}$, $\hat{b}$ and $\hat{\mathbf{F}}$ into blocks as with $\hat{A}$ in equation (15)

$$\boldsymbol{z} = \begin{pmatrix} z_1 \\ z_2 \\ \vdots \\ z_r \end{pmatrix}, \qquad \hat{b} = \begin{pmatrix} \hat{b}_1 \\ \hat{b}_2 \\ \vdots \\ \hat{b}_r \end{pmatrix}, \qquad \hat{F}(\boldsymbol{z}, u) = \begin{pmatrix} \hat{F}_1(\boldsymbol{z}, u) \\ \hat{F}_2(\boldsymbol{z}, u) \\ \vdots \\ \hat{F}_r(\boldsymbol{z}, u) \end{pmatrix}$$

For any real Jordan block $J_{n_k}(\lambda_k)$, $0 < k \leq p$, the eigenvalues are all real and identical to λ_k. If at least one state of the block is controllable (i.e., $\exists 0 < j \leq n_k, \hat{b}_{k,j} \neq 0$), and if the subsystem is stable at the designated bifurcation point $\mu = \mu^0$ (i.e., $\lambda_k(\mu^0) < 0$), we can assign a Hopf bifurcation at this point via a 1-D anti-controller. The procedure is as follows:

1. Find $0 < m \leq n_k$ that $\hat{b}_{k,m} \neq 0$ and $\forall m < j \leq n_k$, $\hat{b}_{k,j} = 0$.
2. Select appropriate damping coefficient $d > 0$ to design a washout filter with state $z_{k,m}$ as its input:

$$\dot{w} = y = z_{k,m} - dw, \qquad d > 0. \tag{18}$$

3. Design the nonlinear control law

$$u = \frac{g(y; \mathbf{K})}{\hat{b}_{k,m}} \triangleq K_l y + K_n y^3. \tag{19}$$

According to equation (6), the linear control gain $K_l = (d - f_1(\mu^0))/\hat{b}_{k,m}$. The nonlinear gain K_n is used to make the bifurcation supercritical.

For the state $j \neq m$ in this Jordan block, if $j > m$, then $\hat{b}_{k,j} = 0$, which means that the input u cannot change the stability of this state. Since the original system is required to be stable, this state is still stable after the controller is applied. If $j < m$, $\hat{b}_{k,j}$ may be nonzero, due to the special structure of Jordan block, the feedback of $z_{k,m}$ does not affect the stability of states above it. Therefore, only state m is destabilized while all the others are untouched. This proved that the controlled system does have a Hopf bifurcation at μ^0.

For any complex Jordan block $C_{n_k}(\alpha_k, \omega_k)$, $p < k \leq r$, the eigenvalues are complex eigenpairs. Similar to the case of real eigenvalues dealt above, we can also introduce a Hopf bifurcation for any controllable state. The procedure also involves selecting largest nonzero term from $\hat{b}_k$. Suppose the state is m, and let the subsystem for this state be written as

$$\dot{z}_{k,m,1} = \hat{F}_{k,m,1}(z_{k,m,1}, z_{k,m,2}; \boldsymbol{z}^0, \mu) + \hat{b}_{k,m,1}u, \tag{20}$$

$$\dot{z}_{k,m,2} = \hat{F}_{k,m,2}(z_{k,m,1}, z_{k,m,2}; \boldsymbol{z}^0, \mu) + \hat{b}_{k,m,2}u. \tag{21}$$

For simplicity, we view all the state variables of $\boldsymbol{z}$, except $z_{k,m,1}$ and $z_{k,m,2}$, as parameters, and use their equilibrium values in the bifurcation analysis.

Next, without lose of generality, we assume $\hat{b}_{k,m,1} \neq 0$. Substituting $z_{k,m,1}$ and $z_{k,m,2}$ with $\nu = z_{k,m,1}/\hat{b}_{k,m,1}$ and $\xi = z_{k,m,2} - \hat{b}_{k,m,2}z_{k,m,1}/\hat{b}_{k,m,1}$ gives

$$\dot{\nu} = \tilde{f}_1(\nu, \xi; \boldsymbol{z}^0, \mu) + u, \tag{22}$$

$$\dot{\xi} = \hat{f}_2(\nu, \xi; \boldsymbol{z}^0, \mu), \tag{23}$$

where $\tilde{f}_1(\nu, \xi) = \hat{F}_{k,m,1}/\hat{b}_{k,m,1}$ and $\tilde{f}_2(\nu, \xi) = \hat{F}_{k,m,2} - \hat{b}_{k,m,2}\hat{F}_{k,m,1}/\hat{b}_{k,m,1}$. This equation has the form of (8)–(9), so we can use the procedure described therein to design the controller as

$$\dot{w} = \xi - dw \triangleq y, \tag{24}$$

$$u = K_l y + K_n y^3, \tag{25}$$

where K_l is used to relocate the bifurcation point and K_n is used to make the bifurcation supercritical. The parameter K_l is actually g_1 in subsection 2.2 whose value can be obtained from equation (13). K_n is much harder to determine, which can be obtained either from analytical formulas or numerical computation of β_2 (e.g. with BIFOR2 [12]). We can also determine it from numerical experiments.

Finally, we should determine the effects of the controller on the original system. In view of the inverse of the transformation (16), where $P = [P_{ij}]$, we know that if a bifurcation is introduced to the state z_j, there is also a bifurcation at the state x_i if and only if $P_{ij} \neq 0$. If we want to induce a bifurcation to x_i, we have to find a j such that both P_{ij} and $\hat{b}_j$ are nonzero. A cross reference table can then be constructed with this condition, so as to facilitate the controller design.

2.4 Anti-control of bifurcations in thermal convection loop

To illustrate the utility of the anti-control laws, we present detailed design studies of a 3-D thermal convection model. This convection loop model is a special case of the well-studied Lorenz equations. In our earlier work [20], bifurcation control techniques have been applied to this system to suppress the various chaotic behaviors over a range of parameters. Here, however, anti-control laws are applied to introduce new Hopf bifurcations and construct stable limit cycles in the system.

The model under consideration is a three-dimensional system of the following form [20]:

$$\dot{x_1} = -px_1 + px_2, \tag{26}$$

$$\dot{x_2} = -x_1x_3 - x_2, \tag{27}$$

$$\dot{x_3} = x_1x_2 - x_3 - R + u. \tag{28}$$

Here $x_i\ (i = 1, 2, 3)$ are real, p and R are positive parameters. The variables x_1, x_2 and x_3 correspond respectively to the cross-sectionally averaged velocity in the loop, the side-to-side temperature difference and the top-to-bottom temperature difference. The parameter R is the Raleigh number, which is proportional to the mean heating rate, and p denotes the Prandtl number which is determined by the fluid property and system geometries. For simplicity, we fix $p = 4.0$. R is used as the bifurcation parameter. The control input u is applied by adding fluctuations to the nominal heating rate R. Fig. 1 shows the bifurcation diagram of the uncontrolled system. In this diagram, a solid line represents a stable equilibrium, a dashed line represents an unstable equilibrium, an open circle represents the maximum amplitude of an unstable periodic orbit, and a solid circle represents the maximum amplitude of a stable periodic orbit. The squares are bifurcation points. An empty square denotes for pitchfork and transcritical bifurcation, while a solid square for Hopf bifurcation.

The bifurcation diagram has two branches of equilibria $\{x_1^0 = x_2^0 = \pm\sqrt{R-1}, x_3^0 = -1\}$ which are stable for $1 \le R < 16$. In this study, we will create a Hopf bifurcation in upper branch to introduce stable limit cycles.

We set the designated bifurcation point as $R = 12$. The eigenvalues of the Jacobian are

$$\lambda_1 = -5.8404, \qquad \lambda_{2,3} = -0.0798 \pm 3.8809i.$$

There are one real eigenvalue and a pair of conjugated eigenvalues. Therefore, we can represent this system as a 1-D subsystem plus a 2-D subsystem.

For the 2-D subsystem, we design the washout filter as

$$\begin{aligned} \dot{w} &= y = \xi - dw = x_1 + 2.17349931942510x_2 - 0.5w, \\ u &= k_l y + k_n y^3, \end{aligned} \tag{29}$$

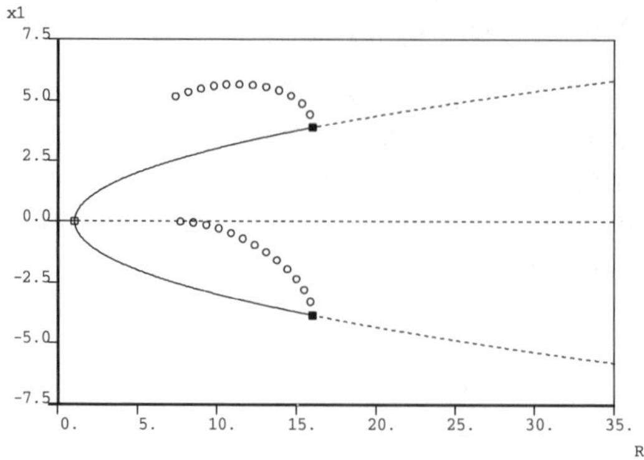

Fig.1. Bifurcation diagram of the uncontrolled convection loop system

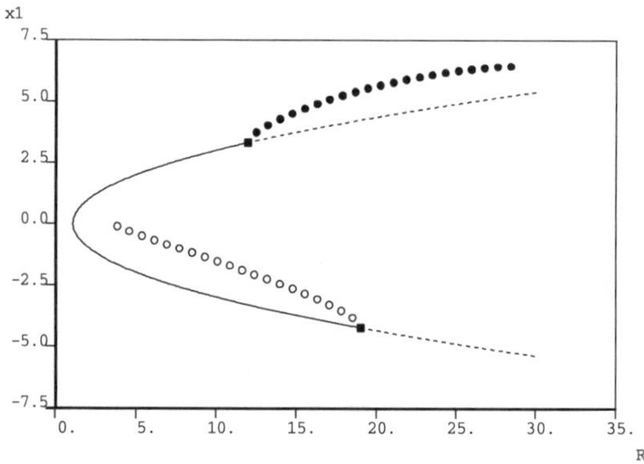

Fig.2. Bifurcation diagram of the controlled convection system with $k_l = -0.5169$ and $k_n = 0.05$

where k_l and k_n are linear and nonlinear control gains. Using equation (13), the linear control gain is found to be $k_l = -0.5169$. k_n can be determined analytically or numerically. Fig. 2 shows the bifurcation diagram of the controlled system for $k_n = 0.05$. It can been seen clearly that a Hopf bifurcation is introduced at $R = 12$.

Recall that for a 2-D system, there is no restriction on the stability of the equilibrium in the development of Section 2.2. We can use the same procedure to assign a Hopf bifurcation at an unstable equilibrium. Suppose we choose $R = 21$, which is at the chaotic region. Similarly, we design a controller

$$\begin{aligned} \dot{w} &= y = x_1 + 1.852794117x_2, \\ u &= 1.47015988335775y + 0.12y^3. \end{aligned} \tag{30}$$

The bifurcation diagram of the controlled system is shown in Fig. 3.

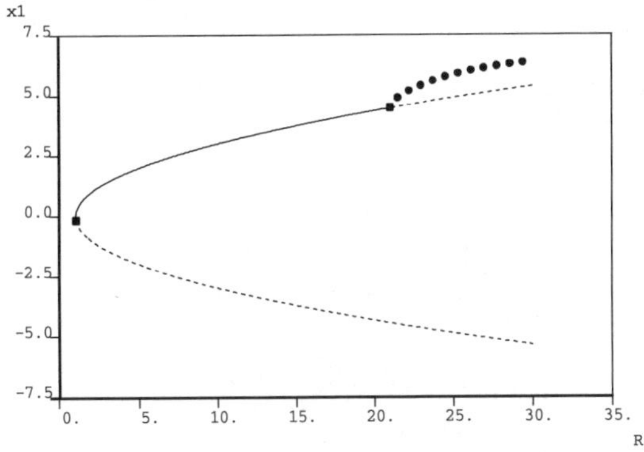

Fig.3. Bifurcation diagram of the controlled convection system with a designated Hopf bifurcation at $R = 21$

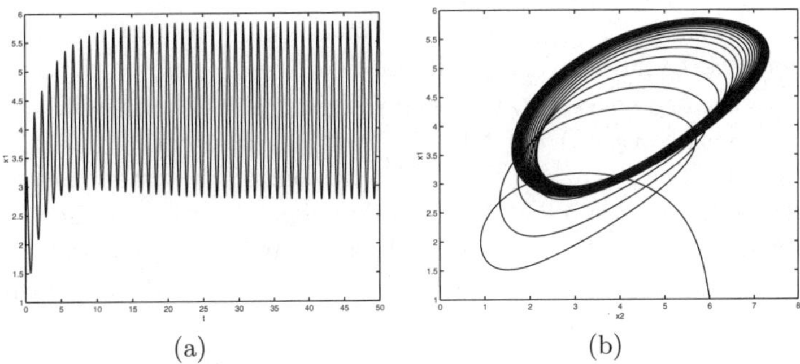

(a) (b)

Fig.4. Trajectory of the controlled convection system under control law (30) at $R = 21.5$

The effects of these controllers can be better appreciated from simulations. In Fig. 4, we show the trajectory of the closed-loop system with $R = 21.5$ under the control law (30).

For the 1-D subsystem, we can also design an anti-controller to assign Hopf bifurcation at a stable equilibrium. In Fig. 5, we show the bifurcation diagram of a controlled system with a controller designed for the 1-D subsystem:

$$\begin{aligned} \dot{w} &= y = x_1 - 0.7502768152x_2 - 0.3388761836x_3, \\ u &= -17.4932y + 1000y^3. \end{aligned} \tag{31}$$

The bifurcation point (HB2) is assigned to be $R = 5$. Notice that this bifurcation takes an opposite direction to the previous two. The other bifurcation (HB1) is so close to HB2 that the stable range of equilibrium is quite small. If we assign the bifurcation at a higher R, the stable range may be too small to be seen. Also note that the control gains are much bigger. To guarantee

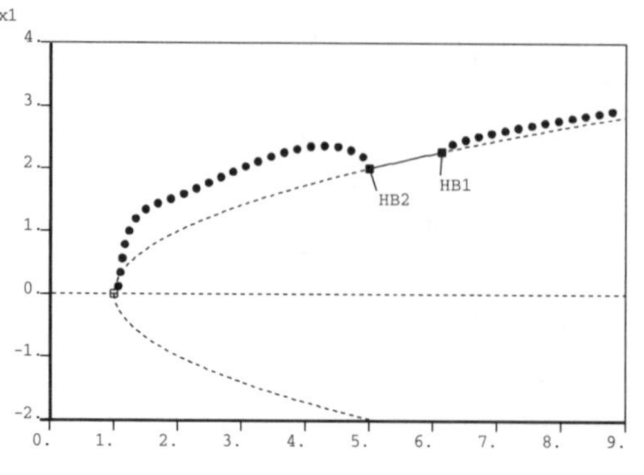

Fig.5. Bifurcation diagram of the controlled convection system under 1-D anti-controller to assign bifurcation at $R = 5$

the stability of limit cycles bifurcated from HB2, the limit cycles from HB1 have been suppressed to have a very small amplitude.

For a 1-D system or subsystem, if the equilibrium point is unstable, we cannot use a 1-D anti-controller directly to create any bifurcation directly. However, a state feedback can be employed to stabilize the system before the applying an anti-controller. For example, at $R = 21$, we can use the linear controller $u = -(x_3 + 1)$ to stabilize the point first. Fig. 6 shows the bifurcation diagram of a system controlled with the combination of a linear feedback and a 1-D anti-controller:

$$\begin{aligned} \dot{w} &= y = x_1 - 0.4018927617x_2 - 0.4199414323x_3, \\ u &= -16.14493693177215y + 50y^3 - (x_3 + 1). \end{aligned} \tag{32}$$

This scheme is more flexible than the original 1-D anti-controller. However, the equilibrium structure of the system has been changed except for the bifurcation point $R = 21$ and the equilibrium points at line $x_1 = 0$. This is due to the use of state feedback which does not preserve the equilibrium structure. An alternative approach is to use an unstable washout filter to ensure the closed-loop stability of the overall system.

3 Fractional Power Control of Bifurcations

In this section, we discuss some new types of border-collision bifurcations, which are introduced by a class of nonsmooth feedback laws consisting of terms with fractional powers, referred to as (nonsmooth) *fractional power control* (FPC), for one-parameter nonlinear systems. These FPC controllers are effective in changing the shape of bifurcations and improving the system behavior. The work begins with the recognition of the effectiveness of

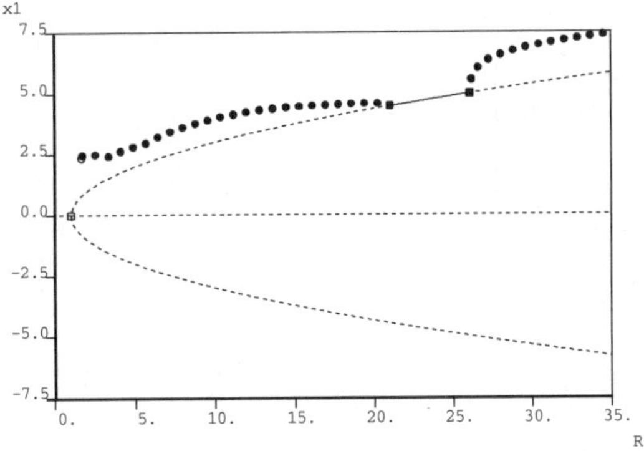

Fig.6. Bifurcation diagram of the controlled convection system under the linear feedback plus 1-D anti-controller

nonsmooth technique applied to local bifurcations in order to control post-instability behavior. Due to space limitation, we concentrate on treatments for one-dimensional systems and two-dimensional systems only. Additional results on three-dimensional systems, in particular, nonsmooth bifurcation control of an axial flow compressor can be found in [24].

3.1 FPC of 1-D systems

We first consider a one-dimensional system with one parameter γ, which will show the essential idea about the tangent property and trumpet bifurcation.

Consider

$$\dot{x} = (\gamma + u)x - Lx^k, \quad L \geq 0 \tag{33}$$

where γ is the bifurcation parameter, $k > 0$ is an odd positive integer, and $u = u(x)$ with $u(0) = 0$ is the feedback control variable. Note that, when $u = 0$, the bifurcation happens at $\gamma = 0$. Namely, the bifurcation critical parameter value is $\gamma^0 = 0$. When $u(x)$ is smooth, its Taylor's expansion can be employed for its analysis as did in the previous section. However, the basic design idea here is to use nonsmooth fractional power control. A simple form of this type of feedback can be taken as follows:

$$u = -K|x|^c, \quad K > 0 \tag{34}$$

where $0 < c < 1$ is a positive constant. Note that when $c \geq 1$, the feedback law is C^1 smooth.

Plugging the nonsmooth factional power control law into system (33), we have

$$\dot{x} = \gamma x - K|x|^{c+1} sign(x) - Lx^k, \tag{35}$$

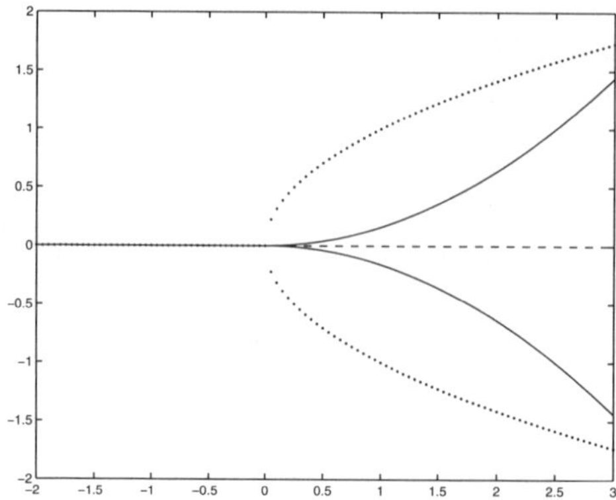

Fig.7. The comparison in two cases: $u = x^2$ and $u = 2.5\sqrt{|x|}$.

where $sign(\cdot)$ is the sign function.

Quite obviously, system (35) does not have bifurcation phenomena when $\gamma < 0$. The fixed point $x = 0$ is an invariant submanifold, unaffected by the value of control input u, which implies that (35) is uncontrollable. However, it is not difficult to check that system (35) still possesses an uncontrollable zero eigenvalue at the stall point. This means that we cannot extend the stable region of the parameter γ crossing 0 for the unstable equilibrium point by any state feedback with $u(0) = 0$ in (35). Even so, we will find that nonsmooth feedback may lead to effective changes when $\gamma > 0$.

Note that $c < 1 < k - 1$, so the term $|x|^{c+1}$, rather than x^k, will be the dominant term. Therefore, locally, after a simple calculation, the stable bifurcation curve has the shape of $|x|^c \sim \gamma/K$, or equivalently,

$$x \sim \pm \left(\frac{\gamma}{K}\right)^b, \quad \text{when} \quad \gamma > 0$$

where $b = 1/c > 1$.

For illustration, we can consider a simple numerical example of system (33) with $L = 0$. Fig. 7 shows a comparison of bifurcation lines between the smooth nonlinear case (with $c = 2$) and the nonsmooth nonlinear case (with $c = \frac{1}{2}$). The dot curve represents locus of locally asymptotically stable equilibrium points of the smooth system, while solid curves correspond to stable points of the fractional power system. Dashed line here stands for unstable equilibrium points.

Note that when the bifurcation occurs near the origin, the shape is determined by the nonsmooth term. Although we cannot move the bifurcation point with the feedback, the critical point seems to have moved because, with γ increasing, the bifurcation line leave the line $x = 0$ very slowly in space (x, γ). In fact, the bifurcation line is tangent to the line $x = 0$ at the stall point because

$$\left.\frac{\partial x}{\partial \gamma}\right|_{x=0,\gamma=0+} = 0$$

(which is the right partial derivative, based on $\gamma - K|x|^c - x^{k-1} = 0$).

Definition 2.1. If there exist several bifurcation solutions, tangent (by changing parameter γ from either side) to one another at the critical point, or calling that the bifurcation solution lines have tangent property, then the bifurcation is called *trumpet*.

From the figure above, we also find that, the trumpet bifurcation lines changes slowly around the critical point as γ changes. This is the significant effect of this type of nonsmooth feedback. Furthermore, when the bifurcation line goes into the region when $|x|$ becomes larger, x^k will be the dominant term. The change is gradual because $\frac{\partial x}{\partial \gamma}$ is small. In fact, the design via nonsmooth but continuous control makes the bifurcation happen in a very mild and gradual way. Note that we can have a subcritical trumpet bifurcation if take $u = K|x|^c$, $K > 0, 0 < c < 1$ with a similar analysis.

We have to admit that the tools for nonsmooth analysis are so scarce, but the main idea of our analysis can be employed in the study of other nonsmooth systems. Since the nonsmooth terms in the dynamics (resulted from feedback laws) become dominant in the beginning of the bifurcation when the parameter changes, by omitting those "higher order terms", we can find that the shape and convexity of the bifurcation solutions are changed, which cannot happen in smooth cases. Then if we would like to consider the global case, the nonsmooth terms influence less when $|x|$ is large. Thus, in this case, the conventional smooth analysis can be used again to understand the bifurcation behaviors. Although we do not present analysis for general nonsmooth systems, the idea can be hopefully applied to the study of higher-dimensional nonlinear systems with fractional power terms.

3.2 2-D systems

With the previous analysis, we begin to recognize the effectiveness of nonsmooth ideas applied to control of local bifurcations. In this subsection, we continue to study the two-dimensional systems, which have more complex behavior. To concentrate on our design idea, we only consider one important case: Hopf bifurcation, though we believe that the idea is able to be used in the analysis for other types of bifurcation of second order systems.

Consider a 2-D system with one-parameter Hopf bifurcation [10]:

$$\begin{cases} \dot{x} = (b_1\gamma + a_1 r^2 + u)x - (b_2\gamma + a_2 r^2)y \\ \dot{y} = (b_2\gamma + a_2 r^2)x + (b_1\gamma + a_1 r^2 + u)y \end{cases} \tag{36}$$

where u is the control variable, $r^2 = x^2 + y^2$, $a_1 \neq 0$, and $b_1 \neq 0$. For simplicity, we take $b_1 < 0$ and $a_1 > 0$ for the following discussion.

According to [10], this is a normal form for Hopf-bifurcation of second order systems, which can also be expressed in polar coordinates:

$$\begin{cases} \dot{r} = (b_1\gamma + a_1 r^2 + u)r \\ \dot{\theta}_0 = (b_2\gamma + a_2 r^2) \end{cases} \tag{37}$$

If $u = 0$, the solutions lie along the parabola $\gamma = -a_1 r^2/b_1$, which implies

$$\left.\frac{\partial\gamma}{\partial r}\right|_{r=0,\gamma=0} = 0.$$

The shape of bifurcation, in fact, is determined by the first equation of (37).

As discussed before, taking

$$u = -K|r|^c, \quad 0 < c < 1,$$

can change the shape of the bifurcation solutions dramatically. It is noted that the significance of the proposed nonsmooth control is not to move the bifurcation point, but make the bifurcation evolve slowly. The reason is that the bifurcation line is tangent to the line $r = 0$ at the critical point because

$$\left.\frac{\partial r}{\partial\gamma}\right|_{r=0,\gamma=0-} = 0.$$

Similar to the previous subsection, we can also find the tangent property and trumpet Hopf bifurcation here. Hence, the proposed controller is also effective in the sense that the quick changes and jumps of the stable system equilibria are prevented. Fig. 8 depicts an illustrative sketch for trumpet Hopf bifurcation resulted from fractional power control, where the dashed line shows the unstable fixed point.

4 Conclusions

In this chapter, we have presented some emerging directions in bifurcation control in continuous-time dynamical systems. Specifically, we have discussed two unconventional bifurcation control problems: anti-control of bifurcations and nonsmooth bifurcation control.

First, we have discussed the problem of anti-control of Hopf bifurcation, that is, a Hopf bifurcation created with desired location and preferable properties by an appropriate dynamic feedback control. In particular, washout-filter-aided dynamic feedback control laws are developed for creation of the intended Hopf bifurcation. A thermal convection loop example is used to illustrate the control procedure. As Hopf bifurcations give rise to limit cycles, anti-control of Hopf bifurcations suggests a new approach to designing limit cycles and oscillations into a system via feedback control when such dynamical behaviors are desirable. The proposed approach also provides a

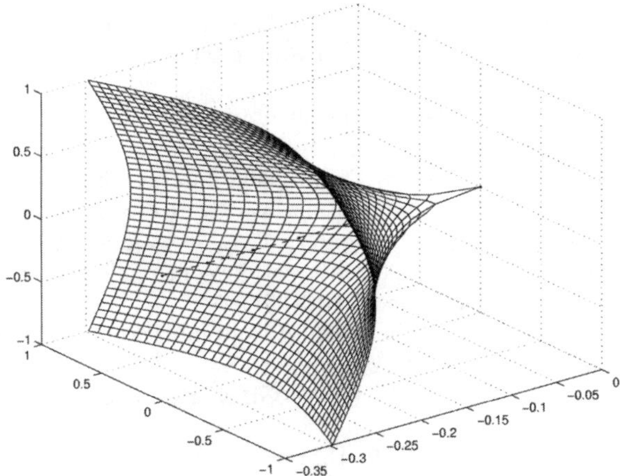

Fig.8. A trumpet Hopf bifurcation

new way of designing warning signals of impending collapse or catastrophe for monitoring and control purposes.

Nonsmooth bifurcation control is yet another important emerging direction of bifurcation control. In this chapter, a new type of bifurcation, namely, the trumpet bifurcation is introduced for several representative cases via nonsmooth fractional power control laws. It is shown that the proposed simple factional power control can result in superior operability to that of smooth feedback. The nonsmooth bifurcation control approach appears to be a viable technique for control of high performance nonlinear system in many situations. Because of the nonsmoothness, many well-known methods effective in smooth cases, such as Taylor series expansion and eigenvalue analysis, may not be employed. Systematic method for analysis and control of nonsmooth bifurcations will be an important topic in our future research.

Acknowledgment.
The authors would like to express their thankfulness to colleagues who have collaborated with the authors in aspects of the research leading to this article, in particular, we would like to express our thanks to Guanrong Chen and Yiguang Hong. This research has been supported in part by the Army Research Office under Grant DAAD19-00-01-0504.

References

1. Abed, E. H., Fu, J. H. (1986) Local feedback stabilization and bifurcation control: Part I. Hopf bifurcation. Syst. Contr. Lett., **7**:11–17
2. Abed, E. H., Fu, J. H. (1987) Local feedback stabilization and bifurcation control: Part II. stationary bifurcation. Syst. Contr. Lett., 8:467–473

3. Abed, E. H., Wang, H. O., Tesi, A. (1995) Control of bifurcations and chaos. The Control Handbook, Levine, W. S. (ed.), Boca Raton, FL: CRC Press & IEEE Press, 951–966
4. Chen, D. S., Wang, H. O., Chen, G. (2001) Anti-control of Hopf bifurcations. IEEE Trans. Circ. Syst.-I, **48**:661–672
5. Chen, D. S., Wang, H. O., Chin, W. (1998) Suppressing cardiac alternans: Analysis and control of a border-collision bifurcation in a cardiac conduction model. In Proc. IEEE Int. Symp. Circ. Syst., Montery, CA, USA, **3**:635–638
6. Chen, G., Moiola, J. L., Wang, H. O. (2000) Bifurcation control: Theories, methods, and applications. Int. J. Bifur. Chaos, **10**(3):511–548
7. Chen, X., Gu, G., Zhou, K. (2002) Local robustness of Hopf bifurcation stabilization. In Proc. Amer. Contr. Conf., 2262–2266
8. Chin, W., Ott, E., Nusse, H. E., Grebogi, C. (1994) Grazing bifurcations in impact oscillators. Phys. Rev. E, **50**:4427–4444
9. Gu, G., Chen, X., Sparks, A., Banda, S. (1999) Bifurcation stabilization with local output feedback. SIAM J. Contr. Optim., **37**:934–956
10. Guckenheimer, J. Holmes, P. (1983) Nonlinear Oscillations, Dynamical Systems, and Bifurcations of Vector Fields. New York: Springer
11. Hamzi, B., Kang, W. (2001) Resonant terms in a class of systems with stationary bifurcations. In Proc. 40th IEEE Conf. Decision Control, 722–727
12. Hassard, B. D., Kazarinoff, N. D., Wan, Y. H. (1981) Theory and applications of hopf bifurcation. Cambridge: Cambridge Univ. Press
13. Hong, Y., Huang, J., Xu, Y. (2001) On an output feedback finite-time stabilization problem. IEEE Trans. Auto. Contr., **46**:305–309
14. Horn, R. A., Johnson, C. R. (1993) Matrix Analysis. Cambridge: Cambridge Univ. Press
15. Iooss, G., Joseph, D. D. (1980) Elementary Stability and Bifurcation Theory. New York: Springer
16. Kang, W. (1998) Bifurcation and normal form of nonlinear control systems – Part I and Part II. SIAM J. Contr. Optim., **36**:193–232
17. Nusse, H. E., Ott, E., Yorke, J. A. (1994) Border-collision bifurcations: An explantation for observed bifurcation phenomena. Phys. Rev. E, **49**:1073–1076
18. Nusse, H. E., Yorke, J. A. (1992) Border-collision bifurcations including 'period two to period three' for piecewise smooth systems. Physica D, **57**:39–57
19. Wang, H. O., Abed, E. H. (1994) Robust control of period doubling bifurcations and implications for control of chaos. In Proc. IEEE Int. Symp. Circ. Syst., Orlando, USA, 3287–3292
20. Wang, H. O., Abed, E. H. (1995) Bifurcation control of a chaotic system. Automatica, **31**(9):1213–1226
21. Wang, H. O., Chen, D. S., Bushnell, L. G. (1997) Control of bifurcations and chaos in heart rhythms. In Proc. 36th IEEE Conf. Decision Control, San Diego, USA, 395–400
22. Wang, H. O., Chen, D. S., Bushnell, L. G. (2001) Dynamic feedback control of bifurcations. Latin American Applied Research, **31**:219–225
23. Wang, H. O., Chen, D. S., Chen, G. (1998) Bifurcation control of pathological heart rhythms. In Proc. IEEE Conf. Contr. Appl., Trieste, Italy, 858–862
24. Wang, H. O., Hong, Y., Bushnell, L. G. (2001), Nonsmooth bifurcation control: from fractional power control to trumpet bifurcation. In Proc. 40th IEEE Conf. Decision Control, Orlando, USA, 2181–2186

Bifurcation Analysis for Control Systems Applications

Mario di Bernardo[1]

Dept. of Engineering Mathematics
University of Bristol, Bristol BS8 1TR, UK
M.diBernardo@bristol.ac.uk

Abstract. In this chapter, the role that bifurcation analysis can play for control systems applications is discussed. It is proposed that bifurcation tools can be used to (i) obtain better models of the system of interest; (ii) design novel control strategies aimed at controlling the bifurcation diagram of a given system rather than its trajectories for a specified parameter value; (iii) assist with the synthesis of traditional controllers such as PID controllers. Some representative examples are discussed in detail to support the arguments presented in the chapter.

1 Introduction

Nonlinear Dynamics has made an enormous progress since Lorenz brought to the attention of the world the existence of seemingly unpredictable behaviour in deterministic dynamical systems. Researchers all over the world have refined their ability to study and characterise the dynamics of nonlinear systems whose describing equations cannot be solved in closed form. An entirely new geometric approach has been developed to cope with complex nonlinear systems. Entire families of bifurcations were studied and classified showing that nonlinear systems can exhibit non trivial transitions from one solution to another as their parameters are varied. New definitions were introduced to describe these phenomena and strategies were developed to classify them into appropriate categories.

Unfortunately, the remarkable theoretical successes of nonlinear dynamics have not translated yet into their successful use in applications. In particular, despite the high degree of overlapping between nonlinear dynamics and control engineering communities and common growing interest for nonlinear systems, the interaction and exchange of information between researchers belonging to these communities has been, and still is, limited.

While nonlinear dynamicists were actively studying bifurcations and chaos, control theorists were busy at developing novel control strategies to deal with complex and nonlinear systems. Robust nonlinear control, geometric control and many others techniques were proposed to deal with nonlinear systems. Surprisingly, in their effort to better describe and control such systems, control engineers kept into little account the methodologies, tools and results of nonlinear dynamics. At the same time (and maybe as a consequence of this

G. Chen, D.J. Hill, and X. Yu (Eds.): Bifurcation Control, LNCIS 293, pp. 249–264, 2003.

attitude!), researchers in nonlinear dynamics seemed to ignore the advances of control theory in characterising, for instance, the stability of nonlinear systems or their controllability and observability.

At the beginning of the nineties when the interest in the so-called control of chaos was at its peak, the scientific community was clearly divided. While physicists and mathematicians discovered (and at times re-discovered) ways to exploit the theory of bifurcations and chaos for control purposes, their efforts were mostly ignored by control theorists and engineers (who had the most relevant experience in the field).

Luckily, over the past few years, this scenario has been quickly turning around as the increasing complexity of systems of relevance in applications is pushing these two communities to exchange know-how and experience. Networks, hybrid systems, neuro-biological systems, large and extended systems are just a few examples of the novel challenges that control theorists are now facing. The classical approaches and tools of control theory are struggling to cope with these new systems showing their limitations. At the same time, nonlinear dynamicists are facing a dangerous information overload and are in urgent need of finding relevant applications for their methodologies. It is becoming increasingly clear that the only successful way forward is the synergy of ideas and results from both communities. The time is ripe for the successful and productive integration of methodologies and techniques from Control Theory and Nonlinear dynamics to tackle the analysis and control of complex nonlinear dynamical systems.

The aim of this chapter is to introduce and discuss ways in which bifurcation analysis and, more generally, bifurcation theory can be used for the analysis and design of control systems in applications. We will discuss the use of bifurcation analysis for control system design, proposing that bifurcation analysis could be used to (i) assess and understand the structural stability of control systems of interest; (ii) design better controllers by exploiting this extra information; (iii) synthesise innovative control strategies where the objective is to use the bifurcation behaviour of the system under investigation. In so doing, we focus on three case studies: the use of bifurcation analysis for identification purposes; the synthesis of so-called bifurcation tailoring laws and the tuning of PID controllers.

The rest of the chapter is organised as follows. The relationship between structural stability and bifurcations is discussed in Sec. 2, together with the possible applications of bifurcation analysis for control systems design. Then, in Sec. 3 the case study of an aircraft rig is presented showing that bifurcation analysis can support the derivation of an appropriate model of the system of interest. A novel scheme for bifurcation control is then presented in Sec. 4, where a strategy is described to "tailor" the bifurcation diagram of the plant to a desired one. Finally, in Sec. 5 the role of bifurcation analysis in assisting the synthesis of traditional PID controllers is discussed, suggesting that continuation methods can be used to tune the control parameters in the case of a nonlinear plant. Conclusions are drawn in Sec. 6.

2 Bifurcations and Structural Stability

An important issue in control is to assess whether the controlled system is structurally stable. Namely, consider a nonlinear control system of the form:

$$\dot{x} = f(x, \mu) \tag{1}$$

where $\mu \in R^k$ is a parameter vector. Then, the question must be addressed of whether the qualitative behaviour of the controlled system remains approximately the same as the vector field $f(x, \mu)$ is perturbed. For example, if the system parameters drift over time, it is important to make sure that the control behaviour remains the same within some acceptable bounds. If this occurs we then say that the system is structurally stable. More precisely, let $f \in C^1(\Omega)$, then the C^1-norm of f on Ω is defined by:

$$\|f\|_1 = \max_{x \in \Omega} |f(x, \mu)| + \sup_{x \in \Omega} \|Df(x)\|. \tag{2}$$

Clearly if K is a compact subset of Ω, then the C^1 norm of f on K is defined as

$$\|f\|_1 = \max_{x \in K} |f(x, \mu)| + \max_{x \in K} \|Df(x)\| < \infty. \tag{3}$$

We can now define structural stability as follows.

System (1) is said to be structurally stable if there exists a $\varepsilon > 0$ such that for all $\tilde{f} \in C^1(\Omega)$ with $\|f - \tilde{f}\| < \varepsilon$, system (1) is topologically equivalent to the system

$$\dot{x} = \tilde{f}(x), \tag{4}$$

i.e. if there exists an orientation preserving homeomorphism $h : \Omega \mapsto \Omega$ which maps trajectories of (1) onto trajectories of (4).

Robustness is an important issue in control. Noise or unwanted perturbations can, in facts, alter the behaviour of the controlled system with unpredictable consequences. Parameter variations are particularly important as parameters are bound to vary with time and/or because of external causes. Control theorists have introduced appropriate robust control strategies, mostly aimed at linear control systems. Examples include H_∞- control, robust multivariable control among many others. An open problem remains that of synthesising robust controllers for nonlinear dynamical systems.

Rather than aiming at controlling the system under investigation but to analyse its behaviour as parameters are varied, nonlinear dynamics has focussed on the study of mechanisms through which structural stability can be lost. It has been observed that given a system of the form (1) at some parameter values qualitative changes are observed in the long term behaviour of the solutions. These points in parameter space are called *bifurcations* and can be easily defined by using the concept of structural stability.

Namely, a value μ^* of system (1) for which the system trajectory of interest is not structurally stable is called a bifurcation value of μ. Examples abound in the literature and bifurcations have been detected in several systems of relevance in applications.

As bifurcations and structural stability of nonlinear systems are undeniably strictly related, it might appear obvious that bifurcation theory could be used in the process of synthesising appropriate controllers for these systems. Despite all appearances, control theorists have not made full use yet of bifurcation analysis in their developments.

As mentioned above, bifurcation analysis can have an impact on several aspects of controller synthesis. In particular, bifurcation theory can support three different aspects of the control synthesis process. Namely, we will discuss how bifurcation diagrams can be used to:

1. support the identification process of the plant to be controlled
2. help designers with the tuning of the control parameters when dealing with nonlinear systems
3. synthesize novel control techniques aimed at changing the entire bifurcation behaviour of the system under investigation

We start by discussing the role that bifurcation analysis can play for model identification. As a representative example of a realistic application, we focus on models related to aircraft flight dynamics.

It is worth emphasizing here that, in our opinion, bifurcation analysis becomes more and more a necessary tool as the complexity of the system of interest grows. Therefore, two application areas are currently the most likely to benefit from the application of bifurcation analysis to control: aerospace engineering and chemical engineering. In both cases unmodelled dynamics, turbulence and highly nonlinear models make the use of traditional controllers more and more inappropriate to deal with the high spec performances required in practical applications. It is therefore expected that bifurcation theory and more generally nonlinear dynamics will play an increasingly important role in control.

3 Bifurcation Analysis for Model Identification

We illustrate the use of bifurcation analysis for identification purposes by describing a representative problem. In the field of aerospace engineering, an important issue is to develop aerodynamic models of flight dynamics. This involves representing the aerodynamic loads in a formulation suitable for use in the equations of motion for arbitrary flight conditions. Typically, the forces and moments are functions of the motion variables and parameters such as control inputs, undercarriage position, etc. The vast majority of such models are quasi-steady and capture only small-amplitude motions.

It has been recognised that unsteady effects can become significant in rapid manoeuvres, or manoeuvres at very low forward speeds. To this end, a substantial body of work has been undertaken on modelling of time-dependent aerodynamic reactions (e.g. [4]–[19]).

The majority of the modelling work has focussed on the phenomenon called 'wing rock'. This is a lateral-directional oscillation, dominant in roll, that occurs in many fighter aircraft configurations as angle of attack is increased. Fewer aircraft exhibit longitudinal pitching limit cycle oscillations ('bucking').

Both wing rock and aerofoil pitching oscillations have been analysed recently in the context of bifurcation theory in [2]. Models incorporating nonlinear damping were evaluated in terms of bifurcation behaviour leading to limit cycles; the nature of the bifurcations (sub- or supercritical) were assessed, as was the rate of growth of the periodic orbits. However, the identification of model structures suitable for representing global bifurcation behaviour over a wide operating envelope has not been addressed.

As with all flight dynamics modelling, a balance must be struck between fine detail in localised regions and the need to represent behaviour over a very large operating envelope. With this respect, bifurcation analysis can be an invaluable tool in understanding the structure and dynamical feature of a flight dynamics model. By providing tools to analyse and characterise different dynamical transitions, it can help to identify the main features to be captured by the model.

In so doing, we need to identify two necessary steps: (i) experimental investigation of the bifurcation structure of the system of interest; (ii) derivation of an appropriate model based on the experimental results.

3.1 Experimental bifurcation analysis

Figure 1 shows the results of two experimental bifurcation runs for an experimental wind-tunnel rig recently studied at the University of Bristol (see [5] for further details). The diagrams show the fixed points and maximum limit-cycle amplitudes for the system as a function of tailplane deflection.

We can clearly see that a stable equilibrium point (corresponding to a certain stable flight attitude) turns into a stable limit cycle at an Hopf bifurcation detected for $\delta_e \approx -10°$. This is precisely the type of wing rock oscillations described earlier.

In Figure 2(a)-(d) the corresponding time histories are plotted for fixed tailplane deflections labelled in Figure 1(a). As expected, points (a) and (c) are fixed equilibria at approximately $\theta = 28°$ and $\theta = 12°$ respectively. Point (b) in Figure 1 is instead a large amplitude limit cycle oscillation, centered at $\theta = 22°$ with an amplitude of approximately $10°$. This is confirmed by

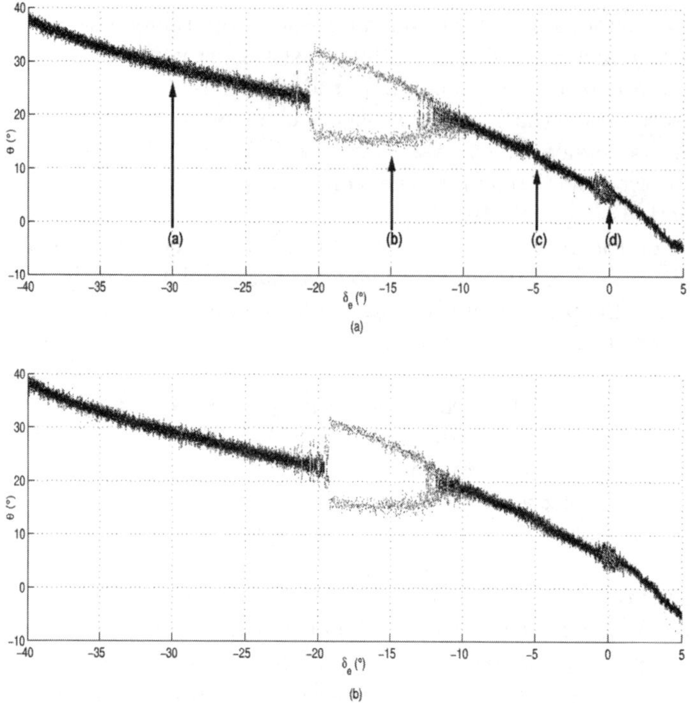

Fig. 1. Experimental bifurcation diagrams: (a) for decreasing and (b) for increasing tailplane deflection.

the experimental time history depicted in Figure 2(b). The experimental bifurcation diagram reveals a region of small-amplitude limit cycles at point (d), centred at approx. $\theta = 5°$.

3.2 Deriving the model

The experimental results obtained by carrying out tests aimed at establishing the bifurcation behaviour of the system can be now manipulated to extract the information needed for the model derivation. In particular, we notice that the diagrams tell us the type of solution to be expected at different tailplane deflections, the amplitude of the oscillatory motion when present and the growth/decay of such oscillations under parameter variations. Moreover, time histories of the system such as those depicted in Fig. 2 show the envelope of the oscillations at specific parameter values and their frequency. It is then possible to use the more traditional time histories with the results of the bifurcation analysis to obtain an appropriate model.

As recently proposed in [5], the starting point is to examine a typical experimental time history of the rig position, such as the one depicted in Fig. 3 where the tailpane deflection is fixed at $-15°$ and the rig is released from a

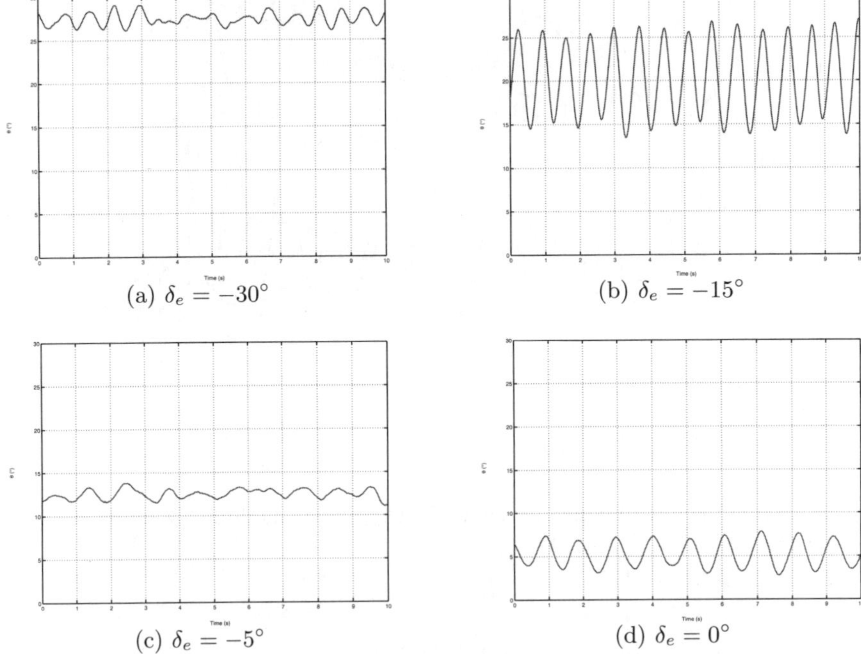

(a) $\delta_e = -30°$ (b) $\delta_e = -15°$

(c) $\delta_e = -5°$ (d) $\delta_e = 0°$

Fig.2. Time histories for points marked in Figure 1.

phase-space point away from the limit cycle. As can be noticed from Figure 3, the oscillations appear approximately sinusoidal with an exponential-type decay (which for practical reasons is better modelled by a tanh function [5]). Also visible are the effects of turbulence on the system (especially at $t \approx 10\,\text{s}$). Assuming the oscillations are symmetrical about the fixed point and neglecting the effects of turbulence, this suggests that the position decays onto the stable period-1 limit cycle according to the equation (as in [6]):

$$\hat{\theta} = (A_u - \Delta(1 + \tanh(Kt)))\sin(\omega t + \phi) \tag{5}$$

where A_u and Δ are extra parameters necessary for defining the tanh function characteristics, and K defines the growth/decay rate. We now have a first equation describing the rig position. This can be appropriately differentiated to get equations for the velocity and acceleration. It can be shown that, through algebraic manipulations, these equations can be used to derive the value of the acceleration (and hence the pitching moment) associated to arbitrary values of the position and velocity. Thus, the model structure can be established [5].

The second stage necessary to complete the derivation of a mathematical model of the experimental system is to find values of all the parameters used in the estimation process.

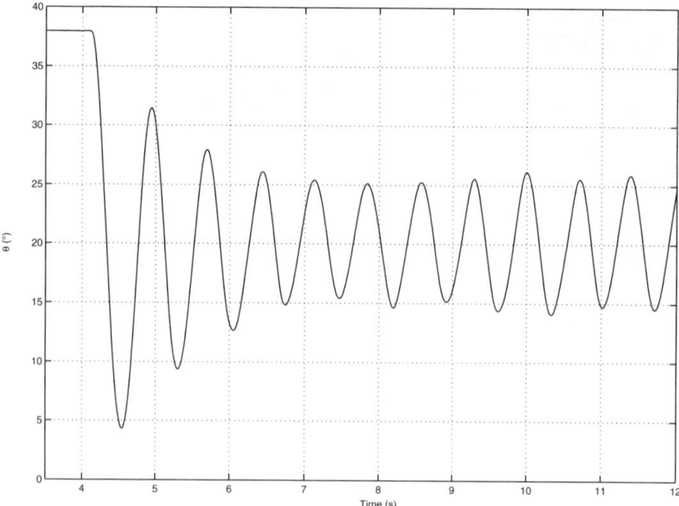

Fig.3. Experimental time history showing decay onto the limit cycle (δ_e fixed at $-15°$).

The parameters we need to find can be broadly grouped in two categories: those that related to the characteristics of the limit cycle oscillations, A and ω and those that are related to the shape of the tanh function, K, A_u and Δ.

As we are interested in modelling the rig behaviour for different values of the tailplane deflection, we need to tabulate these parameters as functions of δ_e. For this purpose, we can make explicit use of the experimental bifurcation analysis carried out on the rig presented above.

The bifurcation diagram produced by AUTO on the model is shown in Figure 4. It can be seen that, as expected, the model bifurcation diagram matches well the experimental one.

4 Bifurcation Tailoring

We have seen sofar that bifurcation analysis can be used to better characterise the system behaviour and, hence, support the model derivation. The next question is whether this extra information can be actually used for control purposes.

Certainly, the model obtained through the procedure described in the previous section can indeed be used to synthesise traditional control strategies to stabilise, for example, unwanted oscillations by traditional control techniques; or by using bifurcation control strategies aimed at moving the occurrence of an unwanted bifurcation at different parameter values.

Without loss of generality, we could assume that the outputs are the first m states of the system which we label with x_I, i.e.,

$$y = x_I = [x_1, x_2, \dots, x_m]^T \tag{8}$$

4.2 Feedback bifurcation tailoring of equilibria

For the sake of clarity, in what follows we focus on the problem where the desired objective is for the controlled system to exhibit a branch of equilibria such that, as the parameter p is varied,

$$x_I(p) = g(p), \qquad p \in [p_a, p_b]$$

where $g : R \mapsto R^m$ is an m-dimensional smooth function of p. This means that, for a given p in the range of interest, the desired output is a point and this point varies smoothly as p varies from p_a to p_b.

Thus, if we say $x_{II} = [x_{m+1}, \dots, x_n]^T$, we want to choose $q = q(x, p)$ such that:

$$f(g(p), x_{II}, p, q) = 0 \equiv f(x_{II}, q, p) = 0, \qquad p \in [p_a, p_b] \tag{9}$$

Let $z = [x_{II} \quad q]$, then according to the Implicit Function Theorem (IFT) [17] if the Jacobian of f with respect to z is invertible then (9) implicitly defines z as a function of p, i.e. $z = z^*(p) = [x^*_{II}(p) \quad q^*(p)]$. This means that if q is varied according to the law $q = q^*(p)$ as p between p_a and p_b, then the system will exhibit the branch of equilibria:

$$x^* = [g(p) \quad x^*_{II}(p)],$$

and therefore the required behaviour of the system output.

Notice that this *open-loop* bifurcation tailoring law only works if x^* is a branch of asymptotically stable equilibria. Moreover, it requires the solution off-line of the implicit equation (9). As discussed in [8], this can be achieved by using an off-line software continuation package such as AUTO [15].

To overcome these problems, the solution recently proposed in [9,10], is to consider the control law:

$$q(p) = q^*(p) + \delta q(x, p) \tag{10}$$

where $q^*(p)$ is the law obtained by solving equation (9) and δq is an additional feedback term which ensures stability of the equilibrium branch as p is varied. As detailed in [8–10], this extra term can be synthesised (i) by using standard state-feedback control based on a sequence of linearisations about the points of interest; (ii) by using a more sophisticated adaptive control action. Moreover to avoid the off-line computation of the term $q^*(p)$, the controller can be equipped with an on-line continuation strategy such as the Newton Flow algorithm (see [9] for further details). A block diagram of the resulting strategy is shown in Fig. 5 The application of the method to several test examples and aircraft flight dynamics model has been recently reported in [10]. Figure 6 shows a numerical example.

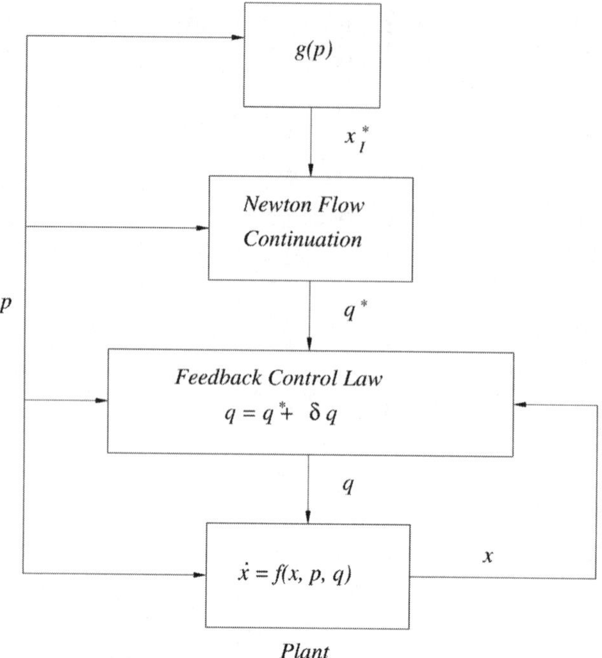

Fig.5. Block Diagram of the Feedback Bifurcation Tailoring control scheme

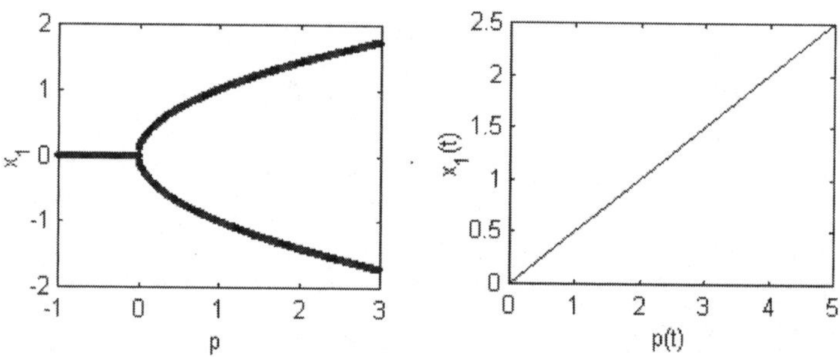

Fig.6. Bifurcation Tailoring of a Duffing Oscillator: diagram before control (left) and bifurcation diagram of the controlled system (right)

5 PID Tuning

Finally, we wish to discuss a third possible application of bifurcation analysis to support control system design. As it is well known in the control community, PID controllers are among the most widely used feedback control strategies in applications. A typical PID controller is based on three separate actions: proportional (P), integral (I) and derivative (D). The contribution of each of these three actions in forming the control signal $u(t)$ is determined by a set of three gains K_P, K_I and K_D that must be appropriately chosen (or tuned) to achieve the control objective. Namely, if $e(t) = y_{\text{ref}} - y$ is the error signal, the PID control law is given by:

$$u(t) = K_P e(t) + K_I \int e(\tau)d\tau + K_D \dot{e}(t). \tag{11}$$

PID control is essentially a linear control strategy and as such can only be applied successfully to control linear systems. The core step of the control design is the *tuning* of the control parameters, i.e. choosing K_P, K_I and K_D so that the controlled output satisfies the desired objective. This can be done by using traditional methods such as the Ziegler-Nichols tuning scheme or more innovative ones (for a review of PID tuning methodologies see [11]). All these techniques share in common the fact that they were developed and tested for linear plants.

In practice, though, because of their advantages in terms of costs and simplicity of operation, PID controllers are used also to control implicitly nonlinear plants, such as airplanes and chemical reactors. In these cases, control engineers carefully isolate the operating regime of interest, obtain the linearised model of the plant and then proceed with tuning the gains of the PID controller using traditional techniques. In general, this implies that the control will be effective as long as the plant remains in a neighborhood of the operating point where the linearisation is valid. Alternatively, the nonlinear disturbances acting on the plant are required to satisfy a set of appropriate conditions (usually extremely conservative).

We propose that bifurcation analysis could offer an effective alternative approach to solve the problem of tuning the gains of PID controllers for nonlinear plants. Namely, from a bifurcation viewpoint, given a plant of the form

$$\dot{x} = f(x, u), \qquad y = g(x) \tag{12}$$

the application of a PID controller such as (11), turns the controlled plant into the system

$$\dot{x} = \tilde{f}(x; K_P, K_I, K_D) \tag{13}$$

where $\tilde{f}$ is the function obtained by substituting (11) into (12) and K_P, K_I, K_D are three free parameters introduced by the PID.

Now, tuning the gains of the PID is equivalent to carry out an extensive bifurcation analysis of the system as the three gains K_P, K_I, K_D are varied (see for example [1]). Currently available continuation software packages such as AUTO97 and CONTENT [16] could be used to uncover the effects of varying these three gains on the controlled plant dynamics. Bifurcation diagrams could be obtained which could assist the designer in tuning the gains of the PID. Namely, we anticipate that locating bifurcation events and coexisting branches of other solutions due to the nonlinear nature of the plant, would be invaluable to better choose the gains and might also suggest solutions which are not necessarily bound by the restrictions brought about by the linearisation procedure. Moreover, other tools from Nonlinear Dynamics such as the computation of the basins of attractions for a given asymptotic solution (and in particular the efficient numerical methods developed to do so) could be used to establish the operating range in which the controller could be used.

Novel methods for PID tuning based on bifurcation analysis based on these ideas are currently under investigation and will be detailed elsewhere [12]. Here we just wish to emphasize that bifurcation analysis seems an invaluable tool to use for the synthesis of controllers for nonlinear systems, even for the simplest type of controller: the PID.

6 Conclusions

In this chapter the role that bifurcation analysis can play for control systems application was discussed. Namely, three applications of bifurcation analysis for control design were presented. Firstly, it was shown that an experimental investigation of the system bifurcations can be used to synthesise better models of its behaviour. A case study was presented of an aircraft wind-tunnel rig recently developed at the University of Bristol. Secondly, the possibility of using bifurcation theory to synthesise novel control strategies was discussed. In particular, so-called bifurcation tailoring schemes were highlighted and shown to provide control strategies aimed at assigning a desired bifurcation diagram to the system of interest. Possible applications of these techniques were discussed. Finally, the possibility was suggested of using bifurcation analysis, and particularly continuation methods, for the synthesis of more conventional control methodologies such as the tuning of PID controllers. It was proposed that bifurcation analysis could be used as an alternative strategy to tune the parameters of such controllers when they are used to tame the dynamics of nonlinear systems.

In conclusion, we anticipate that bifurcation analysis can (and will) play an increasingly important role in the design of better controllers for nonlinear and complex dynamical systems. Bifurcation analysis can provide tools to assist in all phases of the controller synthesis and model identification. It can be used to improve existing methods or to develop entirely new control strategies. Future work should further address the key issue of integrating control expertise with the vast amount of knowledge available today in the field of Nonlinear Dynamics. The challenge will be on one hand to design novel controllers and on the other to spread their use in applications through a rigorous and methodical analysis of their performances.

Acknowledgements.
The author wishes to thank Prof Xiaofan Wang, University of Shangai, China, Mr Guy Charles, Dr Mark Lowenberg and Prof D P Stoten, University of Bristol, U.K., in collaboration with whom the work on bifurcation tailoring was carried out, and Mr Paul Davison, University of Bristol, U.K., who carried out the experiments on the aircraft rig discussed in this chapter and worked on the identification procedure presented in Sec. 3 as part of his Ph.D. under the supervision of the author and Dr. Mark Lowenberg.

References

1. Chang, H.-C., Chen, L.-H. (1984) Bifurcation characteristics of nonlinear systems under conventional PID control. Chem. Eng. Sci., **39**:1127
2. Hui, W. H., Toback, M., (1985) Bifurcation theory applied to aircraft motions. Technical report, NASA
3. Goman, M. G., Khrabrov, A. (1994) State-space representation of aerodynamic characteristics of an aircraft at high angles of attack. J. Aircraft, **31**(5):1109–1115
4. Myatt, H. J., (1996) A nonlinear indicial response model for the rolling 65-degree delta wing. In Proc. AIAA Atmospheric Flight Mechanics Conf., AIAA-96-3406
5. Davison, P., Lowenberg, M. H., di Bernardo, M. (2002) Experimental analysis and modelling of limit cycles and bifurcations in a dynamic wind tunnel rig. Prinprint
6. Hsiao, F. B., Yang, J. S. (1996) The study of wing-rock characteristics on slender delta wings at high angle-of-attack. In Proc. ICAS-96-3.1.1
7. Lowenberg, M. H., Richardson, T. S. (1999) Derivation of non-linear control strategies via numerical continuation. In Proc. AIAA Atmospheric Flight Mechanics Conf., paper no. AIAA-99-4111, 359–369
8. Charles, G. A., Di Bernardo, M., Lowenberg, M. H., Stoten, D. P., Wang, X. F. (2001) Bifurcation tailoring of equilibria: a feedback control approach. Latin American Applied Research, **31**:199–210
9. Wang, X. F., di Bernardo, M., Stoten, D. P., Lowenberg, M. H., Charles, G. (2001) Bifurcation tailoring via Newton flow-aided adaptive control. Preprint
10. Charles, G. A., di Bernardo, M., Lowenberg, M. H., Stoten, D. P., Wang X. F. (2002). On-line bifurcation tailoring applied to a nonlinear aircraft model. In Proc. IFAC World Congress, Barcelona, Spain
11. Aström, K. J., Hagglund, T. (1995) PID Controllers: Theory, Design and Tuning. 2nd ed., pub. by Instrumentation, Systems, and Automation Society
12. di Bernardo, M. (2002) Using bifurcation analysis for PID controller synthesis, in preparation

13. Chen, G., Moiola, J. L., Wang, H. O. (2000) Bifurcation control: theories, methods, and applications. Int. J. Bifur. Chaos, **10**:511–548
14. Chen, G. (ed.) (1999) Controlling Chaos and Bifurcations in Engineering Systems. Boca Raton, FL: CRC Press
15. Doedel, E. J., Wang, X. J. (1995) AUTO94: Software for continuation and bifurcation problems in ordinary differential equations. Technical report, Center for Research on Parallel Computing, California Institute of Technology, Pasadena, CA, 91125 CRPC-95-2
16. Kuznetsov, Y., Levitin, V. (1997) CONTENT: dynamical system software, available at http://www.riaca.win.tue.nl/CAN/SystemsOverview/Special/DiffEqns/Content/GCbody.html
17. Hirsch, M. W., Smale, S. (1974) Differential Equations, Dynamical Systems, and Linear Algebra. New York: Academic Press
18. Kang, W. (1998) Bifurcation and normal form of nonlinear control systems, Part I and II. SIAM J. Contr. Optim., **36**:193-232
19. Tobak, M., Schiff, L. B. (1981) Aerodynamic mathematical modelling basic concepts. In AGARD-LS-114 - AGARD Lecture Series on Dynamic Stability Parameters

Feedback Control of a Nonlinear Dual–Oscillator Heartbeat Model

Michael E. Brandt[1], Guanyu Wang[1], and Hue-Teh Shih[2]

[1] Center for Computational Biomedicine and Neurosignal Analysis Laboratory
University of Texas Health Science Center–SHIS
Houston, TX 77030, USA mbrandt@uth.tmc.edu

[2] Baylor College of Medicine
Department of Medicine
Houston, TX 77030, USA

Abstract. We describe a feedback control method for stabilizing some pathological behaviors of a nonlinear heartbeat model, with and without additive random noise. The controller is discretized to demonstrate how it might be implemented in a practical pacemaker design. Comparisons with some other control methods are discussed.

1 Introduction

In recent years, there has been much attention given to the notion of stabilization of abnormal cardiac rhythms using a nonlinear dynamical systems paradigm. This approach is characterized by the application of low-intensity electrical perturbing stimuli applied with variable timing. The first such attempt at controlling cardiac chaos can be traced back to Garfinkel *et al.* [1], who stabilized drug-induced cardiac arrhythmias in the *in vitro* rabbit ventricle by altering the timing of the interbeat cardiac intervals. It was shown that administering low-intensity electrical shocks directly to the septum at irregular intervals determined by the OGY method [2] could convert the arrhythmia to periodic beating. Despite some success using the OGY method, there remain difficulties in applying it in actual cardiac preparations. It is computationally complex and unintuitive, requiring that the locations of saddle-type UFP's be found using the method of delay-coordinate embedding (to this end a prolonged learning phase is needed). The sensitivity of the state point to changes in parameters, and the timing and the amplitude of the electrical stimulation are difficult to determine.

Another potential drawback of the OGY method is that a constant vigilance must be maintained in case the system veers away from the stable manifold of the unstable target. Several studies (e.g., [7-11]) have overcome some of these limitations by introducing a variety of discrete map models to simulate the dynamics of interbeat intervals and by demonstrating various control strategies using time-delay feedback (for a partial review see [12]).

G. Chen, D.J. Hill, and X. Yu (Eds.): Bifurcation Control, LNCIS 293, pp. 265–273, 2003.

The cardiac conduction system is considered to be a network of self-excitatory pacemakers [3, 4], with the sinoatrial (SA) node having the highest intrinsic rate. Subsidiary pacemakers with slower firing frequencies are located in the atrioventricular (AV) node and the His-Purkinje system. The SA node is the dominant pacemaker of the heart. Electrical impulses travel from it to the ventricles through the AV junction, which is traditionally regarded as a passive conduit. To take advantage of the ever-increasing knowledge of cardiac dynamics we use here a previously detailed mathematical model having a strong correspondence to the physiology of the heart's conduction system [5, 6]. The model is a two-coupled nonlinear oscillator implemented in a set of four ordinary differential equations. We propose a feedback control algorithm which uses the difference in interbeat intervals between the SA and AV nodes. A possible main utility of this method may be as a software-based approach to generating lower chamber pacing stimuli which could replace current pacemaker hardware that utilizes a reel for stimuli timing. Replacing such hardware might allow a pacemaker to become magnet-safe so that patients can undergo magnetic resonance scans without pacemaker removal or damage.

2 Model Description

The coupled nonlinear oscillator model studied here describes in a natural way the interaction between the SA and AV nodes. The generated waveforms of the model resemble the action potentials of cells in both these nodes. The abnormal behavior of the model corresponds well to several pathophysiological symptoms observed in the human heart as detailed in [5, 6]. We use this model to test the method of feedback control to resynchronize the AV and SA nodes when they demonstrate some pathological coupled rhythm pattern such as bigeminy or trigeminy. The continuous model relations are given by

$$\dot{x_1} = \frac{1}{C_{SA}}\, x_2\,, \tag{1}$$

$$\dot{x_2} = -\frac{1}{L_{SA}}\left[x_1 + g(x_2) + R(x_2 + x_4)\right], \tag{2}$$

$$\dot{x_3} = \frac{1}{C_{AV}}\, x_4\,, \tag{3}$$

$$\dot{x_4} = -\frac{1}{L_{AV}}\left[x_3 + f(x_4) + R(x_2 + x_4)\right], \tag{4}$$

where $g(x) = f(x) + h(x)$, $f(x) = x^3/3 - x$, and

$$h(x) = \begin{cases} -x^2 - 0.25\,, & \text{for } |x| < 0.5 \\ -x\,, & \text{for } x > 0.5 \\ \;\;x\,, & \text{for } x < -0.5\,. \end{cases}$$

x_2 and x_4 represent the action potentials of the SA and AV nodes respectively. The interaction between the SA and AV nodes is modelled as two coupled nonlinear oscillators, associated with an equivalent electronic circuit depicted in Fig. 1. To simulate the heartbeat dynamics, Eqs. (1-4) were solved using the fourth order Runge-Kutta algorithm with step size $h = 0.001$.

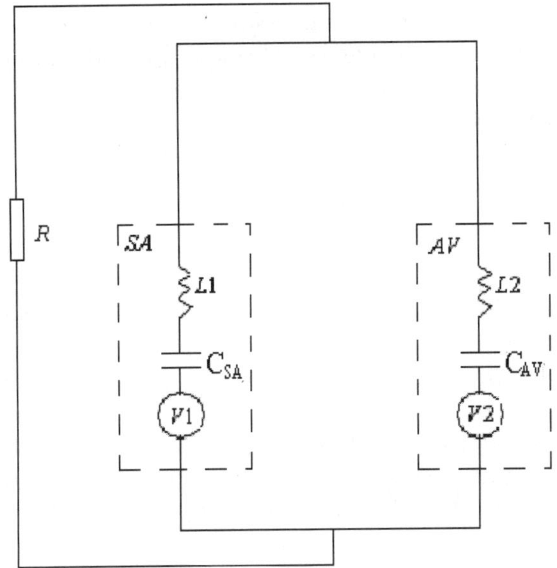

Fig.1. Equivalent electrical circuit describing the model constituted by two coupled nonlinear oscillators representing the SA and AV nodes.

3 Control Method and Results

3.1 Continuous control

In normal sinus rhythm dynamics, the SA and AV nodes are approximately 1:1 phase locked. That is, every time the SA node depolarizes, there is one AV node depolarization. In various arrhythmias, different phase locking (n:m) behaviors occur. Our aim is to use feedback control to resynchronize the depolarization of the SA and AV nodes. That is, to make it 1:1 phase locked again using a gentle control or coaxing strategy (by applying low intensity stimuli to the system). Since an arrhythmia is related to the asynchrony between the SA and AV oscillators, we utilize the difference in heartbeat period between the SA and AV node oscillations as a feedback control signal. The goal is to induce the faster oscillator to run slower and the slow oscillator to run faster. According to [1], the frequency of the SA (AV) beat is proportional to $\frac{1}{C_{SA}}$ $\left(\frac{1}{C_{AV}}\right)$. We thus introduce an (external) control capacitor C whose inverse is updated recursively as

$$\left(\frac{1}{C}\right)_i = \left(\frac{1}{C}\right)_{i-1} + \quad k\,(T_{AV} - T_{SA})_i \qquad \text{for } i = 1,\, 2,\, 3,\, \ldots, \tag{5}$$

where T_{AV} (T_{SA}) is the period between two successive AV (SA) beats, k is a simple gain factor, i is an index for the number of times T_{AV} and T_{SA} are updated.

The controlled model is as follows:

$$\dot{x_1} = \frac{1}{C_{SA}}\,x_2 - \frac{1}{C}\,x_2\,, \tag{6}$$

$$\dot{x_2} = -\frac{1}{L_{SA}}\,[x_1 + g(x_2) + R(x_2 + x_4)]\,, \tag{7}$$

$$\dot{x_3} = \frac{1}{C_{AV}}\,x_4 + \frac{1}{C}\,x_4\,, \tag{8}$$

$$\dot{x_4} = -\frac{1}{L_{AV}}\,[x_3 + f(x_4) + R(x_2 + x_4)]\,, \tag{9}$$

with $1/C$ calculated using the recursive relation given in Eq. (5). Figure 2 is the equivalent electrical circuit of the system with control. The general idea is that if the SA node runs faster than the AV node, then $(T_{AV} - T_{SA}) > 0$,

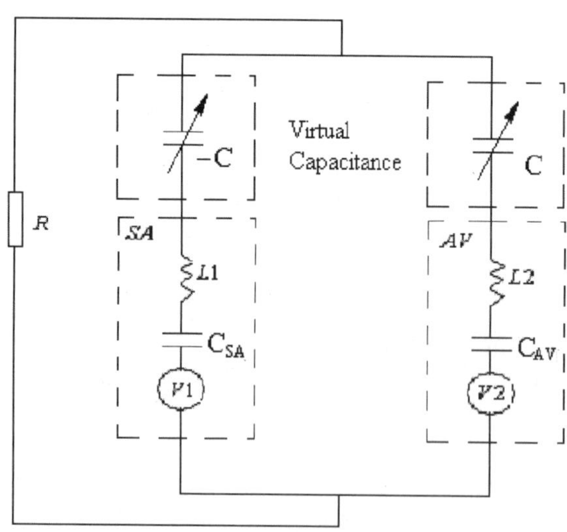

Fig.2. Equivalent electrical circuit for the controlled system.

which will increase the value of $1/C$ from Eq. (5). According to Eqs. (6) and (8), this will cause the SA node to run slower and the AV node faster. The negative feedback of frequency difference is thus realized. Initially, $1/C$ should be set to zero, and then should be self-modified by the feedback error on each iteration until a point in time when the alternating pattern halts and sinus rhythm is restored. The adjustment will not stop as long as the frequency difference remains. In an actual pacemaker implementation one would need to (and would only be able to) speed up the slower node (AV node) in order to keep pace with the faster one (SA node). Therefore the control stimulus would be applied only to the AV node using Eq. (8) and not to the SA node. In the results that follow then, we used Eqs. (1, 5, 7-9) only (and not Eq. (6)) to simulate control of the system.

The values of T_{AV} and T_{SA} are obtained by measuring the time interval between two successive peaks of the waveforms of the AV and SA nodes, respectively. A peak is determined to have occurred if the signal increased through a threshold, passed over a local maximum and then decreased through another threshold.

The value of k can be chosen between 1 and 10 with little difference in performance. The larger k is, the faster synchronization is achieved. Normal heartbeat dynamics can be obtained with

$$C_{SA} = 0.25\,,\; L_{SA} = 0.05\,,\; C_{AV} = 0.675\,,\; L_{AV} = 0.027\,,\; R = 0.11\,. \tag{10}$$

By varying the coupling resistance R while keeping other parameters the same as Eq. (10), we obtain one type of arrhythmia known as 2° AV block of the Wenckebach type. It is usual to describe this arrhythmia with two integers $n : m$, which means the atria contract n times while the ventricles m times. Figure 3(a) shows both the uncontrolled and controlled 3:2 Wenckebach rhythm with $R = 0.018$. The controller is turned on at time 5 and turned off at time 15, with the control gain $k = 2$. By varying C_{SA} while keeping other parameters the same as Eq. (10), we obtain another type of arrhythmia known as $n : 1$ AV block . Figure 3(b) shows both the uncontrolled and controlled 3:1 AV block with $C_{SA} = 0.10$. The controller is turned on at time 5 and turned off at time 10, using the control gain $k = 2$.

Figure 4(a) is analogous to Fig. 3(a) with the exception that zero-mean Gaussian white noise has been added to both x_2 and x_4. The root-mean-square (RMS) amplitude of the white noise (σ) is 0.04 and 0.06, respectively. Similarly, Fig. 4(b) is the "noisy" version of Fig. 4(a) with $\sigma = 0.1$ for both x_2 and x_4.

3.2 Discrete control

In an actual cardiac preparation, control is not normally applied continuously as above. Instead, electrical stimulations are delivered discretely and periodically. From Eq. (8) one sees that the final effect of the control in every time

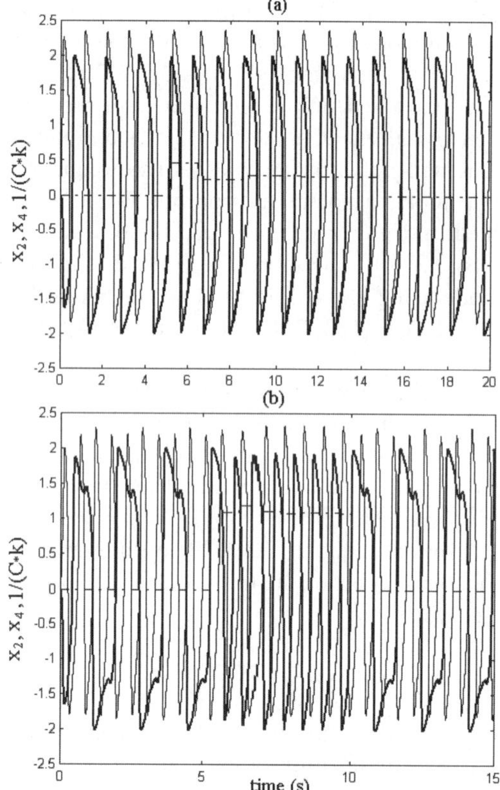

Fig.3. (**a**) 3:2 Wenckebach rhythm with $R = 0.018$, $k = 2$. The controller is turned on at time 5 and turned off at time 15. The thin line is for the SA node, the thick line is for the AV node, the dotted line is the control signal. (**b**) 3:1 AV block. $C_{SA} = 0.10$, $k = 2$. The controller is turned on at time 5 and turned off at time 10. The thin line is for the SA node, the thick line is for the AV node, the dotted line is the control signal $(1/C)$ scaled by $1/k$ for plotting purposes.

step is to change x_4 to $(1 + \frac{C_{AV}}{C})\, x_4$, which is derived from

$$\frac{1}{C_{AV}}\, x_4{}' = \left(\frac{1}{C_{AV}} + \frac{1}{C}\right) x_4\,, \quad \rightarrow \quad x_4{}' = \left(1 + \frac{C_{AV}}{C}\right) x_4\,.$$

To this end, we shall apply a stimulation with strength E proportional to

$$(x_4{}' - x_4) = \left(\frac{C_{AV}}{C}\right) x_4$$

at every time step $i\tau$

$$E(i\tau) \propto \frac{x_4(i\tau)}{C(i\tau)} \quad \text{for } i = 1,\, 2,\, 3,\, \ldots . \tag{11}$$

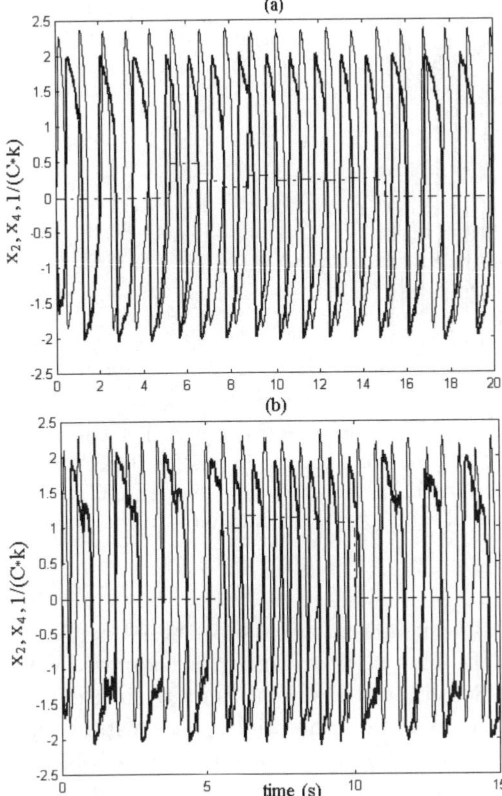

Fig.4. (**a**) 3:2 Wenckebach rhythm with $R = 0.018$, $k = 4$, $\sigma_{SA} = 0.04$, and $\sigma_{AV} = 0.06$. The controller is turned on at time 5 and turned off at time 15. The thin line is for the SA node, the thick line is for the AV node, the dotted line is the control signal. (**b**) 3:1 AV block. $C_{SA} = 0.10$, $k = 2$, $\sigma_{SA} = \sigma_{AV} = 0.1$. The controller is turned on at time 5 and turned off at time 10. The thin line is for the SA node, the thick line is for the AV node, the dotted line is the control signal $(1/C)$ scaled by $1/k$ for plotting purposes.

The constant τ is the time interval between two successive electrical stimulations. Note that any uncertainties are included in the gain factor k (Eq. (5)), thus we can turn Eq. (11) into an equality

$$E(i\tau) = \frac{x_4(i\tau)}{C(i\tau)} \quad \text{for } i = 1, 2, 3, \dots . \tag{12}$$

The discrete and the continuous controls are identical if τ is chosen to be the same as the step size of the Runge-Kutta algorithm. As τ increases, the ability to control the system slowly degrades. However, according to our simulations it can still perform well with τ as large as 0.15. The control is ineffective with τ greater than 0.2. There are several merits of the control strategy Eq. (12).

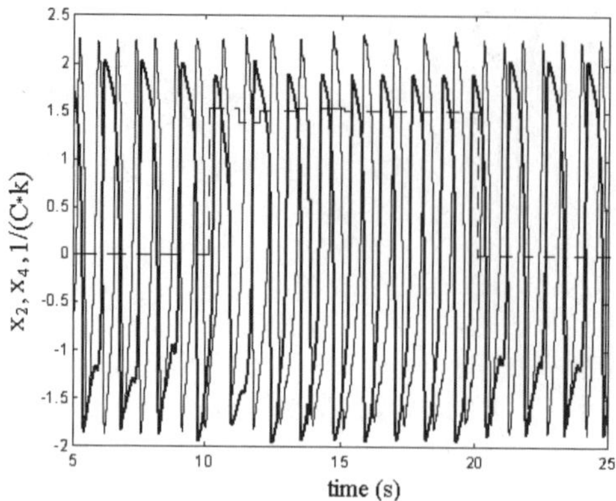

Fig.5. Discrete control. $C_{SA} = 0.15$ (2:1 AV block), $\tau = 0.1$, $k = 8$. The controller is turned on at time 5 and turned off at time 20. The thin line is for the SA node, the thick line is for the AV node, the dotted line is the control signal $(1/C)$ scaled by $1/k$ for plotting purposes (the control signal appears continuous in the plot due to the small τ used).

First, it can be applied directly without any knowledge of the model (even though it is derived from the model). This is an advantage as compared with some methods presented in [7, 8]. Second, the algorithm is quite simple. One needs only to detect the value of the AV action potential x_4 at constant time intervals τ. The control gain k can be chosen over a large range with no essential performance difference. Finally, only one parameter k needs to be adjusted, which is easier as compared with the OGY method. In simulations, the discrete control is realized by altering the action potential of the AV node from x_4 to $x_4{}'$ at every time interval τ. In an actual pacing system discrete electrical stimuli would be applied through an electrode placed near the AV node so as to alter its action potential from x_4 to $x_4{}'$ at every time interval τ. Figure 5 shows both the uncontrolled and controlled 2:1 AV block with $C_{SA} = 0.15$. The controller is turned on at time 5 and turned off at time 20, with $\tau = 0.1$ and $k = 8$.

4 Conclusions

We have presented a simple control method for stabilizing some pathological behaviors of a nonlinear heartbeat model with and without random noise. Comparison of this control method with others (particularly the OGY method) has been discussed. The continuous controller was discretized to demonstrate that it is algorithmically realizable within a digital pacemaking device design by using an external control capacitor C that discharges its energy through a stimulating electrode. This electrode would normally be placed near the AV node in order to speed up its rate to keep pace with the faster SA node. The control algorithm should be able to be implemented in an actual pacing environment at least in the same manner as those of Garfinkel *et al.* [1], Hall *et al.* [13], and Christini *et al.* [14]. As mentioned in the introduction, the control algorithm proposed here might eliminate the need for a timing reel thereby rendering it magnet-safe. This would allow the pacemaker to be used when patients must undergo an MRI scan. The present algorithm does not address the conditions of sinus arrest/prolonged sinus pause and complete atrioventricular block without escape rhythm. However, these problems may be correctable using alternative or adjunctive algorithmic approaches which are beyond the scope of the present chapter.

References

1. Garfinkel, A., Spano, M. L., Ditto, W. L., Weiss, J. N. (1992) Controlling cardiac chaos. Science, **257**:1230–1235
2. Ott, E., Grebogi, C., Yorke, J. A. (1990) Controlling chaos. Phys. Rev. Lett., **64**:1196–1199
3. Van Der Pol M. (1928) The heart-beat considered as a relaxation oscillation and an electrical model of the heart. Phil. Mag., **6**:763–775
4. West, B. J., Goldberger, A. L. (1985) Nonlinear dynamics of the heartbeat. The AV junction: Passive conduit or active oscillator? Physica D, **7**:198–206
5. Signorini, M. G., Cerutti, S. (1998) Simulation of Heartbeat Dynamics: A Nonlinear Model. Int. J. Bifur. Chaos, **8**:1725–1731
6. di Bernardo D., Signorini, M. G., Cerutti, S. (1998) A model of two nonlinear coupled oscillators for the study of heartbeat dynamics. Int. J. Bifur. Chaos, **8**:1975–1985
7. Brandt, M. E., Chen, G. (1996) Feedback control of a quadratic map model of cardiac chaos. Int. J. Bifur. Chaos, **6**:715–723
8. Glass, L., Zeng, W. (1994) Bifurcations in flat-topped maps and the control of cardiac chaos, Int. J. Bifur. Chaos, **4**:1061–1067
9. Brandt, M. E., Chen, G. (1996) Controlling the dynamical behavior of a circle map model of the human heart. Biol. Cybern., **74**:1–8
10. Brandt, M. E., Shih, H. T., Chen, G. (1997) Linear time-delay feedback control of a pathological rhythm in a cardiac conduction model. Phys. Rev. E, **56**:R1334–1337
11. Brandt, M. E., Chen, G. (1997) Bifurcation control of two nonlinear models of cardiac activity. IEEE Trans. Circ. Syst.-I, **44**:1–4
12. Brandt, M. E., Chen, G. (2000) Delay feedback control of cardiac activity models. Controlling Chaos and Bifurcations in Engineering Systems, Chen, G. (ed.), Boca Raton, FL: CRC Press, 325–345
13. Hall, K., Christini, D. J., Tremblay, M., Collins, J. J., Glass, L., Billette, J. (1997) Dynamic control of cardiac alternans. Phys. Rev. Lett., **78**:4518–4521
14. Christini, D. J., Stein, K. M., Markowitz, S. M., *et al.* (2001) Nonlinear-dynamical arrhythmia control in humans. Proc. Nat. Acad. Sci., **98**:5827–5832

Local Robustness of Bifurcation Stabilization with Applications to Jet Engine Control

Xiang Chen[1], Ali Tahmasebi[1], and Guoxiang Gu[2]

[1] Department of Electrical and Computer Engineering
University of Windsor, Windsor, Ontario, Canada N9B 3P4
xchen@uwindsor.ca

[2] Department of Electrical and Computer Engineering
Louisiana State University, Baton Rouge, LA 70803, USA

Abstract. Local robustness of bifurcation stabilization is studied for parameterized nonlinear systems of which the linearized system possesses either a simple zero eigenvalue or a pair of imaginary eigenvalues and the bifurcated solution is unstable at the critical value of the parameter. It is assumed that the unstable mode corresponding to the critical eigenvalue of the linearized system is not linearly controllable by the feedback control, nor linearly affected by the uncertainty signal. Computable conditions are derived to characterize the admissible uncertainty sets for systems with pitchfork, transcritical and Hopf bifurcations. The result for stationary bifurcation is applied to analyzing the robustness of several static stabilizing control laws for axial-flow compressors based on the approximated third-order Moore-Greitzer model.

1 Introduction

Bifurcation stability and stabilization have been widely investigated by researchers in nonlinear control [1,2,4,6,9]. Yet the robustness issue is not so well addressed. In general, engineers need to know *a priori* not only how to design controllers but also how robust the controllers could be, or how much uncertainties could be tolerated. Indeed, the robustness of a controller is very important since any system will be, more or less, subject to uncertainties, either from the modelling process or from external disturbances. Theoretically, there are two approaches to obtain the robustness information of a controller: 1. design a robust controller using the existing robust control methods (for example, the $\mathcal{H}_\infty$ approach for linear systems or the $\mathcal{L}_2$ gain approach for nonlinear systems); 2. design a controller first and then characterize the admissible uncertainty set. The first method is popular because, once the controller is synthesized, the uncertainty set can be characterized in certain form, usually in norm bounds, while the second method is problem dependent, and much less successful in practice. For nonlinear systems, unlike linear ones, direct robust control design is not trivial, and it becomes harder for those involving various bifurcations when the critical modes are not linearly controllable [9]. The difficulties lie in two aspects: first, it is difficult to characterize all the stabilizing controllers – we usually know only a subset of

G. Chen, D.J. Hill, and X. Yu (Eds.): Bifurcation Control, LNCIS 293, pp. 275–300, 2003.

them even for some special nonlinear systems; second, the $\mathcal{L}_2$ gain approach [21] that renders it possible to characterize uncertainties in norm-bounded form yields (robust) stabilizing conditions that require Lyapunov functions. Finding such functions is usually very difficult, because the solutions may not be computable. On the other hand, since nonlinear systems are much more complicated than linear systems, a useful solution normally is the one that takes into account more special details of the underlying nonlinear system.

The theory presented in this chapter finds its application in jet engine control. An axial-flow compressor is the vital part of a turbine-based jet engine. It has been shown that the engine performance could be significantly reduced by rotating stall and surge, which are unsteady aerodynamic instabilities. Modelling post-stall/surge dynamics turns out to be a very difficult job and has been studied in [18,19]. The bifurcation analysis approach has been employed to address the compressor instability in [17] and [3], based on the Moore-Greitzer model from [18]. The original model is a set of PDEs which makes it very difficult to conduct control design based on it. Therefore in terms of rotating stall control using the bifurcation approach, the employed model is usually the finite-dimensional Galerkin projection of the original PDE model. This model is further reduced to a single-mode third order ODE model [17,18], which captures well the rotating stall phenomenon and has been used in various rotating stall control designs using the bifurcation approach as reported in [6,9,11,15,24]. Although the use of the third order model is subjective, the results developed based on this model are instructive, and provide useful and insightful knowledge for understanding the rotating stall control problem. It is interesting to point out that, in [10], a feedback control law based on the third order model is shown to be actually effective for the multi-mode Moore-Greitzer model. It is also noted that the recent work reported in [20] is directly for the multi-mode model. For non-bifurcation approaches tackling rotating stall and surge, it is worth mentioning [12] for more details.

In this chapter, we aim to characterize the robustness of a class of bifurcation stabilizing controllers as obtained in [9]. It is pointed out that, for the specified system, the stabilization conditions obtained in [9] actually give all static feedback stabilizing controllers, and hence it does make sense to examine the robust stability of these controllers. The idea is thus to focus on the stabilizing controllers and characterize the admissible set of uncertainties that can be tolerated by these controllers. For uncertainties, we will assume that it is in the form of smooth state feedback which can be expanded into the Taylor series. We will also present the robustness result of the rotating stall controls developed in [6,9] for axial flow compressors as an application of the robustness result of stationary bifurcation stabilization.

2 Assumptions and Definitions

Consider a nominal parameterized system F subject to some uncertainty Δ as shown in Fig.1, where C is a feedback controller which is designed to stabilize the nominal system F.

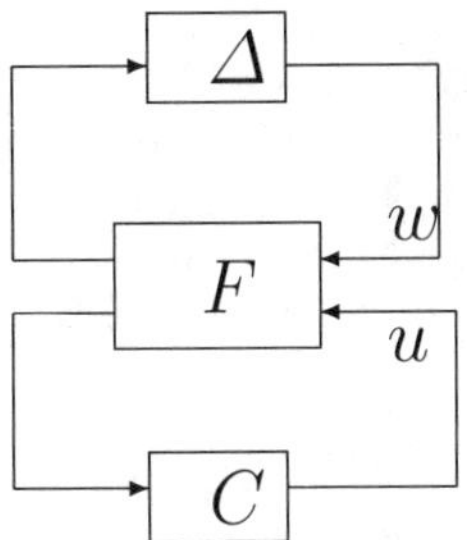

Fig.1. Nonlinear Control System in Robust Consideration

This nonlinear system with uncertainty is governed by the following dynamic equation:

$$\dot{x} = f(\gamma, x) + g(x)u + h(x)w, \tag{1}$$

where γ is a scalar parameter, u is a scalar control, w is a scalar disturbance signal, and

$$f(\cdot,\cdot) : \mathbf{R}^1 \times \mathbf{R}^n \to \mathbf{R}^n, \quad g(\cdot) : \mathbf{R}^n \to \mathbf{R}^n,$$

$$h(\cdot) : \mathbf{R}^n \to \mathbf{R}^n,$$

are all smooth functions with $f(\gamma, 0) = 0$.

This system is assumed to satisfy the following assumption (S) or (H):

(S)

(i) $L(\gamma) = \left.\dfrac{df(\gamma, x)}{dx}\right|_{x=0}$ has a simple eigenvalue (critical mode) $\lambda(\gamma)$ satisfying:

$$\lambda(0) = 0, \quad \lambda'(0) = \left.\frac{d\lambda}{d\gamma}\right|_{\gamma=0} \neq 0,$$

while all other eigenvalues are stable in a neighborhood of $\gamma = 0$.

(ii) The critical mode of $L(0)$ is not linearly controllable by u, nor affected by w, i.e., for all ℓ, $\ell L(0) = 0$, we have $\ell g(0) = 0$, and $\ell h(0) = 0$.

(H)

(i) $L(\gamma) = \left.\frac{df(\gamma,x)}{dx}\right|_{x=0}$ possesses a pair of complex eigenvalues $\lambda(\gamma) = \alpha(\gamma) + j\beta(\gamma)$, $\bar{\lambda}(\gamma) = \alpha(\gamma) - j\beta(\gamma)$, which are called 'critical modes', satisfying:

$$\alpha(0) = 0, \quad \beta(0) = \omega_c \neq 0, \quad \alpha'(0) = \frac{d\alpha}{d\gamma}(0) \neq 0,$$

while all other eigenvalues are stable in a neighborhood of $\gamma = 0$.

(ii) This pair of critical modes are not linearly controllable by u in the sense that, for all $\ell \neq 0$, $\ell L(0) = 0$ and $\bar{\ell}L(0) = 0$, we have $\ell g(0) = 0$, $\bar{\ell}g(0) = 0$.

(iii) This pair of critical modes are not linearly affected by w in the sense that, for all $\ell \neq 0$ with $\ell L(0) = 0$ and $\bar{\ell}L(0) = 0$, we have $\ell h(0) = 0$ and $\bar{\ell}h(0) = 0$.

The assumption (S) corresponds to the stationary bifurcations and the assumption (H) to the Hopf bifurcation. It is noted that, for both bifurcations, the critical mode of the linearized system is not only uncontrollable by u but also not linearly affected by w. If w can move the critical mode, for example, if the critical mode can be shifted into the right-half complex plane by some disturbance w at $\gamma = 0$, then the perturbed system will no longer be stable and hence will no longer be a critical system. Actually, in this case, since F is not linearly controllable by u, it is not feasible to find any control action to stabilize the whole system [5]. Therefore, we only consider the case that w does not affect the critical mode.

To accommodate our presentation, we provide definitions of *bifurcation stability, bifurcation stabilizing control* and *admissible set*, respectively, as follows.

Definition 1. Consider a parameterized autonomous system

$$\dot{x} = f(\gamma, x), \quad f(\gamma, 0) = 0, \tag{2}$$

where $f(\gamma, x)$ satisfies the assumption (S) (or (H)). This system is said to have local bifurcation stability if the bifurcated solution $x_e \neq 0 : f(\gamma, x_e) = 0$ (or the bifurcated periodic solution of $f(\gamma, x_e) = 0$) is locally asymptotically stable in a sufficiently small neighborhood of $\gamma = 0$.

Definition 2. Consider the nominal parameterized control system

$$\dot{x} = f(\gamma, x) + g(x)u, \quad f(\gamma, 0) = 0, \tag{3}$$

where $f(\gamma, x)$ satisfies either the assumption (S) or (H). A feedback $u = C(x), C(0) = 0$, is called a bifurcation stabilizing control law if the closed-loop system $\dot{x} = f(\gamma, x) + g(x)C(x)$ possesses bifurcation stability.

For the uncertainty, we restrict w to be smooth state feedbacks which possess Taylor series expansions. The admissible set of w is given in the next definition.

Definition 3. Consider the system (1). Let u be a stabilizing control. The following set U

$$U = \{w : \ w = \Delta(x) = P_1 x + P_2[x, x] + P_3[x, x, x] + \cdots\}, \tag{4}$$

is called the local admissible set of u if $\forall w \in U$ the system (1) has bifurcation stability, where x is the state vector of the system (1) and $P_1 x + P_2[x, x] + P_3[x, x, x] + \cdots$ is the Taylor series expansion of $w = \Delta(x)$ around $x = 0$.

Therefore, the local robustness problem for bifurcation stabilization can be stated as: *given a u which is a bifurcation stabilizing control for the nonlinear system (3), find the admissible set U.*

3 Projection Method and Bifurcation Stabilization

In this section, the projection method, as developed in [13] and advocated in [2], is presented, followed by the bifurcation stabilization theorems [9].

3.1 Projection method

Consider the nonlinear system (2) which satisfies the assumption (S) or (H). Let S_e be the eigen-space spanned by the critical eigenvector(s). In [13], it is shown that, for both stationary bifurcations and Hopf bifurcation, the stability of (2) can be concluded by analyzing the projected dynamics in S_e. The results are stated as the so-called projection method.

Stationary bifurcation: Let ℓ and r denote the left row and right column eigenvectors of $L(0)$, corresponding to the critical eigenvalue $\lambda(0) = 0$. Then $\ell r = 1$ by suitable normalization. Denote $\varepsilon = \ell x_e$, where $x_e(\gamma) \neq 0$ satisfying $f(\gamma, x_e) = 0, x_e(0) = 0$, is a bifurcated solution in a sufficiently small neighborhood of $\gamma = 0$. Then there exists a series expansion

$$\begin{bmatrix} x_e(\varepsilon) \\ \gamma(\varepsilon) \end{bmatrix} = \sum_{k=1}^{\infty} \begin{bmatrix} x_{ek} \\ \gamma_k \end{bmatrix} \varepsilon^k.$$

Since $f(\gamma, x)$ is sufficiently smooth, there exists a Taylor expansion near the origin of $\mathbf{R}^n$ of the form

$$\dot{x} = f(\gamma, x) = L(\gamma)x + Q(\gamma)[x, x] + C(\gamma)[x, x, x] + \cdots \tag{5}$$

Note that $L(\gamma)x$, $Q(\gamma)[x, x]$, and $C(\gamma)[x, x, x]$ can each be expanded into

$$L(\gamma)x = L_0 x + \gamma L_1 x + \gamma^2 L_2 x + \cdots, Q(\gamma)[x, x] = Q_0[x, x] + \gamma Q_1(x, x) + \cdots,$$

and $C(\gamma)[x, x, x] = C_0[x, x, x] + \gamma C_1[x, x, x] + \cdots$, where L_0, L_1, and L_2 are $n \times n$ constant matrices.

Let $\tilde{\lambda}(\gamma)$ be the critical eigenvalue of the linearized system matrix at the new (bifurcated) equilibrium close to the origin. Then $\tilde{\lambda}(0) = \lambda(0) = 0$ at $\gamma = 0$. There also exists a series expansion [13]

$$\tilde{\lambda}(\varepsilon) = \sum_{i=1}^{\infty} \tilde{\lambda}_i \varepsilon^i = \tilde{\lambda}_1 \varepsilon + \tilde{\lambda}_2 \varepsilon^2 + \cdots .$$

The computation of the first two coefficients of $\tilde{\lambda}$ can proceed as follows [1,13]:

- Step 1: Calculate $\lambda'(0) = \ell L_1 r$ where λ is a function of γ.
- Step 2: Set $x_{e1} = r$, and calculate $\gamma_1 = -\ell Q_0[r,r]/\lambda'(0)$.
- Step 3: Compute x_{e2} from equations $\ell x_{e2} = 0$ and $L_0 x_{e2} = -Q_0[r,r]$, and γ_2 from

$$\gamma_2 = -\frac{1}{\lambda'(0)} \left(\gamma_1 \ell L_1 x_{e2} + \gamma_1^2 \ell L_2 r + 2\ell Q_0[r, x_{e2}] + \gamma_1 \ell Q_1[r,r] + \ell C_0[r,r,r] \right).$$

- Step 4: Set $\tilde{\lambda}_1 = -\gamma_1 \lambda'(0)$ and $\tilde{\lambda}_2 = -2\gamma_2 \lambda'(0)$.

Theorem 1. *Suppose that all eigenvalues of L_0 are stable, except one critical eigenvalue. For the case $\gamma_1 \neq 0$, the branch of the bifurcated equilibrium solution is locally stable for γ sufficiently close to 0 if $\ell Q_0[r,r]\varepsilon < 0$, and unstable if $\ell Q_0[r,r]\varepsilon > 0$. For the case $\gamma_1 = 0$, the bifurcated solution is locally stable for γ sufficiently close to 0 if $\tilde{\lambda}_2 < 0$, and unstable if $\tilde{\lambda}_2 > 0$, where*

$$\tilde{\lambda}_2 = 2\ell \left(2Q_0[r, x_{e2}] + C_0[r,r,r] \right), \quad x_{e2} = -(\ell^T \ell + L_0^T L_0)^{-1} L_0^T Q_0[r,r].$$

It should be clear that the local bifurcation for the case $\gamma_1 \neq 0$ is transcritical. Thus the branch of the bifurcated solution at $\varepsilon > 0$ has the opposite stability property as the one at $\varepsilon < 0$. The local bifurcation for the case $\gamma_1 = 0$, and $\gamma_2 \neq 0$ is pitchfork, where both branches of the bifurcated solution share the same stability property.

Hopf bifurcation: The Hopf Bifurcation Theorem asserts the existence of a one-parameter family $\{p_\varepsilon\}$, where $0 < \varepsilon \leq \varepsilon_0$, of non-constant periodic solutions of (2) emerging from the zero solution at $\gamma = 0$. ε is a measure of the amplitude of the periodic solution and ε_0 is sufficiently small. The periodic solution $p_\varepsilon(t)$ has a period near $2\pi/\omega_c$ and occurs for the parameter value $\gamma(\varepsilon)$. Exactly one of the characteristic exponents of p_ε is near zero, and is given by

$$\tilde{\lambda}(\varepsilon) = \tilde{\lambda}_2 \varepsilon^2 + \tilde{\lambda}_4 \varepsilon^4 + \cdots = \sum_{i=1}^{\infty} \tilde{\lambda}_{2i} \varepsilon^{2i}. \tag{6}$$

Let the Taylor series expansion of $f(\gamma, x)$ be of the form in (5), where $L_0 = L(0)$. An algorithm to compute $\tilde{\lambda}_2$ is quoted from [1].

- Step 1: Compute left row eigenvector ℓ and right column eigenvector r of L_0 corresponding to the critical eigenvalue of $\lambda(0) = j\omega_c$, normalized as $\ell r = 1$.
- Step 2: Solve column vectors μ and ν from the equations

$$-L_0\mu = \frac{1}{2}Q_0[r, \bar{r}], \quad (2j\omega_c I - L_0)\nu = \frac{1}{2}Q_0[r, r], \quad j = \sqrt{-1}.$$

- Step 3: The coefficient $\tilde{\lambda}_2$ is given by

$$\tilde{\lambda}_2 = 2\mathrm{Re}\left\{2\ell Q_0[r, \mu] + \ell Q_0[\bar{r}, \nu] + \frac{3}{4}\ell C_0[r, r, \bar{r}]\right\}.$$

Local stability of a Hopf bifurcation is given by the following lemma [1].

Theorem 2. *Suppose all eigenvalues of L_0 are stable in a neighborhood of $\gamma = 0$ except for the critical pair of complex eigenvalues. Then the Hopf bifurcation is stable if $\tilde{\lambda}_2 < 0$ and unstable if $\tilde{\lambda}_2 > 0$.*

3.2 Bifurcation stabilization

We shall only consider the smooth state feedback control. Suppose that a smooth state feedback control $u = K(x) = K_1 x + K_Q[x, x] + K_C[x, x, x] + \cdots, \ K(0) = 0$ is applied to the nominal control system (3). Let the Taylor series expansion of (3) be

$$\dot{x} = L_0 x + \gamma L_1 x + \gamma^2 L_2 x + B_1 u + u\tilde{L}_1 x + Q_0[x, x] + u\tilde{L}_{12}[x, x] + B_2 u^2$$
$$+\gamma Q_1[x, x] + C_0[x, x, x] + \cdots.$$

The bifurcation stabilizing control laws are given in the following theorems (see [9] for proofs).

Theorem 3. *Consider the nonlinear system (3) satisfying the assumption (S). For the case $\tilde{\lambda}_1 > 0$, i.e., $\ell Q_0[r, r] > 0$, there exists a nonlinear state feedback control law $u = K(x)$ that stabilizes the given branch of the bifurcated solution at $\varepsilon > 0$ if and only if there exists a linear state feedback control law that stabilizes the given branch of the bifurcated solution. Moreover there exists a nonsingular matrix $T \in \mathbf{R}^{n\times n}$ such that*

$$TL_0T^{-1} = \begin{bmatrix} L_{00} & 0 \\ 0 & 0 \end{bmatrix}, \quad TB_1 = \begin{bmatrix} B_{11} \\ 0 \end{bmatrix}. \tag{7}$$

Let ℓ_i be the ith element of ℓ, and $r^T Q_{0k} r$ be the kth element of $Q_0[r, r]$ with $Q_{0k} = Q_{0k}^T$ for $k = 1, 2, \cdots, n$. Partition

$$(T^{-1})^T \left(\sum_{k=1}^{n} \ell_k Q_{0k}\right) T^{-1} = \begin{bmatrix} \tilde{Q}_{00} & \tilde{Q}_{01} \\ \tilde{Q}_{10} & \tilde{Q}_{11} \end{bmatrix},$$

where $\tilde{Q}_{10} = \tilde{Q}_{01}^T$ *and* $\tilde{Q}_{00} \in \mathbf{R}^{(n-1)\times(n-1)}$. *Then* $\tilde{Q}_{11} = \tilde{\lambda}_1 > 0$ *and the existence of a stabilizing state feedback control law is equivalent to the existence of a* $K_1 = \begin{bmatrix} K_{11} \; K_{12} \end{bmatrix} T \neq 0$ *such that*

$$\text{(i)} \quad (1 + K_{11}L_{00}^{-1}B_{11}) \left(\tilde{\lambda}_1(1 + K_{11}L_{00}^{-1}B_{11}) + aK_{12}\right) + bK_{12}^2 < 0,$$

$$\text{(ii)} \quad \lambda'(0)\left(\lambda'(0) + (K_{11}\lambda'(0) - K_{12}d)L_{00}^{-1}B_{11}\right) > 0, \quad \text{and}$$

$$\text{(iii)} \quad L_{00}^* = L_{00} + B_{11}K_{11} \quad \text{is stable.}$$

where $a = \tilde{d}_0 - 2\tilde{Q}_{10}L_{00}^{-1}B_{11}$, $b = \ell B_2 + \left((L_{00}^{-1}B_{11})^T\tilde{Q}_{00} - \tilde{d}\right) L_{00}^{-1}B_{11}$, $\begin{bmatrix} \tilde{d} \; \tilde{d}_0 \end{bmatrix} = \ell\tilde{L}_1T^{-1}$ *with* $\tilde{d}_0$ *a scalar,* $d^T = \begin{bmatrix} I_{n-1} \; 0 \end{bmatrix} (\ell L_1 T^{-1})^T$, *and* $\lambda(\gamma)$ *is the critical eigenvalue.*

Theorem 4. *Consider the nonlinear system (3) satisfying the assumption (S). For the case* $\tilde{\lambda}_1 = \ell Q_0[r, r] = 0$, *there exists a feedback control law* $u = K_1x + K_2[x, x] + K_3[x, x, x] + \cdots$, *where* $K_1r = 0$; K_1x , $K_2[x, x]$ *and* $K_3[x, x, x]$ *are of linear, quadratic and cubic terms. This control law stabilizes the bifurcated solution if and only if:*

$$\tilde{\lambda}_2^* = \tilde{\lambda}_2 + K_{11}L_{00}^{-1}(B_{11}\tilde{\lambda}_2 + \rho\beta) - \rho K_2[r, r] < 0 \tag{8}$$

and that $L_{00} + B_{11}K_{11}$ *is stable for some* $K_{11} \neq 0$,
where:

$$\tilde{\lambda}_2 = 2\ell(2Q_0[r, x_{e2}] + C_0[r, r, r]), \quad x_{e2} = -(\ell^T\ell + L_0^TL_0)^{-1}L_0^TQ_0[r, r],$$

$$\rho = 4\ell Q_0[r, (\ell^T\ell + L_0^TL_0)^{-1}L_0^TB_1] - 2\ell\tilde{L}_1r, \quad \beta = \begin{bmatrix} I_{n-1} \; 0 \end{bmatrix} TQ_0[r, r],$$

and T *is a nonsingular matrix such that:*

$$TL_0T^{-1} = \begin{bmatrix} L_{00} & 0 \\ 0 & 0 \end{bmatrix}, \quad TB_1 = \begin{bmatrix} B_{11} \\ 0 \end{bmatrix}, \quad \begin{bmatrix} K_1 \\ \ell \end{bmatrix} T^{-1} = \begin{bmatrix} K_{11} & 0 \\ 0 & 1 \end{bmatrix}.$$

Note that, in this case, the right column eigenvector r' of $L_0' = L_0 + B_1K_1$ corresponding to the critical eigenvalue can be calculated as:

$$r' = T^{-1}\begin{bmatrix} \eta \\ 1 \end{bmatrix}, \quad \eta = -\frac{L_{00}^{-1}B_{11}K_{12}}{1 + K_{11}L_{00}^{-1}B_{11}},$$

while the left raw eigenvector ℓ remains unchanged and $\ell r' = 1$.

Theorem 5. *Consider the nonlinear system (3) satisfying the assumption (H). Let* $\ell(\bar{\ell})$ *and* $r(\bar{r})$ *be the left row and right column vectors of* L_0 *corresponding to the complex critical eigenvalues and* μ, ν *be the vectors solved from* $-L_0\mu = \frac{1}{2}Q_0[r, \bar{r}]$, $(2j\omega_cI - L_0)\nu = \frac{1}{2}Q_0[r, r]$. *Then there exists a feedback control law* $u = K_2[x, x] + K_3[x, x, x] + \cdots$ *if and only if:*

$$\tilde{\lambda}_2^* = \tilde{\lambda}_2 - \mathrm{Re}\{K_2[r, \bar{r}]\ell(2Q_0[r, L_0^{-1}B_1] - \tilde{L}_1r) +$$

$$+K_2[r,r]\ell(Q_0[\bar{r},(2j\omega_c I - L_0)^{-1}B_1] - 0.5\tilde{L}_1\bar{r})\} < 0,$$

where

$$\tilde{\lambda}_2 = 2Re\left\{2\ell Q_0[r,\mu] + \ell Q_0[\bar{r},\nu] + \frac{3}{4}\ell C_0[r,r,\bar{r}]\right\}.$$

Theorem 6. *Consider the nonlinear system (3) satisfying the assumption (H). Let $\ell(\bar{\ell})$ and $r(\bar{r})$ be the left row and right column vectors of L_0 corresponding to the complex critical eigenvalues. Then there exists a stabilizing controller $u = K_1 x$, if $K_1 r = K_1 \bar{r} = 0$ and*

$$\tilde{\lambda}_2^* = \tilde{\lambda}_2 + \mathrm{Re}(\Theta + \Phi) < 0,$$

where

$$\Theta = \frac{K_1 L_0^{-1} Q_0[r,\bar{r}]\theta}{1 + K_1 L_0^{-1} B_1}, \quad \Phi = \frac{K_1 (L_0 - 2j\omega_c I)^{-1} Q_0[r,r]\phi}{1 + K_1 (L_0 - 2j\omega_c I)^{-1} B_1},$$

$$\theta = -\left(\ell\tilde{L}_1 r - 2\ell Q_0[r, L_0^{-1} B_1]\right),$$

$$\phi = \frac{1}{2}\left(\ell\tilde{L}_1\bar{r} - 2\ell Q_0[\bar{r},(L_0 - 2j\omega_c I)^{-1}B_1]\right).$$

$$\tilde{\lambda}_2 = 2Re\left\{2\ell Q_0[r,\mu] + \ell Q_0[\bar{r},\nu] + \frac{3}{4}\ell C_0[r,r,\bar{r}]\right\}.$$

and μ, ν are the vectors solved from

$$-L_0\mu = \frac{1}{2}Q_0[r,\bar{r}], \ \ (2j\omega_c I - L_0)\nu = \frac{1}{2}Q_0[r,r].$$

4 Local Robustness of Bifurcation Stabilization

As formulated in Section 1, the robustness problem for bifurcation stabilization requires us to quantify the admissible set U of uncertainty signal w, which is assumed to take the form of smooth state feedback $w = \Delta(x) = P_1 x + P_2[x,x] + P_3[x,x,x] + \cdots$, given a stabilizing control law for the nominal system (3). Therefore, the characterization of U can be done by determination of all coefficients in the Taylor series expansion. To derive our results, we need the Taylor series expansion for the system (1):

$$\begin{aligned}\dot{x} = {} & L_0 x + \gamma L_1 x + \gamma^2 L_2 x + B_1 u + h_0 w + u\tilde{L}_1 x + w h_1 x + Q_0[x,x] \\ & + u\tilde{L}_2[x,x] + w h_2[x,x] + \gamma Q_1[x,x] + C_0[x,x,x] + \cdots . \end{aligned} \tag{9}$$

The following theorems provide robustness results for both stationary and Hopf bifurcations (see [7,8]).

4.1 Pitchfork bifurcation

Suppose the system (1) yields a pitchfork bifurcation and (3) is its nominal control system. So the assumption (S) is satisfied and $\tilde{\lambda}_1 = 0$. Then according to Theorem 4, a controller $u = K_1x + K_2[x,x]$ with $K_1r = 0$, can be obtained to stabilize the nominal system (3). Substituting u into the system equation (1), we obtain

$$\dot{x} = f(\gamma, x) + g(x)(K_1x + K_2[x,x]) + h(x)w = f'(\gamma, x) + h(x)w,$$

where $f'(\gamma,x) = f(\gamma,x) + g(x)(K_1x + K_2[x,x])$. In this case the Taylor series expansion becomes:

$$\begin{aligned} \dot{x} = &L'_0x + \gamma L'_1x + \gamma^2 L'_2x + h_0w + wh_1x + Q'_0[x,x] \\ &+wh_2[x,x] + \gamma Q'_1[x,x] + C'_0[x,x,x] + \cdots, \end{aligned} \tag{10}$$

with

$$L'_0 = L_0 + B_1K_1, \quad L'_1 = L_1, \quad L'_2 = L_2, \quad Q'_1[x,x] = Q_1[x,x],$$

$$Q'_0[x,x] = Q_0[x,x] + B_1K_2[x,x] + K_1x\tilde{L}_1x,$$

$$C'_0[x,x,x] = C_0[x,x,x] + \tilde{L}_1xK_2[x,x] + K_1x\tilde{L}_2[x,x].$$

The following lemma is needed.

Lemma 1. *For the system (1) satisfying the assumption (S), let ℓ and r be the left row and right column eigenvectors of L_0 corresponding to the critical eigenvalue such that $\ell r = 1$. Then there exists a nonsingular matrix T' such that*

$$T'(L_0 + B_1K_1)T'^{-1} = \begin{bmatrix} L'_{00} & 0 \\ 0 & 0 \end{bmatrix}, \quad T'h_0 = \begin{bmatrix} h_{01} \\ 0 \end{bmatrix}, \quad \ell T'^{-1} = \begin{bmatrix} 0 & 1 \end{bmatrix}.$$

Proof: Note that $B_1 = g(0)$, $\ell(L_0 + B_1K_1) = \ell L_0 = 0$, $(L_0 + B_1K_1)r = 0$. The proof of this Lemma can be done using Kalman decomposition (see [14]). ∎

Now we consider the uncertainties in U. It is noted that, although the control law $u = K_1x + K_2[x,x]$ does not change the pitchfork bifurcation, the uncertainty w may change the pitchfork bifurcation into transcritical bifurcation if the linear term $P_1r \neq 0$. If this is the case, only one branch of the bifurcated solution can be robustly stabilized instead of all branches.

Theorem 7. *Suppose that the system (1) satisfies the assumption (S). Let $u = K_1x + K_2[x,x]$ be a stabilizing controller synthesized using Theorem 4. Let T' be a nonsingular matrix given in Lemma 1. Let l_i be the ith element of ℓ, $r^TQ_{0k}r$ be the kth element of $Q_0[r,r]$ and partition*

$$(T'^{-1})^T\left(\sum_{k=1}^{n} \ell_k Q_{0k}\right)T'^{-1} = \begin{bmatrix} q_{00} & q_{01} \\ q_{10} & q_{11} \end{bmatrix}, \quad q_{10} = q_{01}^T, \quad q_{00} \in \mathbf{R}^{(n-1)\times(n-1)}.$$

Then the controller can robustly stabilize the system for uncertainties $w = P_1x+P_2[x,x]+P_3[x,x,x]+\cdots$, *if and only if* $P_1 = \begin{bmatrix} P_{11} \ P_{12} \end{bmatrix} T' \neq 0$ *satisfies:*

(i) $\left(1 + P_{11}L_{00}'^{-1}h_{01}\right) aP_{12} + bP_{12}^2 < 0, \;\; P_1 r \neq 0$

(ii) $\frac{d\lambda'}{d\gamma}(0)\left(\frac{d\lambda'}{d\gamma}(0) + (P_{11}\frac{d\lambda'}{d\gamma}(0) - P_{12}d)L_{00}'^{-1}h_{01}\right) > 0,$

(iii) $L_{00}'^{*} = L_{00}' + h_{01}P_{11}$ is stable,

where:

$$a = \tilde{d}_0 - 2q_{10}L_{00}'^{-1}h_{01}, \;\; b = \left((L_{00}'^{-1}h_{01})^T q_{00} - \tilde{d}\right) L_{00}'^{-1}h_{01},$$

$\begin{bmatrix} \tilde{d} \ \tilde{d}_0 \end{bmatrix} = \ell h_1 T'^{-1}$, $\tilde{d}_0$ *is a scalar*, $d^T = \begin{bmatrix} I_{n-1} \ 0 \end{bmatrix} (\ell L_1 T'^{-1})^T$, $\lambda'(\gamma)$ *is the critical eigenvalue of* L_0' *and* $\frac{d\lambda'}{d\gamma}(0) = \ell L_1 r$.

Proof: Substituting $u = K_1x + K_2[x,x]$ into (1), we get :

$$\dot{x} = f(\gamma,x) + g(x)(K_1x + K_2[x,x]) + h(x)w = f'(\gamma,x) + h(x)w, \tag{11}$$

where $f'(\gamma,0) = f(\gamma,0) = 0$. Clearly $f'(\gamma,x)$ possesses the same critical properties as those of $f(\gamma,x)$ since $\ell B_1 = 0$ and :

$$\ell L_0' = \ell L_0 + \ell B_1 K_1 = 0, \quad L_0' r = L_0 r + B_1 K_1 r = 0.$$

Hence the critical eigenvalue of L_0 is also that of L_0' and the system

$$\dot{x} = f'(\gamma,x) + h(x)w$$

satisfies the assumption (S). On the other hand

$$Q_0'[r,r] = Q_0[r,r] + B_1K_2[r,r] + K_1 r\tilde{L}_1 r = Q_0[r,r] + B_1K_2[r,r].$$

So $\tilde{\lambda}_1' = \ell Q_0'[r,r] = \ell Q_0[r,r] + \ell B_1 K_2[r,r] = 0$. Now let $K_2[r,r] = r^T Q_{K_2} r$, $B_1 = [b_1,\cdots,b_n]^T$, $r^T Q_{0k}' r$ be the kth element of $Q_0'[r,r]$ and ℓ_i be the ith element of ℓ. Again, by $\ell B_1 = 0$, we have

$$(T'^{-1})^T\left(\sum_{k=1}^n \ell_k Q_{0k}'\right)T'^{-1} = (T'^{-1})^T\left(\sum_{k=1}^n \ell_k Q_{0k} + Q_{K_2}\sum_{k=1}^n \ell_k b_k\right)T'^{-1}$$

$$= (T'^{-1})^T\left(\sum_{k=1}^n \ell_k Q_{0k}\right)T'^{-1} = \begin{bmatrix} q_{00} & q_{01} \\ q_{10} & q_{11} \end{bmatrix}, \quad q_{10} = q_{01}^T, \;\; q_{00} \in \mathbf{R}^{(n-1)\times(n-1)}.$$

Applying Theorem 3 to the system

$$\dot{x} = f'(\gamma,x) + h(x)w,$$

we can see that the control law $u = K_1x + K_2[x,x]$ stabilizes this perturbed system for uncertainties $w = P_1x + P_2[x,x] + P_3[x,x,x] + \cdots$ if and only if $P_1 = \left[P_{11}\ P_{12}\right]T' \neq 0$ satisfying:

$$\left(1 + P_{11}L_{00}'^{-1}h_{01}\right)aP_{12} + bP_{12}^2 < 0, \quad P_1r \neq 0,$$

$$\frac{d\lambda'}{d\gamma}(0)\left(\frac{d\lambda'}{d\gamma}(0) + (P_{11}\frac{d\lambda'}{d\gamma}(0) - P_{12}d)L_{00}'^{-1}h_{01}\right) > 0,$$

and

$$L_{00}'^{*} = L_{00}' + h_{01}P_{11} \text{ is stable,}$$

where

$$a = \tilde{d}_0 - 2q_{10}L_{00}'^{-1}h_{01}, \quad \left[\tilde{d}\ \tilde{d}_0\right] = \ell h_1 T'^{-1}, \quad b = \left((L_{00}'^{-1}h_{01})^T q_{00} - \tilde{d}\right)L_{00}'^{-1}h_{01},$$

$\tilde{d}_0$ is a scalar, $d^T = \left[I_{n-1}\ 0\right](\ell L_1 T'^{-1})^T$, $\lambda'(\gamma)$ is the critical eigenvalue of L_0' and $\frac{d\lambda'}{d\gamma}(0) = \ell L_1 r$.

∎

It is noted that only the linear term P_1 determines the admissible uncertainty set.

If $P_1r = 0$, then the admissible set is different and is characterized in the next theorem.

Theorem 8. *Suppose that the system (1) satisfies the assumption (S). Let $u = K_1x + K_2[x,x]$ be a stabilizing controller synthesized using Theorem 4. Let T' be given as that in Lemma 1. Denote I_{n-1} as an $(n-1)\times(n-1)$ identity matrix, and*

$$\rho' = 4\ell Q_0'[r, (\ell^T\ell + L_0'^T L_0')^{-1}L_0'^T h_0] - 2\ell h_1 r, \quad \beta' = [I_{n-1}\ 0]T'Q_0'[r,r].$$

Then the controller can robustly stabilize the system for uncertainties $w = P_1x + P_2[x,x] + P_3[x,x,x] + \cdots$, $P_1r = 0$, if and only if

$$\tilde{\lambda}_2^* + P_{11}L_{00}'^{-1}(h_{01}\tilde{\lambda}_2^* + \rho'\beta') - \rho' P_2[r,r] < 0,$$

where $P_1 = \left[P_{11}\ P_{12}\right]T'$ and $\tilde{\lambda}_2^*$ is calculated as that in Theorem 4.

Proof: Because the procedures to prove this theorem are very similar to that of Theorem 7, we only give the outline of the proof.
Note that for the control system subject to uncertainty w

$$\dot{x} = f(\gamma,x) + g(x)u + h(x)w = f'(\gamma,x) + h(x)w,$$

where $u = K_1x + K_2(x,x)$ and $f'(\gamma,x) = f(\gamma,x) + g(x)(K_1x + K_2(x,x))$ are given in Theorem 4, $f'(\gamma,x)$ continues to satisfy the assumption (S) and the controller does not change ℓ and r of L_0 since the critical eigenvalue is neither controllable nor observable. If we consider only those uncertainties with $P_1 r = 0$, then we can treat the robust analysis problem as another stabilizing control problem for the pitchfork bifurcation when the critical mode is not observable. Thus, applying Theorem 4 and using Lemma 1, we obtain the result of this theorem. ∎

4.2 Transcritical bifurcation

Suppose that the system (1) satisfies the assumption (S) and yields a transcritical bifurcation, i.e., $\tilde{\lambda}_1 \neq 0$. Then Theorem 3 provides a class of linear stabilizing controllers. Let $u = Kx$ be such a stabilizing controller. Substituting it into the system, we get:

$$\dot{x} = f(\gamma, x) + g(x)Kx + h(x)w = f'(\gamma, x) + h(x)w, \tag{12}$$

where

$$f'(\gamma, x) = f(\gamma, x) + g(x)Kx, \quad f'(\gamma, 0) = 0.$$

In this case the Taylor series expansion becomes:

$$\begin{aligned} \dot{x} &= L_0'x + \gamma L_1'x + \gamma^2 L_2'x + h_0 w + wh_1 x + Q_0'[x, x] \\ &\quad + wh_2[x, x] + \gamma Q_1'[x, x] + C_0'[x, x, x] + \cdots, \end{aligned} \tag{13}$$

where

$$L_0' = L_0 + B_1 K, \quad L_1' = L_1, \quad L_2' = L_2, \quad Q_0'[x, x] = Q_0[x, x] + \tilde{L}_1 x K x,$$

$$Q_1'[x, x] = Q_1[x, x], \quad C_0'[x, x, x] = C_0[x, x, x] + \tilde{L}_2[x, x]Kx.$$

Let ℓ' and r' be the left row and right column eigenvectors of L_0' and ℓ and r be the ones of L_0, both corresponding to the critical eigenvalues and $\ell' r' = 1$, $\ell r = 1$. Then $\ell' = \ell$ and r' can be calculated as stated before. We introduce some matrices and constants as follow:

Let T be given as that in Theorem 3 and T' be a nonsingular matrix given in Lemma 1(well, replacing K_1 by K) such that:

$$T' L_0' T'^{-1} = \begin{bmatrix} L_{00}' & 0 \\ 0 & 0 \end{bmatrix}, \quad T' h_0 = \begin{bmatrix} h_{01} \\ 0 \end{bmatrix}, \quad \ell T'^{-1} = \begin{bmatrix} 0 \cdots 0 \; 1 \end{bmatrix}.$$

Let l_i be the ith element of ℓ, $r'^T Q_{0k}' r'$ be the kth element of $Q_0'[r', r']$ and $r^T Q_{0k} r$ be the kth element of $Q_0[r, r]$. Partition

$$(T'^{-1})^T \left(\sum_{k=1}^{n} \ell_k Q_{0k}' \right) T'^{-1} = \begin{bmatrix} q_{00}' & q_{01}' \\ q_{10}' & q_{11}' \end{bmatrix}, \quad q_{10}' = q_{01}'^T, \quad q_{00}' \in \mathbf{R}^{(n-1)\times(n-1)}.$$

The admissible set of uncertainties $w = P_1 x + P_2[x, x] + P_3[x, x, x] + \cdots$ is characterized in the next two theorems.

Theorem 9. *Suppose that the system (1) satisfies the assumption (S). Let* $u = Kx = \begin{bmatrix} K_{11} & K_{12} \end{bmatrix} Tx$ *be a stabilizing controller obtained from Theorem 3. Then this controller will robustly stabilize the system for all uncertainties* $w = P_1 x + P_2[x,x] + P_3[x,x,x] + \cdots,\ \ P_1 r' \neq 0$, *if and only if* $P_1 = \begin{bmatrix} P_{11} & P_{12} \end{bmatrix} T'$ *satisfies:*

1.

$$(1 + P_{11} L'^{-1}_{00} h_{01}) \left(\tilde{\lambda}'_1 (1 + P_{11} L'^{-1}_{00} h_{01}) + a' P_{12} \right) + b' P_{12}^2 < 0,\ P_1 r' \neq 0,$$

2.

$$\frac{d\lambda'}{d\gamma}(0) \left(\frac{d\lambda'}{d\gamma}(0) + (P_{11} \frac{d\lambda'}{d\gamma}(0) - P_{12} d') L'^{-1}_{00} h_{01} \right) > 0,$$

3.

$$L'^{*}_{00} = L'_{00} + h_{01} P_{11} \ \ \textit{is stable,}$$

where

$$\tilde{\lambda}'_1 = \ell Q_0[r,r] + a(1 + K_{11} L^{-1}_{00} B_{11})^{-1} K_{12} + b(K_{12})^2 (1 + K_{11} L^{-1}_{00} B_{11})^{-2} < 0,$$

$$a = \tilde{d}_0 - 2q_{10} L^{-1}_{00} B_{11},\ \begin{bmatrix} \tilde{d} & \tilde{d}_0 \end{bmatrix} = \ell \tilde{L}_1 T^{-1},\ b = \left((L^{-1}_{00} B_{11})^T q_{00} - \tilde{d} \right) L^{-1}_{00} B_{11},$$

$$a' = \tilde{d}'_0 - 2q'_{10} L'^{-1}_{00} h_{01},\ \begin{bmatrix} \tilde{d}' & \tilde{d}'_0 \end{bmatrix} = \ell h_1 T'^{-1},\ b' = \left((L'^{-1}_{00} h_{01})^T q'_{00} - \tilde{d}' \right) L'^{-1}_{00} h_{01},$$

$\tilde{d}'_0$ *is a scalar,* $d'^T = \begin{bmatrix} I_{n-1} & 0 \end{bmatrix} (\ell L_1 T'^{-1})^T$, $\lambda'(\gamma)$ *is the critical eigenvalue of* L'_0 *and* $\dfrac{d\lambda'}{d\gamma}(0) = \ell L_1 r'$.

Proof: Let $u = Kx = \begin{bmatrix} K_{11} & K_{12} \end{bmatrix} Tx$ be the stabilizing controller synthesized from Theorem 3. Substituting it into (1), we obtain (12). Note that $f'(\gamma, x)$ possesses all critical properties of $f(\gamma, x)$ and we can calculate $\tilde{\lambda}'_1$ for the controlled system as (see [9]):

$$\begin{aligned} \tilde{\lambda}'_1 = \ell Q'_0[r', r'] = \ell Q_0[r', r'] + \ell \tilde{L}_1 r' K r' = \ell Q_0[r,r] + \\ + a(1 + K_{11} L^{-1}_{00} B_{11})^{-1} K_{12} + b(K_{12})^2 (1 + K_{11} L^{-1}_{00} B_{11})^{-2} < 0, \end{aligned}$$

where

$$a = \tilde{d}_0 - 2q_{10} L^{-1}_{00} B_{11},\ \begin{bmatrix} \tilde{d} & \tilde{d}_0 \end{bmatrix} = \ell \tilde{L}_1 T^{-1},\ b = \left((L^{-1}_{00} B_{11})^T q_{00} - \tilde{d} \right) L^{-1}_{00} B_{11}.$$

We introduce T' as stated before and let $\lambda'(\gamma)$ be the critical eigenvalue of L'_0. The uncertainty characterization becomes a transcritical bifurcation stabilization problem. Therefore, by applying Theorem 3, the uncertainty set can be characterized as given. ∎

Theorem 10. *Suppose that all conditions in Theorem 9 are satisfied except that* $P_1 r' = 0$. *Let* $u = Kx = \begin{bmatrix} K_{11} & K_{12} \end{bmatrix} Tx$ *be a stabilizing controller obtained from Theorem 3. Then this controller will robustly stabilize the system for all uncertainties* $w = P_1 x + P_2[x, x] + P_3[x, x, x] + \cdots$ *with* $P_1 r' = 0$.

Proof: Note that, since $u = Kx = \begin{bmatrix} K_{11} & K_{12} \end{bmatrix} Tx$ is a stabilizing controller given in Theorem 3, the control system $\dot{x} = f(\gamma, x) + g(x)Kx$ is stable so $\tilde{\lambda}_1' = \ell Q_0'[r', r'] < 0$. For the perturbed system $\dot{x} = f(\gamma, x) + g(x)Kx + h(x)w = f'(\gamma, x) + h(x)w$, since $P_1 r' = 0$, by Theorem 3.1 in [9], it is claimed that the uncertainty $w(x)$ will not affect $\tilde{\lambda}_1'$. Therefore, the perturbed system will remain to be stable so $u = Kx$ robustly stabilize the system for the uncertainties

$$w = P_1 x + P_2[x, x] + P_3[x, x, x] + \cdots, \quad P_1 r' = 0.$$

∎

Clearly, if $P_1 = 0$ then the robust control problem is equivalent to stabilizing control problem. This suggest that the stabilizing control law designed by Theorem 3 can tolerate all uncertainties of higher terms in absence of the linear term.

4.3 Hopf bifurcation

Suppose that the nonlinear system (1) satisfies the assumption (H). That is, it yields a Hopf bifurcation. We can use a quadratic controller $u = K_2[x, x]$ to stabilize the system as that in Theorem 5 or a linear feedback $u = K_1 x, \ K_1 r = K_1 \bar{r} = 0$ as that in Theorem 6. To characterize the uncertainty $w = P_1 x + P_2[x, x] + P_3[x, x, x] + \cdots$, we need to classify it into two categories in terms of $P_1 = 0$ and $P_1 \neq 0$.

Let us consider $u = K_2[x, x]$ first. Substitute this controller into (1), we obtain

$$\dot{x} = f^*(\gamma, x) + h(x)w, \tag{14}$$

where $f^*(\gamma, x) = f(\gamma, x) + g(x)K_2[x, x]$. For uncertainty w with $P_1 \neq 0$, we have the following result.

Theorem 11. *For the system (1) satisfying the assumption* (H), *let the stabilizing controller* $u = K_2[x, x]$, $\tilde{\lambda}_2^*$, μ *and* ν *be given as that in Theorem 5. Then this controller will stabilize the system for any* $w = P_1 x + P_2[x, x] + \cdots$, *as long as* P_1 *satisfies the following conditions*

1. $P_1 r = P_1 \bar{r} = 0$,
2. $\tilde{\lambda}_2^* + \mathrm{Re}(\Theta^* + \Phi^*) < 0$, *where* $Q_0^*[x, x] = Q_0[x, x] + B_1 K_2[x, x]$,

$$\Theta^* = \frac{P_1 L_0^{-1} Q_0^*[r, \bar{r}]\theta^*}{1 + P_1 L_0^{-1} h_0}, \quad \Phi^* = \frac{P_1 (L_0 - 2j\omega_c I)^{-1} Q_0^*[r, r]\phi^*}{1 + P_1 (L_0 - 2j\omega_c I)^{-1} h_0},$$

$$\theta^* = -\left(\ell h_1 r - 2\ell Q_0^*[r, L_0^{-1}h_0]\right),$$

$$\phi^* = \frac{1}{2}\left(\ell h_1 \bar{r} - 2\ell Q_0^*[\bar{r}, (L_0 - 2j\omega_c I)^{-1}h_0]\right),$$

Proof: After substituting $u = K_2[x, x]$ into the system (1), it becomes:

$$\dot{x} = f^*(\gamma, x) + h(x)w,$$

where

$$f^*(\gamma, x) = f(\gamma, x) + g(x)K_2[x, x].$$

The Taylor series expansion is:

$$\dot{x} = L_0^* x + \gamma L_1^* x + \gamma^2 L_2^* x + h_0 w + w h_1 x + Q_0^*[x, x] +$$

$$+ w h_2[x, x] + \gamma Q_1^*[x, x] + C_0^*[x, x, x] + \cdots .$$

Note that the following relations hold:

$$L_0^* = L_0, Q_0^*[x, x] = Q_0[x, x] + B_1 K_2[x, x],$$

$$C_0^*[x, x, x] = C_0[x, x, x] + \tilde{L}_1 x K_2[x, x].$$

So $f^*(\gamma, x)$ possesses the same critical properties (H) as that of $f(\gamma, x)$ and. Moreover the effect of w can be treated as another Hopf bifurcation stabilization problem synthesized in Theorem 6 with $\tilde{\lambda}_2^* < 0$ (playing the role of $\tilde{\lambda}_2$ in Theorem 6). Therefore, the perturbation on the $\tilde{\lambda}_2^*$ by the uncertainty w (playing the role of u in Theorem 6) can be obtained, by applying Theorem 6, as

$$\tilde{\lambda}_2^* + \mathrm{Re}(\Theta^* + \Phi^*) < 0,$$

where

$$\Theta^* = \frac{P_1 L_0^{-1} Q_0^*[r, \bar{r}]\theta^*}{1 + P_1 L_0^{-1} h_0}, \quad \Phi^* = \frac{P_1 (L_0 - 2j\omega_c I)^{-1} Q_0^*[r, r]\phi^*}{1 + P_1 (L_0 - 2j\omega_c I)^{-1} h_0},$$

$$\theta^* = -\left(\ell h_1 r - 2\ell Q_0^*[r, L_0^{-1}h_0]\right),$$

$$\phi^* = \frac{1}{2}\left(\ell h_1 \bar{r} - 2\ell Q_0^*[\bar{r}, (L_0 - 2j\omega_c I)^{-1}h_0]\right).$$

Here we use the fact that the left row and right column eigenvectors of L_0^* are the same as those of L_0. ∎

For the uncertainty w with $P_1 = 0$, i.e., $w = x^T P_2 x + P_3[x, x, x] + \cdots$, we have the following result:

Theorem 12. *Consider the nonlinear system equation (1), which satisfies the assumption (H). Let $u = K_2[x,x]$ be a stabilizing controller derived from Theorem 5. Then it robustly stabilizes the system for all uncertainties $w = x^T P_2 x + P_3[x,x,x] + \cdots$ if P_2 satisfies :*

$$\tilde{\lambda}_2^* - \mathrm{Re}\{r^T P_2 \bar{r}\ell(2Q_0^*[r, L_0^{-1}h_0] - h_1 r)$$
$$+ r^T P_2 r\ell(Q_0^*[\bar{r}, (2j\omega_c I - L_0)^{-1}h_0] - 0.5h_1\bar{r})\} < 0$$

Proof: Similarly, let $u = K_2[x,x]$ be a controller synthesized in Theorem 5 and substitute it into the system equation (1):

$$\dot{x} = f(\gamma, x) + g(x)K_2[x,x] + h(x)w = f^*(\gamma, x) + h(x)w.$$

Still, we have the following relations for the Taylor series:

$$L_0^* = L_0, \quad Q_0^*[x,x] = Q_0[x,x] + B_1 K_2[x,x],$$
$$C_0^*[x,x,x] = C_0[x,x,x] + \tilde{L}_1 x K_2[x,x].$$

So $f^*(\gamma, x)$ possesses the same critical properties (H) as that of $f(\gamma, x)$. Since $P_1 = 0$, the effect of w on the system can be treated as another Hopf bifurcation stabilization problem as solved in Theorem 5 with $\tilde{\lambda}_2^*$ playing the role of $\tilde{\lambda}_2$ and w playing the role of u. Therefore, by applying Theorem 5, the perturbation on $\tilde{\lambda}_2^*$ caused by w can be computed as

$$\tilde{\lambda}_2^* - \mathrm{Re}\{r^T P_2 \bar{r}\ell(2Q_0^*[r, L_0^{-1}h_0] - h_1 r)$$
$$+ r^T P_2 r\ell(Q_0^*[\bar{r}, (2j\omega_c I - L_0)^{-1}h_0] - 0.5h_1\bar{r})\} < 0,$$

which has to be satisfied by P_2 in order to make the system asymptotically stable. ■

Next, we consider the controller obtained from Theorem 6: $u = K_1 x$, $K_1 r = K_1 \bar{r} = 0$. Substitute u into the system equation (1). The Taylor series expansion (9) becomes:

$$\dot{x} = L_0^* x + \gamma L_1^* x + \gamma^2 L_2^* x + h_0 w + w h_1 x + Q_0^*[x,x] +$$
$$w h_2[x,x] + \gamma Q_1^*[x,x] + C_0^*[x,x,x] + \cdots,$$

where

$$L_0^* = L_0 + B_1 K_1, \quad Q_0^*[x,x] = Q_0[x,x] + K_1 x \tilde{L}_1 x,$$
$$C_0^*[x,x,x] = C_0[x,x,x] + K_1 x \tilde{L}_2[x,x].$$

However, it is noted that the left row and right column eigenvectors of L_0^* corresponding to the critical modes are the same as those of L_0, and that L_0^* continues to satisfy the assumption (H).

For the uncertainty w with $P_1 \neq 0$, we have the following theorem:

Theorem 13. *For the system (1), let the stabilizing controller* $u = K_1 x$, $K_1 r = K_1 \bar{r} = 0$, $\tilde{\lambda}_2^*$, μ *and* ν *be given as that in Theorem 6. Then this controller will stabilize the system for any* $w = P_1 x + P_2[x,x] + \cdots$, *as long as* P_1 *satisfies the following conditions*

1. $P_1 r = P_1 \bar{r} = 0$,
2. $\tilde{\lambda}_2^* + \mathrm{Re}(\Theta^* + \Phi^*) < 0$, *where*

$$\Theta^* = \frac{P_1 L_0^{*-1} Q_0[r,\bar{r}]\theta^*}{1 + P_1 L_0^{*-1} h_0}, \quad \Phi^* = \frac{P_1 (L_0^* - 2j\omega_c I)^{-1} Q_0[r,r]\phi^*}{1 + P_1 (L_0^* - 2j\omega_c I)^{-1} h_0},$$

$$\theta^* = -\left(\ell h_1 r - 2\ell Q_0^*[r, L_0^{*-1} h_0]\right),$$

$$\phi^* = \frac{1}{2}\left(\ell h_1 \bar{r} - 2\ell Q_0^*[\bar{r}, (L_0^* - 2j\omega_c I)^{-1} h_0]\right).$$

Next theorem states the local robustness regarding the uncertainty w with $P_1 = 0$:

Theorem 14. *Consider the nonlinear system equation (1), which satisfies all assumptions in Theorem 6. Let* $u = K_1 x$, $K_1 r = K_1 \bar{r} = 0$ *be a stabilizing controller derived from Theorem 6. Then it robustly stabilizes the system for all uncertainties* $w = x^T P_2 x + P_3[x,x,x] + \cdots$ *if* P_2 *satisfies :*

$$\tilde{\lambda}_2^* - \mathrm{Re}\{r^T P_2 \bar{r} \ell (2Q_0^*[r, L_0^{*-1} h_0] - h_1 r)$$

$$+ r^T P_2 r \ell (Q_0^*[\bar{r}, (2j\omega_c I - L_0^*)^{-1} h_0] - 0.5 h_1 \bar{r})\} < 0$$

The proofs of Theorem 13 and 14 can be similarly derived.

5 Robustness of Rotating Stall Control for Axial Flow Compressors

In this section, we shall apply the results in Section 4 to deriving the robustness of some rotating stall control laws obtained in [6,9,15,24]. As stated in [9], bifurcation analysis shows that, depending on the state variable used in the third order Moore-Greitzer model, the rotating stall problem can be characterized as either a transcritical bifurcation or a pitchfork bifurcation. Correspondingly, rotating stall control laws can be derived based on the theorems in Section 3 and the robustness of these control laws can be derived by applying the results in Section 4.

5.1 Moore-Greitzer model for post-stall dynamics

The third order Moore-Greitzer model with control u for post-stall dynamics is of the form

$$\dot{\Psi} = \frac{1}{\beta^2}\left(\Phi - (\gamma + u)\sqrt{\Psi} + 1\right), \tag{15}$$

$$\dot{\Phi} = -\Psi + \psi_c(\Phi) + 6c_3\Phi R, \quad \psi_c(\Phi) = c_0 + c_1\Phi + c_3\Phi^3, \tag{16}$$

$$\dot{R} = \sigma R(1 - \Phi^2 - R), \tag{17}$$

where Φ is the average flow rate, Ψ the pressure rise, R the amplitude square of the disturbance flow ($R = A^2$), and u the actuating signal implemented with throttle, which are all non-dimensionalized. If the amplitude A is used as a state variable, then the third equation can be replaced by

$$\dot{A} = 0.5\sigma A(1 - \Phi^2 - A^2). \tag{18}$$

An obvious equilibrium (Ψ_e, Φ_e, R_e) for $u = 0$ satisfies

$$R_e = 0, \quad \Psi_e = \psi_c(\Phi_e), \quad \Psi_e = \frac{1}{\gamma}(1 + \Phi_e)^2. \tag{19}$$

It can be easily shown that there exists $\gamma_c > 0$ such that the above equilibrium is stable for $\gamma > \gamma_c$, but unstable for $\gamma < \gamma_c$ [17]. Denote

$$x = \begin{bmatrix} x_1 \\ x_2 \\ x_3 \end{bmatrix} = \begin{bmatrix} \Psi - \Psi_e \\ \Phi - \Phi_e \\ R \end{bmatrix}, \quad g(x) = \begin{bmatrix} -1 \\ 0 \\ 0 \end{bmatrix} \frac{\sqrt{\Psi}}{\beta^2},$$

with $\Psi = x_1 + \Psi_e$. Then we have

$$\dot{x} = f(\delta\gamma, x) + g(x)u, \quad \delta\gamma = \gamma - \gamma_c. \tag{20}$$

Thus the equilibrium in (19) is the local zero solution for $u = 0$, and both Ψ_e and Φ_e are functions of γ. Moreover the linearized system at the origin possesses exactly one zero eigenvalue at $\gamma = \gamma_c$, which implies that γ_c is the critical value. The equilibrium in (19) at the critical value of γ is determined as (see [15,24])

$$R_c = 0, \quad \Phi_c = 1, \quad \Psi_c = \Psi_c(\Phi_c) = c_0 + c_1 + c_3, \quad \gamma_c = \frac{2}{\sqrt{\Psi_c}},$$

where $c_0 = 8/3$, $c_1 = 1.5$, and $c_3 = -0.5$. It is shown in [9] that, for state variables (Ψ, Φ, R), this model yields a transcritical bifurcation with

$$\tilde{\lambda}_1 = \ell Q_0[r, r] = -\sigma\left(1 + \frac{6c_3\gamma_c}{\sqrt{\Psi_c}}\right) = -\sigma\left(1 + \frac{12c_3}{\Psi_c}\right) > 0,$$

$$\lambda'(0) = -2\sigma\sqrt{\Psi_c} < 0,$$

and that, for state variables (Ψ, Φ, A), this model yields a subcritical pitchfork bifurcation with

$$\tilde{\lambda}_1 = 0, \quad \tilde{\lambda}_2 = -\sigma(\Psi_c - 6)/\Psi_c > 0.$$

5.2 Rotating stall control laws

Several feedback rotating stall control laws have been synthesized based on the third order Moore-Greitzer model by applying the bifurcation stabilization result in Section 3 [6,9,15,24].

Clearly the nonlinear system (20) can be expanded into the Taylor series form with

$$L_0 = \begin{bmatrix} -\frac{\gamma_c\beta^{-2}}{2\sqrt{\Psi_c}} & \beta^{-2} & 0 \\ -1 & c_1 + 3c_3\Phi_c^2 & 6c_3 \\ 0 & 0 & 0 \end{bmatrix}, \quad Q_0[x,x] = \begin{bmatrix} \frac{\gamma_c}{8\beta^2\Psi_c^{\frac{3}{2}}} x_1^2 \\ 3c_3\Phi_c x_2^2 + 6c_3x_2x_3 \\ -\sigma x_3^2 - 2\sigma\Phi_c x_2x_3 \end{bmatrix},$$

$$B_1 = \begin{bmatrix} -1 \\ 0 \\ 0 \end{bmatrix} \frac{\sqrt{\Psi_c}}{\beta^2}, \quad \tilde{L}_1 = \begin{bmatrix} -\frac{1}{2\beta^2\sqrt{\Psi_c}} & 0 & 0 \\ 0 & 0 & 0 \\ 0 & 0 & 0 \end{bmatrix}, C_0[x,x,x] = \begin{bmatrix} -\frac{\gamma_c}{16\beta^2\Psi_c^{\frac{5}{2}}} x_1^3 \\ c_3x_2^3 \\ -\sigma x_2^2x_3 \end{bmatrix},$$

and ℓ and r can be chosen as

$$\ell = \begin{bmatrix} 0 & 0 & 1 \end{bmatrix}, \quad r = \begin{bmatrix} 6c_3 \\ \frac{3c_3\gamma_c}{\sqrt{\Psi_c}} \\ 1 \end{bmatrix}.$$

Applying Theorem 3, a control law $u = Kx = \begin{bmatrix} k_\Psi & k_\Phi & k_R \end{bmatrix} x$ can be obtained as

$$\frac{6-\Psi_c}{\Psi_c^{3/2}} < -6k_\Psi - k_\Phi + 2k_R, \quad k_\Phi < \Psi_c^{-1/2}, \quad k_\Psi > -\Psi_c^{-3/2}, \tag{21}$$

with

$$T = \begin{bmatrix} 1 & 0 & -6c_3 \\ 0 & 1 & -\frac{3c_3\gamma_c}{\sqrt{\Psi_c}} \\ 0 & 0 & 1 \end{bmatrix}, \quad L_{00} = \begin{bmatrix} -\frac{\gamma_c\beta^{-2}}{2\sqrt{\Psi_c}} & \beta^{-2} \\ -1 & c_1 + 3c_3\Phi_c^2 \end{bmatrix}, \quad B_{11} = \begin{bmatrix} -\frac{\sqrt{\Psi_c}}{\beta^2} \\ 0 \end{bmatrix}$$

and $a = -2\sqrt{\Psi_c}\sigma$, $b = 0$, $d = \tilde{d} = 0$.

Taking $k_R = k_\Phi = 0$ yields

$$-\frac{3}{11}\sqrt{\frac{3}{11}} = -\Psi_c^{-3/2} < k_\Psi < \frac{1}{6\sqrt{\Psi_c}} - \Psi_c^{-3/2} = -\frac{7}{66}\sqrt{\frac{3}{11}}. \tag{22}$$

Taking $k_\Psi = k_\Phi = 0$ yields

$$k_R > 3\Psi_c^{-3/2} - \frac{1}{2\sqrt{\Psi_c}} = 0.1175 \tag{23}$$

which implies the control law with $k_R = 0.5$ as developed in [15,24].

Besides, a nonlinear control law is also synthesized in[6] as $u = k_n \frac{x_1}{\sqrt{x_1 + \Psi_e}}$ with

$$-\frac{1}{c_0 - 2c_3} = -\frac{3}{11} < k_n < -\frac{1}{12c_3} - \frac{1}{c_0 - 2c_3} = -\frac{7}{66}. \tag{24}$$

5.3 Robustness of rotating stall control

We shall present the robustness results of the rotating stall control laws (22) – (24) (see also [22,23]). For simplicity, the disturbance signal is added into the pressure rise equation only:

$$\begin{aligned}\dot{\Psi} &= \frac{1}{\beta^2}\left(\Phi - (\gamma+u)\sqrt{\Psi}+1\right) + n_1 w(x),\\ \dot{\Phi} &= -\Psi + \psi_c(\Phi) + 6c_3\Phi R,\\ \dot{R} &= \sigma R(1-\Phi^2 - R).\end{aligned}$$

This actually yields the robust control system

$$\dot{x} = f(\delta\gamma, x) + g(x)u + h(x)w(x), \quad \delta\gamma = \gamma - \gamma_c, \tag{25}$$

with $h(x) = [n_1\ 0\ 0]^T$.

For the control law $u = k_\Psi x_1$, the robustness result is summarized in the following theorem.

Theorem 15. *For the robust rotating stall control problem stated in* (25), *let the stabilizing control be $u = k_\Psi x_1$, then u will also achieve rotating stall control for any uncertainty $w = P_1 x + P_2[x,x] + P_3[x,x,x] + \cdots$ satisfying*

$$n_1 p_\Psi < \frac{1}{\beta^2}\left(\frac{3}{11} + k_\Psi\sqrt{\frac{11}{3}}\right), \quad -p_\Phi n_1 \beta^2 < 1,$$

$$n_1(2p_R - 6p_\Psi - p_\Phi) < -\frac{1}{\beta^2}\left(\frac{7}{11} + 6k_\Psi\sqrt{\frac{11}{3}}\right).$$

In particular, if $n_1 = 1$, then we have

$$p_\Psi < \frac{1}{\beta^2}\left(\frac{3}{11} + \sqrt{\frac{11}{3}}k_\Psi\right), \quad p_\Phi > -\frac{1}{\beta^2},$$

$$2p_R - 6p_\Psi - p_\Phi < -\frac{1}{\beta^2}\left(\frac{7}{11} + 6\sqrt{\frac{11}{3}}k_\Psi\right),$$

where $P_1 = \left[p_\Psi\ p_\Phi\ p_R\right]$.

The proof of this theorem is a straightforward application of Theorem 9 plus tedious computation.

Corollary 1. *The controller $u = k_\Psi x_1 = -\frac{12.5}{66}\sqrt{\frac{3}{11}}x_1$ will achieve stabilizing rotating stall control for any uncertainty $w = P_1x + P_2[x,x] + P_3[x,x,x] + \cdots$:*

$$n_1 p_\Psi < \frac{1}{12\beta^2}, \quad -p_\Phi n_1\beta^2 < 1, \quad 2p_R n_1 \beta^2 - 6p_\Psi n_1\beta^2 - p_\Phi n_1 \beta^2 < \frac{1}{2}.$$

In particular, if $n_1 = 1$, *then we have*

$$p_\Psi < \frac{1}{12\beta^2}, \ p_\Phi > -\frac{1}{\beta^2}, \ 2p_R - 6p_\Psi - p_\Phi < \frac{1}{2\beta^2}.$$

where $P_1 = \begin{bmatrix} p_\Psi \ p_\Phi \ p_R \end{bmatrix}$.

The result of Corollary 1 is illustrated in Figure 2 for the special case $k_\Psi = -0.0989$ and $n_1 = 1$. The gray area in this normalized graph shows the acceptable set of uncertainty signals tolerated by the controller.

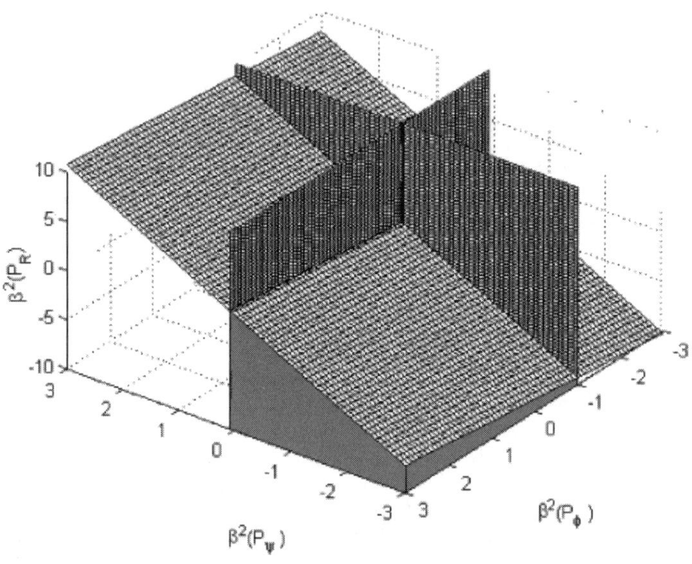

Fig.2. Acceptable uncertainty region for control law $u = k_\Psi x_1 = -0.0989x_1$

For the control law $u = k_R x_3$, the robustness result is summarized in the following theorem.

Theorem 16. *For the robust rotating stall control problem stated in* (25), *let the stabilizing control law be* $u = k_R x_3$, *then* u *will also achieve rotating stall control for any uncertainty* $w = P_1 x + P_2[x, x] + P_3[x, x, x] + \cdots$ *with*

$$n_1 p_\Psi < \frac{3}{11\beta^2}, \quad -p_\Phi n_1 \beta^2 < 1,$$

$$2p_R n_1 \beta^2 - 6p_\Psi n_1 \beta^2 - p_\Phi n_1 \beta^2 < 2k_R \sqrt{\frac{11}{3}} - \frac{7}{11}.$$

In particular, if $n_1 = 1$*, then we have*

$$p_\Psi < \frac{3}{11\beta^2}, \; p_\Phi > -\frac{1}{\beta^2}, \; 2p_R - 6p_\Psi - p_\Phi < -\frac{1}{\beta^2}\left(\frac{7}{11} - 2\sqrt{\frac{11}{3}}k_R\right),$$

where $P_1 = \left[p_\Psi \; p_\Phi \; p_R\right]$.

Once again, the proof of this theorem is a straightforward application of Theorem 9 plus tedious computation.

Corollary 2. *The controller* $u = k_R x_3 = 0.5 x_3$ *will achieve rotating stall control for any uncertainty* $w = P_1 x + P_2[x,x] + P_3[x,x,x] + \cdots$ *with*

$$n_1 p_\Psi < \frac{3}{11\beta^2}, \; -p_\Phi n_1 \beta^2 < 1,$$

$$2p_R n_1 \beta^2 - 6p_\Psi n_1 \beta^2 - p_\Phi n_1 \beta^2 < \sqrt{\frac{11}{3}} - \frac{7}{11}.$$

In particular, if $n_1 = 1$*, we have:*

$$p_\Psi < \frac{3}{11\beta^2}, \; p_\Phi > -\frac{1}{\beta^2}, \; 2p_R - 6p_\Psi - p_\Phi < \frac{1}{\beta^2}\left(\sqrt{\frac{11}{3}} - \frac{7}{11}\right),$$

where $P_1 = \left[p_\Psi \; p_\Phi \; p_R\right]$.

The result of Corollary 2 is illustrated in Figure 3 for the special case $n_1 = 1$. The gray area in this normalized graph shows the acceptable set of uncertainty tolerated by the controller.

Finally, we prove that the nonlinear control law (24) is equivalent to some linear control law from (22) and, hence, possesses same robustness as that linear control law.

Theorem 17. *Let* $u_n = k_n \dfrac{\Psi - \Psi_e}{\sqrt{\Psi}}$ *be a nonlinear controller. Then it is a stabilizing controller if and only if the linear controller* $u = k_\Psi(\Psi - \Psi_e)$ *with* $k_\Psi = \frac{k_n}{\sqrt{\Psi_c}}$ *is stabilizing.*

Proof: First, if u_n is a stabilizing controller, then we have:

$$-\frac{1}{c_0 - 2c_3} = -\frac{3}{11} < k_n < -\frac{1}{12c_3} - \frac{1}{c_0 - 2c_3} = -\frac{7}{66}.$$

Substituting $k_n = k_\Psi \sqrt{\Psi_c} = \sqrt{\frac{11}{3}} k_\Psi$, we obtain

$$-\frac{3}{11}\sqrt{\frac{3}{11}} = -\Psi_c^{-3/2} < k_\Psi < \frac{1}{6\sqrt{\Psi_c}} - \Psi_c^{-3/2} = -\frac{7}{66}\sqrt{\frac{3}{11}}.$$

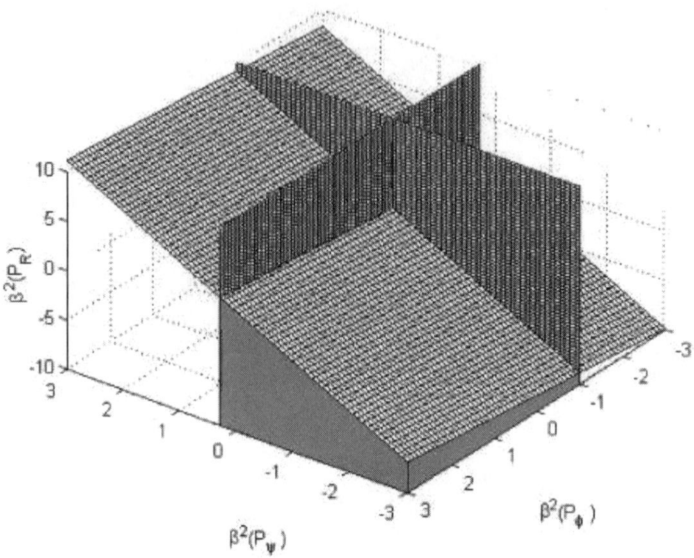

Fig.3. Acceptable uncertainty region for control law $u = 0.5x_3$

Hence, $u = k_\Psi(\Psi - \Psi_e)$ is a stabilizing controller as given in (22).

On the other hand, if $u = k_\Psi(\Psi - \Psi_e)$ is a stabilizing controller, then

$$-\frac{3}{11}\sqrt{\frac{3}{11}} < k_\Psi < -\frac{7}{66}\sqrt{\frac{3}{11}}.$$

Now taking $u_n = k_n \frac{\Psi - \Psi_e}{\sqrt{\Psi}}$ with $k_n = k_\Psi \sqrt{\Psi_c}$, we have

$$-\frac{3}{11}\sqrt{\frac{3}{11}} = -\Psi_c^{-3/2} < \frac{k_n}{\sqrt{\Psi_c}} < \frac{1}{6\sqrt{\Psi_c}} - \Psi_c^{-3/2} = -\frac{7}{66}\sqrt{\frac{3}{11}},$$

or

$$-\frac{3}{11} < k_n < -\frac{7}{66},$$

which means that u_n is also a stabilizing controller. ∎

6 Remarks

Not surprisingly, the results obtained in this chapter show that only the dominant terms (linear and/or quadratic terms) in the series expansion are of importance. This is true for characterization of both the stabilizing control laws and the admissible uncertainty set. We can make the following remarks:

1. For the transcritical bifurcation stabilization, a linear control is good enough to achieve the goal so a nonlinear control is not necessary. For the pitchfork bifurcation stabilization, a linear plus quadratic control is better used to yield more flexibility of choosing the feedback gain. For Hopf bifurcation, both linear and quadratic controls are presented.
2. Up to quadratic terms in the Taylor series expansion of disturbance signal w are shown in the characterization of w so higher order terms can be robustly tolerated by the relevant stabilizing control.
3. The significance of the approach described in this chapter is that it can be applied to selecting a robust control from a class of stabilizing controls. For example, in Corollary 1, we obtain the robustness of the central controller $u = -\frac{12.5}{66}\sqrt{\frac{3}{11}}x_1$ and in Corollary 2 we obtain the robustness of the controller $u = 0.5x_3$, both for $n_1 = 1$. Since the second inequalities are exactly same and for the first and third inequalities we have

$$\frac{1}{12\beta^2} < \frac{3}{11\beta^2}, \quad \frac{1}{2\beta^2} < \frac{1}{\beta^2}\left(\sqrt{\frac{11}{3}} - \frac{7}{11}\right),$$

clearly, the controller in Corollary 2 is more robust than the one in Corollary 1 so it is a better choice for robustness consideration. The same conclusion can be drawn by observing Figure 2 and Figure 3.
4. For robustness of rotating stall control, the nonlinear control law is not superior or inferior to a linear control law, which mirrors the conclusion for the bifurcation stabilization.
5. The future research work should focus on developing dynamical feedback control laws for multi-mode Moore-Greitzer model and on deriving the robustness of dynamical feedback control laws. This should be both a research and an application problem and hence deserves further efforts from the control community.

Acknowledgment.
This research was supported in part by the Research Grant from NSERC (Grant Number: #217351).

References

1. Abed, E. H., Fu, J-H. (1986) Local feedback stabilization and bifurcation control, I. Hopf bifurcation. Syst. Contr. Lett., **7**:11–17
2. Abed, E. H., Fu, J-H. (1987) Local feedback stabilization and bifurcation control, II. Stationary bifurcation. Syst. Contr. Lett., **8**:467–473
3. Abed, E. H., Houpt, P. K., Hosny, W. M. (1993) Bifurcation analysis of surge and rotating stall in axial flow compressors. J. Turbomachinery, **115**(10):817–824

4. Baillieul, J., Dahlgren, S., Lehman, B. (1995) Nonlinear control design for systems with bifurcations with applications to stabilization and control of compressors. In Proc. 34th IEEE Conf. Decision Control, 3062–3067
5. Brokett, R. W. (1983) Asymptotic stability and feedback stabilization. Differential Geometry Control Theory, Brokett, R. W., Millman, R. S., Sussmann, H. J. (eds.), Boston: Birkhäuser, 181–191
6. Chen, X., Gu, G., Martin, P., Zhou, K. (1998) Rotating stall control via bifurcation stabilization. Automatica, **34**(4):437–443
7. Chen, X., Gu, G., Zhou, K. (2002) Local robustness of Hopf bifurcation control. In Proc. Amer. Contr. Conf., 2263–2266
8. Chen, X., Sparks, A. G., Zhou, K., Gu, G. (2001) Local robustness of stationary bifurcation control. Latin American Applied Research, **31**(3):163–170
9. Gu, G., Chen, X., Sparks, A. G., Banda, S. S. (1999) Bifurcation stabilization with local output feedback. SIAM J. Contr. Optim., **37**(3):934–956
10. Gu, G., Sparks, A. G. (2000) Bifurcation stabilization with applications in jet engine control. Chapter 16, Controlling Chaos and Bifurcations in Engineering Systems, Chen, G. (ed.), Boca Raton, FL: CRC Press
11. Gu, G., Sparks, A. G., Banda, S. S. (1997) Bifurcation based nonlinear feedback control for rotating stall in axial flow compressors. Int. J. Contr., **6**:1241–1257
12. Gravdahl, J. T., Egeland, O. (1999) Compressor Surge and Rotating Stall: Modeling and Control. New York: Springer
13. Iooss, G., Joseph, D. D. (1990) Elementary Stability and Bifurcation Theory. New York: Springer-Verlag
14. Kailath, T. (1980) Linear Systems. Englewood Cliffs, NJ: Prentice-Hall
15. Liaw, D.-C., Abed, E. H. (1996) Active control of compressor stall inception: A bifurcation-theoretical approach. Automatica, **32**:109–116
16. Lu, W.-M. (1995) Control of Uncertain Systems: State Space Characterizations. Ph.D. Thesis, CalTech, USA
17. McCaughan, F. E. (1990) Bifurcation analysis of axial flow compressor stability. SIAM J. Appl. Math., **20**:1232–1253
18. Moore, F. K., Greitzer, E. M. (1986) A theory of post-stall transients in axial compressors: Part I – development of the equations. ASME J. Engr. for Gas Turbines and Power, **108**:68–76
19. Paduano, J. D., Valavani, L., Epstein, A. H., Greitzer, E. M., Guenette, G. R. (1994) Modeling for control of rotating stall. Automatica, **30**(9):1357–1373
20. Xiao, M.-Q., Basar, T. (2000) Analysis and control of multi-mode axial flow compression system models. ASME J. Dynam. Syst., Measure. Contr., **122**:393–401
21. van der Schaft, A. (1996) $\mathcal{L}_2$-Gain and Passivity Techniques in Nonlinear Control. New York: Springer
22. Tahmasebi, A., Chen, X. (2002) Robustness of rotating stall control for axial-flow compressors subject to both pressure and flow perturbation. In Proc. Amer. Contr. Conf., 3295–3300
23. Tahmasebi, A., Chen, X. (2001) Robustness of control design for axial flow compressors. In Proc. IEEE Canadian Conf. Electr. Comput. Eng., 35–40
24. Wang, H. O., Adomatis, R. A., Abed, E. H. (1994) Nonlinear analysis and control of rotating stall in axial flow compressors. In Proc. Amer. Contr. Conf., 2317–2321

Bifurcations and Chaos in Turbo Decoding Algorithms

Zarko Tasev[1], Petar Popovski[2], Gian Mario Maggio[3], and Ljupco Kocarev[1]

[1] Institute for Nonlinear Science
University of California, San Diego
La Jolla, CA 92093-0402, USA
lkocarev@ucsd.edu

[2] CPK, Aalborg University
Niels Jernes Vej 12
A5-211 9220 Aalborg, Denmark

[3] STMicroelectronics and
Center for Wireless Communications
University of California, San Diego
La Jolla, CA 92093-0407, USA

Abstract. The turbo decoding algorithm is a high-dimensional dynamical system parameterized by a large number of parameters (for a practical realization the turbo decoding algorithm has more than 10^3 variables and is parameterized by more than 10^3 parameters). In this chapter we treat the turbo decoding algorithm as a dynamical system parameterized by a single parameter that closely approximates the signal-to-noise ratio (SNR). A whole range of phenomena known to occur in nonlinear systems, like the existence of multiple fixed points, oscillatory behavior, bifurcations, chaos and transient chaos are found in the turbo-decoding algorithm. We develop a simple technique to control transient chaos in turbo decoding algorithm and improve the performance of the standard turbo codes.

1 Introduction

Recently, it has been recognized that two classes of codes, namely *turbo-codes* [1] and *low-density parity-check* (LDPC) codes [2–4], perform at rates extremely close to the Shannon limit imposed by the noisy channel coding theorem [5]. Both codes are based on a similar philosophy: constrained random code ensembles, described by some fixed parameters plus randomness, decoded using iterative decoding algorithms (or message passing decoders). Iterative decoding algorithms may be viewed as a complex nonlinear dynamical system. The aim of the present work is to contribute to the in-depth understanding of these families of error-correction codes, based on the well developed theory of nonlinear dynamical systems [6].

Turbo codes were discovered by Berrou *et al.* in 1993 [1]. On the other hand, LDPC codes were originally introduced by Gallager [7] in 1962. The crucial innovation of LDPC codes being the introduction of iterative decoding algorithms. LDPC codes were rediscovered by MacKay *et al.* [2] in 1996.

G. Chen, D.J. Hill, and X. Yu (Eds.): Bifurcation Control, LNCIS 293, pp. 301–320, 2003.

Moreover, iterative decoding of turbo codes was recognized as instances of sum-product algorithms for codes defined on general graphs [8]. The past few years have seen many new developments in the area of iterative decoding algorithms for both turbo and LDPC codes. We now briefly mention some of these achievements. The complexity, introduced by the interleaver in turbo coding, makes a rigorous analysis of the distance spectrum difficult, if not impossible, for an arbitrary interleaver instance. However, Benedetto and his co-workers [9] introduced the concept of "uniform interleaver", which proved to be a very useful tool for investigating the average distance spectrum of turbo-codes. Moreover, they generalized the original parallel concatenation of convolutional codes (*i.e.* turbo-codes) to the case of serially and hybridly concatenated codes, showing that serial concatenation results in a larger interleaver gain and, therefore, in a better average distance spectrum. Although turbo-codes, on average, have poor free distance, their extraordinary performance was explained from the distance spectrum perspective by the phenomenon of spectral thinning, observed by Perez *et al.* [10].

Very recently, in a pioneering paper [11], Richardson has presented a geometrical interpretation of the turbo-decoding algorithm and formalized it as a discrete-time dynamical system defined on a continuous set. This approach clearly demonstrates the relationship between turbo-decoding and maximum-likelihood decoding. The turbo-decoding algorithm appears as an iterative algorithm aimed at solving a system of $2n$ equations in $2n$ unknowns, where n is the block-length size. If the turbo-decoding algorithm converges to a certain codeword, then the later constitutes a solution to this set of equations. Conversely, solutions to these equations provide fixed points of the turbo-decoding algorithm, seen as a nonlinear mapping. In a follow-up by Agrawal and Vardy [12] a rigorous bifurcation analysis of the iterative decoding process as a dynamical system parameterized by SNR has been carried out. These works open new research directions for analyzing and designing random coding schemes.

In this chapter, we will consider the iterative decoding algorithm as a nonlinear dynamical system, where the codewords—to which the algorithm converges—correspond to fixed points in the symbol state space. We emphasize that in general the iterative decoding algorithm, being a nonlinear dynamical system, may exhibit a whole range of phenomena known to occur in nonlinear systems [13,14]. These include the existence of multiple fixed points, oscillatory behavior, and even chaos.

The outline of this chapter is as follows. In Section 2 we recall Richardson's formulation of the turbo decoding algorithm as a dynamical system [11]. To this aim, we consider a classical turbo code with parallel concatenation of identical recursive convolutional codes generated by the polynomials $\{D^4 + D^3 + D^2 + D^1 + 1, D^4 + 1\}$, resulting in a rate-1/3 turbo code. The codewords are transmitted over an AWGN (additive white Gaussian noise) channel using BPSK (binary phase shift keying) modulation. With an interleaver length of

$n = 1024$, the turbo decoding algorithm may be described as a dynamical system of dimension $n(= 1024)$, parameterized by $3n(= 3072)$ parameters.

In Section 3, we analyze the character of the fixed points in the turbo decoding algorithm. Simulations show that at low SNRs, the turbo decoding algorithm often converges to an *indecisive* fixed point, which corresponds to many erroneous decisions on the information bits. On the other hand, at slightly higher SNRs, after the waterfall region, the turbo decoding algorithm converges to an *unequivocal* fixed point that corresponds to correct decisions on information bits.

In Section 4, we treat the turbo decoding algorithm as a dynamical system parameterized by a single parameter that closely approximates the SNR. By varying this parameter, we analyze the turbo decoding algorithm as a function of SNR. In each instance of the turbo decoding algorithm that we analyzed, an unequivocal fixed point was found in a wide range of SNRs: we found that this point is stable even for SNR ≈ -1.5 dB. However, at low SNR, before and at the waterfall region, the decoding algorithm often fails to converge to this fixed point, while spending time, instead, onto another attracting (chaotic) invariant set. The reason why the turbo decoding algorithm is unable to find the unequivocal fixed point for low SNRs, even when the fixed point is stable, is due to the fact the basin of attraction of this fixed point can be very small. In our simulations we found that the indecisive fixed point looses its stability at low SNR, typically in the range of -7 dB to -5 dB. The region -5 dB to 0 dB is characterized by chaotic behavior: the turbo decoding algorithm as a dynamical system possesses a chaotic attractor. In the waterfall region, the turbo decoding algorithm converges either to the chaotic invariant set or to the unequivocal fixed point, after a long transient behavior. The later indicates the existence of a chaotic non-attracting invariant set in the vicinity of the unequivocal fixed point.

Section 5 considers an application of the theory developed here. We use a simple technique for controlling transient chaos, thereby reducing the number of iterations needed by the turbo decoding algorithm to reach the unequivocal fixed point.

2 Dynamics of Iterative Decoding Algorithms

2.1 Preliminaries

Let $\mathcal{H}$ be the set of all ordered binary strings of length n. We use $\boldsymbol{b}^0 = (0, 0, \ldots, 0)^T$, $\boldsymbol{b}^1 = (1, 0, \ldots, 0)^T$, $\boldsymbol{b}^2 = (0, 1, \ldots, 0)^T$, $\ldots$, $\boldsymbol{b}^n = (0, 0, \ldots, 1)^T$, $\boldsymbol{b}^{n+1} = (1, 1, \ldots, 0)^T$, $\ldots$, $\boldsymbol{b}^{2^n-1} = (1, 1, \ldots, 1)^T$ to denote the elements of $\mathcal{H}$ sorted in the increasing order of Hamming weight, and within each weight class, sorted in the reverse lexicographical order. A *density* on $\mathcal{H}$ is a positive real function defined over $\mathcal{H}$. A density f on $\mathcal{H}$ induces a *probability measure*

Pr_f on the set of all subsets of $\mathcal{H}$, $\mathcal{P}(\mathcal{H})$, as follows:

$$Pr_f(A) := \frac{\sum_{\boldsymbol{b} \in A} f(b)}{\sum_{\boldsymbol{b} \in \mathcal{H}} f(b)},$$

for all $A \in \mathcal{P}(\mathcal{H})$.

A density p is called *probability density* if $\sum_{\boldsymbol{b} \in \mathcal{H}} p(\boldsymbol{b}) = 1$. A density p is normalized with respect to the all zero binary string, $\boldsymbol{b}^0 = \mathbf{0}$, if $p(\mathbf{0}) = 1$. We say that the densities p and q are equivalent if they determine the same probability density. In each equivalence class, there is a unique probability density and a unique density normalized with respect to $\mathbf{0}$. For brevity, a density normalized with respect to $\mathbf{0}$ will simply be called a normalized density.

It is useful to represent densities in the logarithmic domain. Given a density f, let $F = log \circ f$ be its logarithmic representation. We say F is a *log-density* on $\mathcal{H}$. A log-density F is a real valued function on $\mathcal{H}$, taking both positive and negative values. Let Φ denote the set of all log densities that correspond to the normalized densities, that is, $F \in \Phi$ if and only if $F(\mathbf{0}) = 0$.

Let $\mathcal{H}_i \subset \mathcal{H}$ be the set of all binary strings whose i-th bit is 1. A density f is referred to as a *product density* if according to the induced probability measure Pr_f, all bits are independent of each other. For a normalized product density f, $f(\boldsymbol{b}^i)$, $i = 1, \ldots, n$, is the likelihood ratio of the $i-$th bit according to the density f,

$$f(\boldsymbol{b}^i) = \frac{Pr_f(\mathcal{H}_i)}{Pr_f(\mathcal{H}_i^c)} = \frac{\sum_{\boldsymbol{b} \in \mathcal{H}_i} f(\boldsymbol{b})}{\sum_{\boldsymbol{b} \in \mathcal{H}_i^c} f(\boldsymbol{b})}.$$

It is clear that for a product log-density we have

$$f(\boldsymbol{b} = b_1 b_2 \ldots b_n) = \prod_{i: b_i = 1} f(\boldsymbol{b}^i).$$

We refer to a log density that corresponds to a product density as a *product log-density*. Let Π be the set of all product log-densities in Φ. Using the last expression, for a product long-density $F \in \Pi$, we can write

$$F(\boldsymbol{b} = b_1 b_2 \ldots b_n) = \sum_{i: b_i = 1} F(\boldsymbol{b}^i).$$

Therefore, $F(\boldsymbol{b}^i)$ is the log-likelihood ratio of the i-th bit according to the density f and densities in Π are completely specified by their values on $\boldsymbol{b}^1$, $\boldsymbol{b}^2, \ldots, \boldsymbol{b}^n$. Furthermore, Π is an n-dimensional linear subspace of Φ. A basis for Π is given by a $2^n \times n$ matrix B, having $(\boldsymbol{b}^i)^T$ as its i-th row.

We say that two densities p and q have the same bitwise marginal distributions if

$$Pr_p(\mathcal{H}_i) = Pr_q(\mathcal{H}_i)$$

for $i = 1, \ldots, n$. For a log-density P, we define a projection map $\pi_P : \Pi \rightarrow \Pi$ by setting $\pi_P(Q)$ to be the unique normalized product log-density that has the same bitwise marginals as $P + Q$. In another words,

$$\pi_P(Q)(\boldsymbol{b}^i) = \log \frac{\sum_{\boldsymbol{b} \in \mathcal{H}_i} p(\boldsymbol{b})q(\boldsymbol{b})}{\sum_{\boldsymbol{b} \in \mathcal{H}_i^c} p(\boldsymbol{b})q(\boldsymbol{b})} \tag{1}$$

for $i = 1, \ldots, n$.

2.2 The turbo decoding algorithm as a dynamical system

A classical turbo code is a parallel concatenation of two recursive systematic binary convolutional codes, C_1 and C_2 [1]. Let $\boldsymbol{i}$ be the information bit sequence of length n at the input to the turbo encoder, and let $\boldsymbol{c}_1(\boldsymbol{i})$ (respectively $\boldsymbol{c}_2(\boldsymbol{i})$) be the parity bits produced by the first (respectively second) encoder. The information bit sequence $\boldsymbol{i}$ along with the parity bit sequences $\boldsymbol{c}_1(\boldsymbol{i})$ and $\boldsymbol{c}_2(\boldsymbol{i})$ form a turbo codeword $(\boldsymbol{i}, \boldsymbol{c}_1(\boldsymbol{i}), \boldsymbol{c}_2(\boldsymbol{i}))$.

We assume that the turbo code is transmitted over a noisy binary-input memoryless channel. Let $\tilde{\boldsymbol{i}}$, $\tilde{\boldsymbol{c}}_1$ and $\tilde{\boldsymbol{c}}_2$ be the channel outputs corresponding to the input sequences $\boldsymbol{i}$, $\boldsymbol{c}_1$ and $\boldsymbol{c}_2$, respectively. Ideally, we would like to compute the *posterior* probability density, $p(\boldsymbol{b}|\tilde{\boldsymbol{i}}, \tilde{\boldsymbol{c}}_1, \tilde{\boldsymbol{c}}_2)$, where $\boldsymbol{b} \in \mathcal{H}$. Let us assume that the input bits are independent of each other and are equally likely to be either 0 or 1. Under this standard assumption, $p(\boldsymbol{b}|\tilde{\boldsymbol{i}}, \tilde{\boldsymbol{c}}_1, \tilde{\boldsymbol{c}}_2)$ is equivalent to the density $p_{ML}(\boldsymbol{b}) = p(\tilde{\boldsymbol{i}}|\boldsymbol{b})p(\tilde{\boldsymbol{c}}_1|\boldsymbol{b})p(\tilde{\boldsymbol{c}}_2|\boldsymbol{b})$. A direct computation of p_{ML} requires taking both sets of parity bits, $\tilde{\boldsymbol{c}}_1$ and $\tilde{\boldsymbol{c}}_2$, simultaneously into account by constructing a joint trellis of two convolutional encoders, which is computationally prohibitive.

The turbo decoder consists of two components: a decoder D_1 for the convolutional code C_1 and a decoder D_2 for the code C_2. These decoders use the BCJR [15] algorithm to compute the a posteriori probabilities of the information bits. Let q_1 be the a priori product density used by decoder D_2 and let q_2 be the a priori product density used by decoder D_1. We assume that both densities q_1 and q_2 are initialized to the uniform density. The decoding begins with the decoder D_1 computing the posterior likelihood ratios of the information bits based on the prior density q_2 and the observations $\tilde{\boldsymbol{i}}$ and $\tilde{\boldsymbol{c}}_1$. The posterior likelihood ratio of the i-th information bit is given by

$$\frac{\sum_{\boldsymbol{b} \in \mathcal{H}_i} p(\boldsymbol{b}|\tilde{\boldsymbol{i}}, \tilde{\boldsymbol{c}}_1)}{\sum_{\boldsymbol{b} \in \mathcal{H}_i^c} p(\boldsymbol{b}|\tilde{\boldsymbol{i}}, \tilde{\boldsymbol{c}}_1)} = \frac{\sum_{\boldsymbol{b} \in \mathcal{H}_i} p(\tilde{\boldsymbol{i}}|\boldsymbol{b})p(\tilde{\boldsymbol{c}}_1|\boldsymbol{b})q_2(\boldsymbol{b})}{\sum_{\boldsymbol{b} \in \mathcal{H}_i^c} p(\tilde{\boldsymbol{i}}|\boldsymbol{b})p(\tilde{\boldsymbol{c}}_1|\boldsymbol{b})q_2(\boldsymbol{b})}. \tag{2}$$

Let p_0, p_1 and p_2 be the normalized densities equivalent to $p(\tilde{\boldsymbol{i}}|\boldsymbol{b})$, $p(\tilde{\boldsymbol{c}}_1|\boldsymbol{b})$ and $p(\tilde{\boldsymbol{c}}_2|\boldsymbol{b})$, respectively. The first decoder D_1 uses the normalized density $p_0 p_1 q_2$ to compute posterior likelihood ratios. In the logarithmic domain, $p_0 p_1 q_2$ corresponds to $P_0 + P_1 + Q_2$. Since $(P_0 + Q_2) \in \Pi$, it follows from (1) that the posterior log-likelihood ratio of the i-th information bit is given by $\pi_{P_1}(P_0+Q_2)(\boldsymbol{b}^i)$. Next, the extrinsic information for the i-th information bit is obtained by dividing its posterior likelihood ratios, computed with (2), by the product of its channel and prior likelihood ratios. This extrinsic information is then passed from the decoder D_1 to the decoder D_2 by setting the prior density q_1 in such a way that the likelihood ratios of the information bits according to q_1 equal their extrinsic information, that is, we set

$$\frac{q_1(\boldsymbol{b}^i)}{q_1(\boldsymbol{b}^0)} = \frac{\sum_{\boldsymbol{b}\in\mathcal{H}_i} p(\boldsymbol{b}|\tilde{\boldsymbol{i}},\tilde{c}_1)}{\sum_{\boldsymbol{b}\in\mathcal{H}_i^c} p(\boldsymbol{b}|\tilde{\boldsymbol{i}},\tilde{c}_1)} \frac{p(\boldsymbol{b}^0|\tilde{\boldsymbol{i}})}{p(\boldsymbol{b}^i|\tilde{\boldsymbol{i}})} \frac{q_2(\boldsymbol{b}^0)}{q_2(\boldsymbol{b}^i)}.$$

In the logarithmic domain, the last equation can be rewritten as

$$Q_1(\boldsymbol{b}^i) = \pi_{P_1}(P_0+Q_2)(\boldsymbol{b}^i) - (P_0+Q_2)(\boldsymbol{b}^i)$$

for $i = 1,\ldots,n$. Recall that q_1 and q_2 are initialized to induce the uniform probability distribution on $\mathcal{H}$. In the logarithmic domain, this corresponds to setting $Q_1^{(0)} = Q_2^{(0)} = \mathbf{0}$. Since q_1 is a product density, the likelihood ratios $q_1(\boldsymbol{b}^i)/q_1(\boldsymbol{b}^0)$ determine the density q_1 uniquely. Therefore, we have:

$$Q_1^{(l+1)} = \pi_{P_1}(P_0+Q_2^{(l)}) - (P_0+Q_2^{(l)}) \tag{3}$$

The second decoder D_2 performs a similar operation and compute the modified prior log-density Q_2:

$$Q_2^{(l)} = \pi_{P_2}(P_0+Q_1^{(l)}) - (P_0+Q_1^{(l)}) \tag{4}$$

The decoding algorithm iteratively performs the operations indicated by (3) and (4): $l = 0,1,2,\ldots$.

Equations (3) and (4) may be considered as a discrete-time dynamical system. The log-densities P_0, P_1 and P_2 are completely specified by the channel likelihood ratios of the codeword bits. Consequently, the turbo decoding algorithm is parameterized by $3n$ parameters. The iterated variables, Q_1 and Q_2, are product log-densities, and each of them can be specified by n log-likelihood ratios. Hence, in the above formulation, the turbo-decoding algorithm is a n-dimensional dynamical system depending on $3n$ parameters. As shown in [11], this mapping depends smoothly on its variables and parameters.

What are the parameters of the turbo-decoding algorithm? We assume that the turbo codewords are transmitted over an AWGN channel using BPSK modulation. Let $\mathbf{s}(\boldsymbol{b})$ be the Euclidean-space representation of the binary string $\boldsymbol{b}$ under the BPSK map, and let $\mathbf{s}_1 = \mathbf{s}\circ\mathbf{c}_1$ and $\mathbf{s}_2 = \mathbf{s}\circ\mathbf{c}_2$, where "$\circ$" denotes the composition of two functions. Without loss of generality, we consider the case when the vector $(\mathbf{s}(\boldsymbol{b}^0),\mathbf{s}_1(\boldsymbol{b}^0),\mathbf{s}_2(\boldsymbol{b}^0))$ is transmitted and the vector $(\boldsymbol{x},\boldsymbol{y},\boldsymbol{z})$ is received. The normalized posterior density p_1, induced on the information bits by $\boldsymbol{y}$, is given by:

$$p_1(\boldsymbol{b}) = \frac{Pr(\boldsymbol{b}|\boldsymbol{y})}{Pr(\boldsymbol{b}^0|\boldsymbol{y})} = \exp\frac{||\boldsymbol{y}-\mathbf{s}_1(\boldsymbol{b}^0)||^2 - ||\boldsymbol{y}-\mathbf{s}_1(\boldsymbol{b})||^2}{2\sigma^2},$$

where σ^2 is the noise variance and $||.||^2$ denotes the squared Euclidean distance in $\mathbb{R}^n$. The corresponding normalized log-density P_1 is given by

$$P_1(\boldsymbol{b}) = \frac{||\boldsymbol{y} - \mathbf{s}_1(\boldsymbol{b}^0)||^2 - ||\boldsymbol{y} - \mathbf{s}_1(\boldsymbol{b})||^2}{2\sigma^2} = -\frac{2}{\sigma^2} \sum_{j:s_{1j}(\boldsymbol{b}^0) \neq s_{1j}(\boldsymbol{b})} y_j$$

where s_{1j} and y_j are the j-th components of $\mathbf{s}_1$ and $\boldsymbol{y}$, respectively. In a similar manner, log densities P_0 and P_2 are induced by the received vectors $\boldsymbol{x}$ and $\boldsymbol{z}$, respectively. Therefore, the turbo-decoding algorithm has $3n$ parameters: $x_1, \ldots, x_n, y_1, \ldots y_n, z_1, \ldots, z_n$.

3 Fixed Points in the Turbo-Decoding Algorithm

3.1 Basic concepts of dynamical system theory

The basic goal of the theory of dynamical systems is to understand the asymptotic behavior of the system itself. If the process is described by a differential equation whose independent variable is time, then the theory attempts to predict the ultimate behavior of the solutions of the equation in either the distant future ($t \to \infty$) or the distant past ($t \to -\infty$). If, on the other hand, the process is a discrete-time process such as iteration of the function $G : \mathbb{R}^m \to \mathbb{R}^m$, then the theory attempts to understand the eventual behavior of the set of the points $\{\boldsymbol{x}, G(\boldsymbol{x}), G^2(\boldsymbol{x}), \ldots\}$, called a trajectory (or orbit) of $\boldsymbol{x}$. Functions which determine discrete-time dynamical systems are also called *mappings*, or *maps*. Trajectories of points can be quite complicated sets, even for very simple nonlinear mappings. However, there are some trajectories which are especially simple and which play a central role in the study of the entire system, as described in the following.

The point $M \in \mathbb{R}^m$ is a *fixed point* for $G : \mathbb{R}^m \to \mathbb{R}^m$ if: $G(M) = M$. The point M is a *periodic point* of period k if: $G^k(M) = M$. The least positive k for which $G^k(M) = M$ is called the prime period of M. The set of all iterates of a periodic point form a periodic trajectory (orbit). A fixed point M for $G : \mathbb{R}^m \to \mathbb{R}^m$ is called hyperbolic if $DG(M)$ has no eigenvalues on the unit circle, where $DG(M)$ is the Jacobian matrix of G computed at M. If M is a periodic point of period k, then M is hyperbolic if $DG^k(M)$ has no eigenvalues on the unit circle. There are three types of hyperbolic periodic points: sinks, sources and saddles. M is a sink (attracting periodic point) if all of the eigenvalues of $DG^k(M)$ are less than one in absolute value. M is a source (repelling periodic point) if all of the eigenvalues of $DG^k(M)$ are greater than one in absolute value. M is a saddle point if some of the eigenvalues of $DG^k(M)$ are larger and some are less than one in absolute value. Suppose that G admits an attracting fixed point at M. Then there is an open set about M in which all points tend to M under forward iterations of G. The largest such open set is called the stable set or basin (domain) of attraction of M and is denoted by $W^s(M)$.

We now consider a discrete-time dynamical system parameterized by a single parameter α, $G(\boldsymbol{x}, \alpha)$, $\alpha \in I\!R$. We assume that G is a smooth function. Since the system is smooth, the fixed point as well as the Jacobian evaluated at this point is a continuous function of the system parameter. As the system parameter α is changed, the magnitudes of the eigenvalues may also change, and it is possible that one or more eigenvalues cross the unit circle. This non-hyperbolic behavior usually indicates the occurrence of a bifurcation. In a generic system, by changing the system parameter, either a single eigenvalue will cross the unit circle through ± 1 or two complex conjugate eigenvalues will cross the unit circle together. Therefore, when a fixed point changes continuously with the parameter, it can bifurcate and loose its stability by one of the following three mechanisms: i) *tangent* (or *fold*) bifurcation: an eigenvalue approaches $+1$; ii) *flip* (or *period doubling*) bifurcation: an eigenvalue approaches -1; iii) *Neimark-Sacker* bifurcation: a pair of complex conjugate eigenvalues crosses the unit circle. These mechanisms will be illustrated in details on the turbo-decoding algorithm, in Sec. 4.2.

3.2 Indecisive and unequivocal fixed points

In our analysis we considered the classical turbo code [1] with identical constituent recursive convolutional codes generated by the polynomials $\{D^4 + D^3 + D^2 + D^1 + 1, D^4 + 1\}$, producing a rate-1/3 turbo code. The codewords were transmitted over an AWGN channel using BPSK modulation. The length of the interleaver was $n = 1024$. Figure 1 shows the performance of the turbo code.

Assume that the log-densities P_1 and P_2 are product log-densities. We know that if P is a product log-density then $\pi_P(Q) = P + Q$. Therefore, if P_1 and P_2 are product log-densities, the turbo decoding algorithm converges to the fixed point $(Q_1^*, Q_2^*) = (P_1, P_2)$ in a single iteration, regardless of the initial conditions. The continuity of the fixed points with respect to the parameters (P_1, P_2) implies that if (P_1, P_2) is close enough to a pair of product log-densities (P_1^*, P_2^*), then the turbo decoding algorithm has a unique fixed point close to (P_1^*, P_2^*). Moreover, we expect that the unbiased initialization $(\mathbf{0}, \mathbf{0})$ will be in the domain of attraction, since for product densities, the domain of attraction is $\Pi \times \Pi$ which includes the unbiased initialization.

The turbo decoding algorithm has two types of fixed points: indecisive and unequivocal [12]. For asymptotically low SNRs, P_1 and P_2 converge to the product log-density $\mathbf{0}$, and therefore, the turbo decoding algorithm should have a fixed point close to $(\mathbf{0}, \mathbf{0})$. Simulations show that not only is this true but the signal-to-noise ratio required for the existence of this fixed point is not extremely low. In fact, for low SNR, the turbo-decoding algorithm converges and most of the extrinsic log-likelihood ratios, $Q_1^*(\boldsymbol{b}^i)$ and $Q_2^*(\boldsymbol{b}^i)$, for $i = 1, \ldots, n$, are close to 0. In this case, the probability measures induced by Q_1^* and Q_2^* are close to 0.5. We refer to a fixed point with these characteristics as an *indecisive* fixed point [12]. At such fixed points the turbo decoding

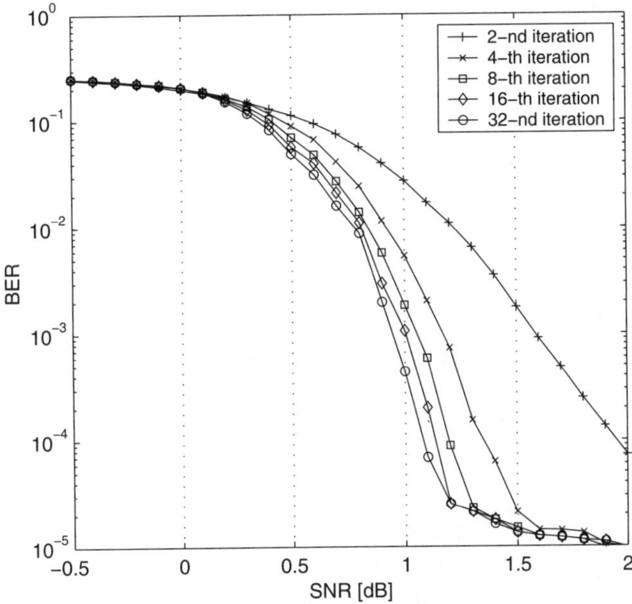

Fig.1. The performance of the classical turbo code with interleaver length 1024 and rate 1/3, for increasing number of iterations. Note that the waterfall region of the turbo code corresponds approximately to the SNR region spanning 0.25 dB to 1.25 dB.

algorithm is relatively ambiguous regarding the values of the information bits. Therefore, hard decisions corresponding to these fixed points, typically will not form a codeword. For high signal-to-noise ratios, the log-densities Q_1^* and Q_2^* are concentrated on the information sequence that corresponds to the codeword closest to the received vector. In other words, with high probability, $Pr_{q_1^*}(b_i = 0)$ and $Pr_{q_2^*}(b_i = 0)$ are either approximately 0 or 1, depending upon whether the i-th information bit of the codeword closest to the received vector is 1 or 0. Since the final log-likelihood ratios computed by the decoding algorithm are strong indicators of the information bits, we refer to a fixed point with these characteristics as an *unequivocal* fixed point [12]. Hard decisions corresponding to unequivocal fixed points, will usually form a codeword.

Since the turbo-decoding algorithm is a high dimensional dynamical system, we suggest the following representation of its trajectories in the state space. At each iteration l, the turbo-decoding algorithm computes $2n$ log-densities Q_1 and Q_2. From these log-densities one can calculate, for each iteration, the probabilities $p_i^l(0)$ and $p_i^l(1)$ that i-th bit is 0 or 1. Let us define

$$E(l) = -\frac{1}{n}\sum_{i=1}^{n} p_i^l(0)\ln p_i^l(0) + p_i^l(1)\ln p_i^l(1).$$

Thus, E represents the a posteriori average entropy, which is in a way a measure of the reliability of bit decisions of an information block with size $n = 1024$. When all bits are detected correctly or almost correctly, p_i is close to 1 for all i and, therefore, the unequivocal fixed point is represented by the point close to $E = 0$. On the other hand, for an indecisive fixed point, when all bits are equally probable, which is the case when SNR goes to $-\infty$, $E = 1$. In the following we present two types of figures: $E(l)$ versus l, and $E(l+1)$ versus $E(l)$.

4 Bifurcation Analysis of the Turbo-Decoding Algorithm

The turbo-decoding algorithm is an n-dimensional system with $3n$ parameters. This is a complex dynamical system with a large number of variables and parameters and, therefore, is not readily amenable for analysis. As a discrete-time dynamical system, the turbo decoding algorithm is parameterized by the log-densities P_0, P_1 and P_2. Given the transmitted codeword, P_0, P_1 and P_2 are completely specified by the noise values $x_1, x_2, \ldots, x_n$, $y_1, y_2, \ldots$, and $y_n, z_1, z_2, \ldots, z_n$. To study the dynamics of the turbo decoding algorithm we would like to parameterize it by the SNR, that is essentially $1/\sigma^2$ [12]. For large enough values of n, as typical for turbo codes, the noise variance σ^2 is approximately equal to: $\tilde{\sigma}^2 = \sum(x_j^2 + y_j^2 + z_j^2)/3n$. In this work, we focus on bifurcations in the turbo decoding algorithm for a single parameter, namely parameter $\tilde{\sigma}^2$ [12]. In particular, we fix the $3n-1$ noise ratios $x_1/x_2, x_2/x_3, \ldots, z_{n-1}/z_n$ and treat the turbo decoding algorithm as it was depending on a single parameter $\tilde{\sigma}^2$, which is closely related to the SNR. By varying this parameter, we were able to analyze the turbo decoding algorithm as a function of SNR. Figure 2 schematically summarizes our results.

4.1 Bifurcation diagram

We have performed many simulations changing the parameter SNR from $-\infty$ to $+\infty$ with different realizations of the noise (different noise ratios x_1/x_2, $x_2/x_3, \ldots, z_{n-1}/z_n$). In each instance of the turbo decoding algorithm that we analyzed, an unequivocal fixed point existed for all values of SNR from $-\infty$ to $+\infty$: this point is always represented with the point corresponding to the average entropy $E = 0$. This fixed point becomes stable at around -1.5 dB. However, the algorithm "cannot see" this point until 0–0.5 dB, when, in some cases, the initial point of the algorithm (which is always at $E = 1$) is within the basin of the attraction of the unequivocal fixed point.

The algorithm has another fixed point: the indecisive fixed point. For low values of SNR, when SNR goes to $-\infty$, this fixed point is represented by $E = 1$. In our simulations we found that the indecisive fixed point moves

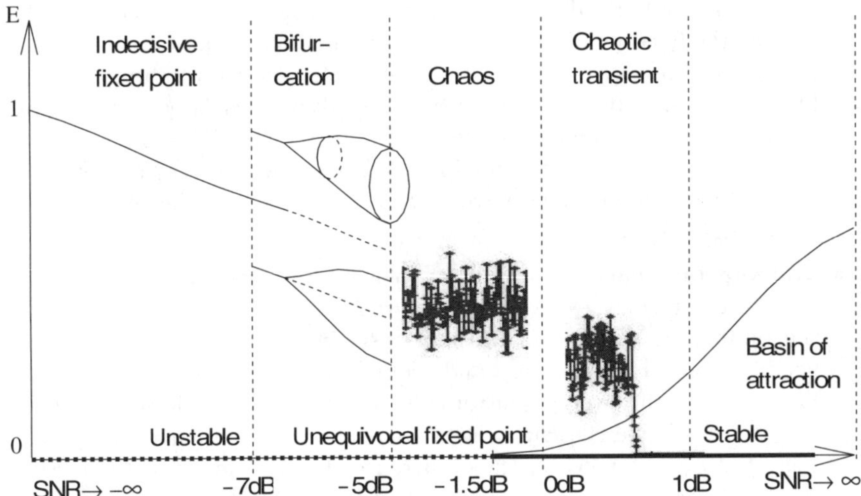

Fig.2. Schematic bifurcation diagram of the turbo-decoding algorithm.

toward smaller values of E with increasing SNR, and looses its stability (or disappears) at low SNRs, typically in the range of -7 dB to -5 dB. The mechanisms responsible for the instability/disappearance of the indecisive fixed point of the turbo-decoding algorithm are the following: tangent bifurcation, flip (or period doubling) bifurcation and Neimark-Sacker bifurcation, as illustrated in Sec. 4.2. Correspondingly, a transition from indecisive to unequivocal fixed points occurs in the turbo-decoding algorithm. In Fig. 2 this transition corresponds to a large region of SNRs: from -7– -6 dB to 0.5–1 dB. On the other hand, the region -5 dB–0 dB is characterized by chaos: the turbo decoding algorithm as a dynamical system has a chaotic attractor. In the region 0 dB–1 dB chaotic transients occur. These aspects are addressed in more details in Sec. 4.3.

We note here, by comparison with Fig. 1, that the waterfall region of the turbo code corresponds to the transient chaos behavior. Also, in this region the unequivocal fixed point is stable and the size of its basin of attraction gradually grows for increasing SNRs.

4.2 Bifurcations of fixed points

As explained above, the transition from indecisive to unequivocal fixed points in the turbo-decoding algorithm is due to bifurcations of the indecisive fixed point. There are three ways in which a fixed point of a discrete-time dynamical system may fail to be hyperbolic: when the Jacobian matrix evaluated at the fixed point has a pair of complex eigenvalues crossing the unit circle, an eigenvalue at +1, or an eigenvalue at -1. In what follows we describe such bifurcations as they were observed in the classical turbo-decoding algorithm:

- **Neimark-Sacker bifurcation** In this case the Jacobian matrix evaluated at the fixed point undergoing the bifurcation admits a pair of complex conjugate eigenvalues on the unit circle. After the bifurcation, the fixed point goes unstable and is surrounded by an isolated, stable, close invariant curve, topologically equivalent to a circle. This bifurcation is illustrated in Figs. 3a)-b) and Figs. 4a)-b), respectively. Note that the Fig. 3 shows the average entropy E versus time l, while in Fig. 4 we report $E(l+1)$ versus $E(l)$.
- **Tangent bifurcation** Tangent (or fold) bifurcations are associated with a real eigenvalue at +1. After the bifurcation, the fixed point undergoing the bifurcation disappears, without resulting in an invariant set in its neighborhood. Figures 5a-b), and 6 illustrate this bifurcation. Note that Fig. 5 shows the average entropy E versus time l, while on Fig. 6 we report $E(l+1)$ versus $E(l)$. Figure 6d) shows an enlargement of part of Fig. c); it clearly indicates the occurrence of a tangent bifurcation. After the bifurcation the trajectory spends a long time in the vicinity of the disappeared fixed point, then approaches a chaotic attractor.
- **Flip bifurcation** Flip (or period doubling) bifurcation are associated with an eigenvalue at -1. As a result of the flip bifurcation, a stable fixed point becomes unstable, and an asymptotically stable period-two orbit appears in the neighborhood of the resulting unstable fixed point. Figures 7a)–d) illustrate this bifurcation. Figures 7a) and b) show the average entropy E versus time l, while in Fig. 7c) and d) we report $E(l+1)$ versus $E(l)$.

4.3 Chaotic behavior

From the bifurcation diagram in Fig. 2 it can be seen that the turbo-decoding algorithm exhibits chaotic behavior for a relatively large range of SNR values. Our analysis indicates three routes to chaos: period-doubling, intermittent and torus breakdown. A torus breakdown route to chaos is evident, for example, in Fig. 4, where a fixed point undergoes a Neimark-Sacker bifurcation giving rise to a periodic orbit; the later forms a torus which bifurcates leading to a chaotic attractor. Some chaotic time series are visible in Fig. 5. The same route to chaos can be observed in Fig. 7 where, this time, a period doubling cascade is interrupted by the occurrence of Neimark-Sacker bifurcations. The largest Lyapunov exponent of the chaotic attractor from Fig. 4f) was computed to be equal to 0.051. This chaotic attractor exists for all values of SNRs in the interval [-6.1,0.5]. The values of the largest Lyapunov exponent for some parameter values are: 0.63 for SNR=-4dB, 1.28 for SNR=-2dB, 1.68 for SNR=0dB, and 1.73 for SNR=0.5dB.

In the waterfall region, the turbo decoding algorithm converges either to the chaotic invariant set or to the unequivocal fixed point, after a long transient behavior indicating an existence of a chaotic non-attracting invariant set

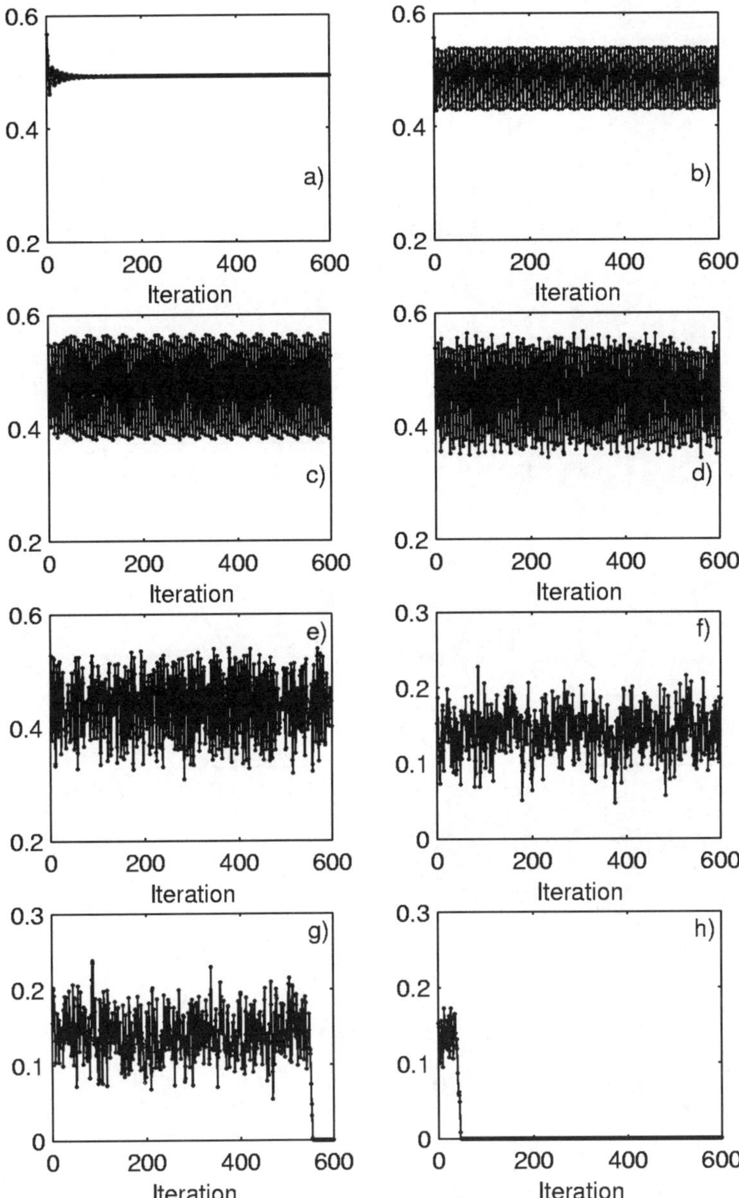

Fig.3. Transition to unequivocal fixed point via Neimark-Sacker bifurcation: average entropy vs. time ($E(l+1)$ vs. l). The values of SNR are: a) -6.7 dB; b) -6.5 dB; c) -6.3 dB; d) -6.1 dB; e) -5.9 dB; f) 0.75 dB; g) 0.80 dB; h) 0.85 dB. Figures a) and b) indicate the occurrence of Neimark-Sacker bifurcation. Note also the chaotic transients in Figs. g) and h).

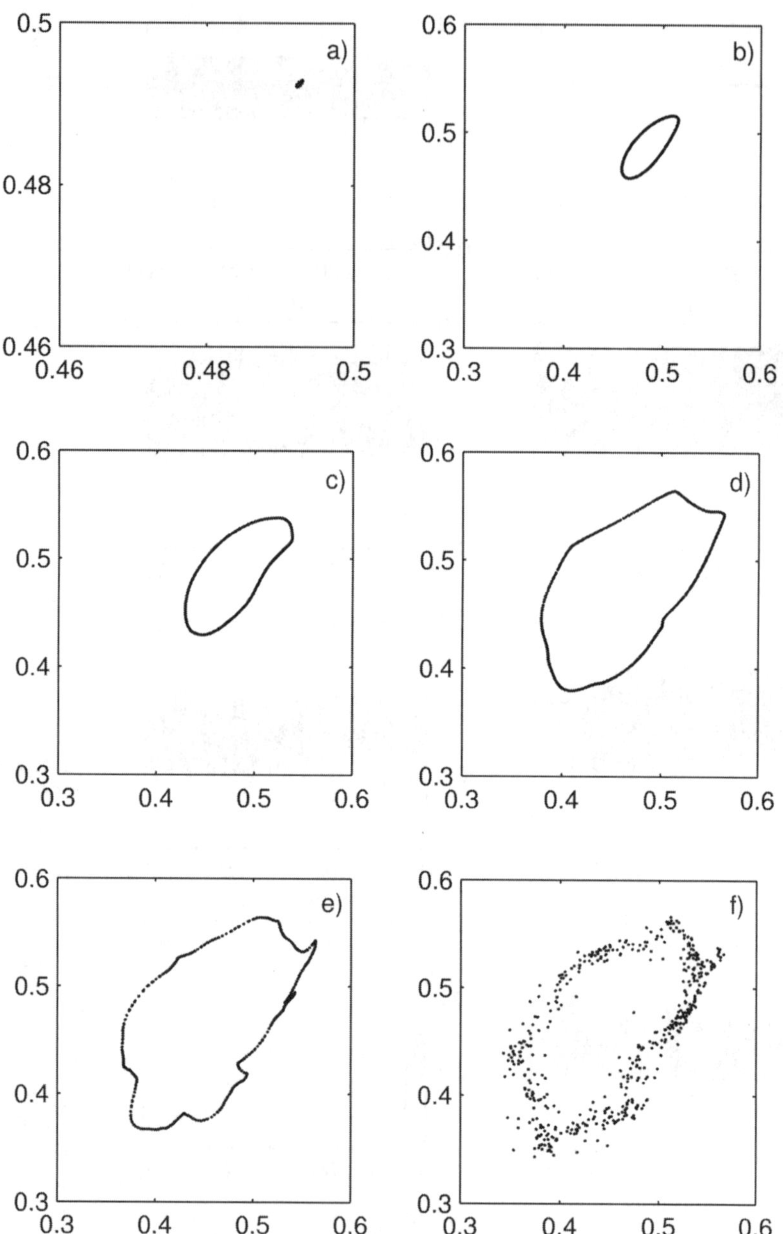

Fig.4. Neimark-Sacker bifurcation and transition to chaos: $E(l+1)$ vs $E(l)$. The values of SNR are: a) -6.7 dB; b) -6.6 dB; c) -6.5 dB; d) -6.3 dB; e) -6.2 dB; f) -6.1 dB. Figures a) and b) indicate the occurrence of a Neimark-Sacker bifurcation. Note also the torus-breakdown route to chaos in Figs. d) through f).

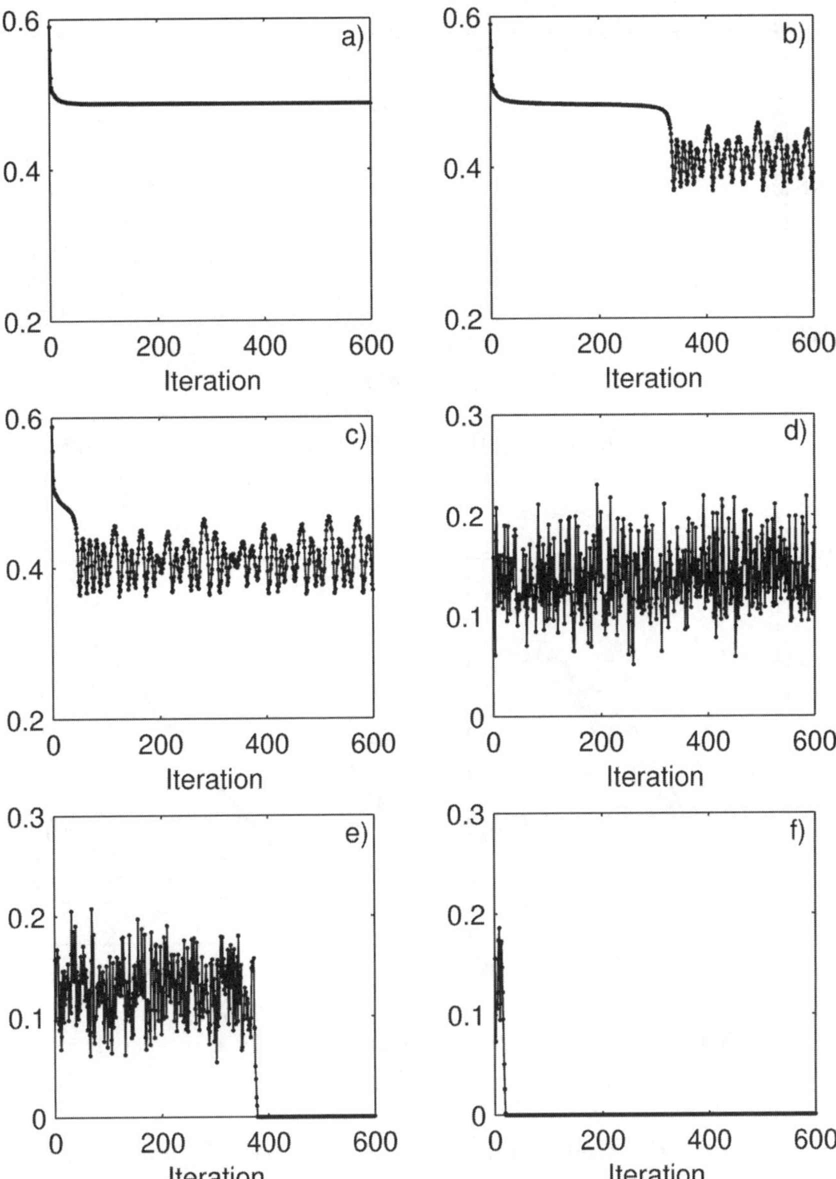

Fig.5. Transition to unequivocal fixed point via tangent bifurcation: average entropy $E(l)$ vs. time l. The values of SNR are: a) -7.65 dB; b) -7.645 dB; c) -7.6 dB; d) 0.30 dB; e) 0.35 dB; f) 0.4 dB. The indecisive fixed point a) looses its stability via tangent bifurcation, and b) the corresponding trajectory in state space approaches a chaotic attractor. Figure e) indicates transient chaos. In f) the algorithm reaches the fixed point solution in a few iterations.

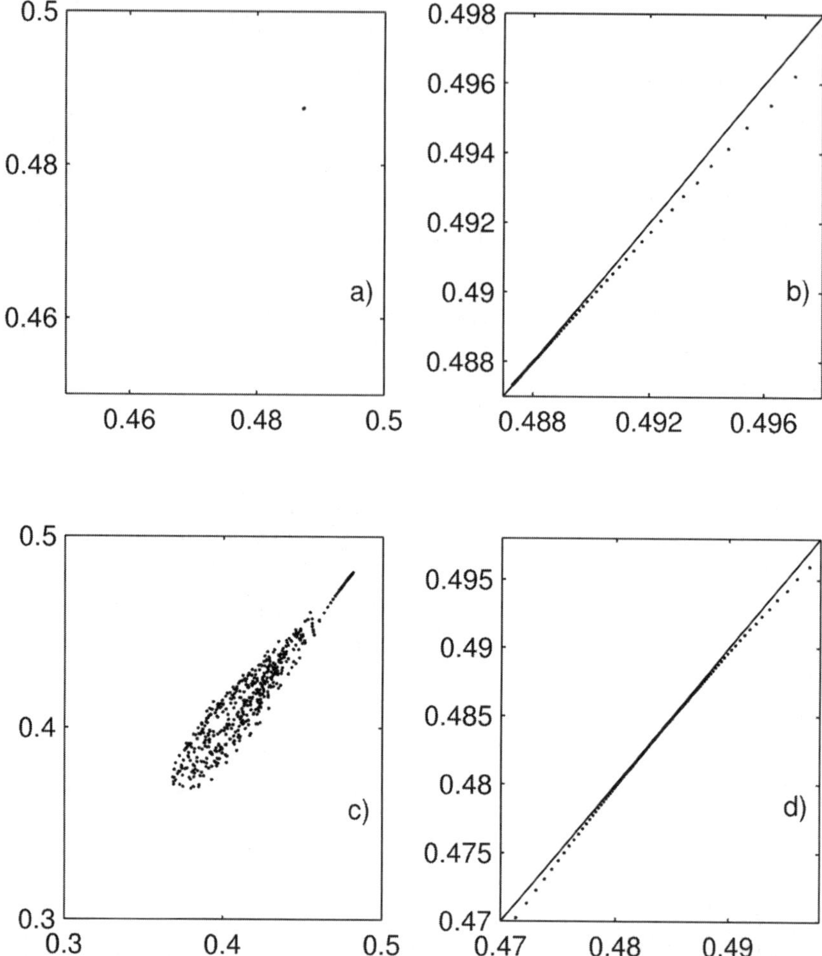

Fig.6. Tangent bifurcation: $E(l+1)$ vs $E(l)$. The values of SNR are: a)-b) -7.65 dB; c)-d) -7.64 dB. Figures b) and d) are nothing but zooms of Figs. a) and c), respectively.

in the vicinity of the unequivocal fixed point. In some cases, the algorithm spends a few thousand iterations before reaching the fixed point solution. As SNR increases, the number of iterations decreases, as can be seen, for example, from Fig. 3g)-h), and Fig. 5e)-f).

5 Control of Transient Chaos

In this section we consider an application of nonlinear control theory in order to speed-up the convergence of the turbo-decoding algorithm. First of all, we

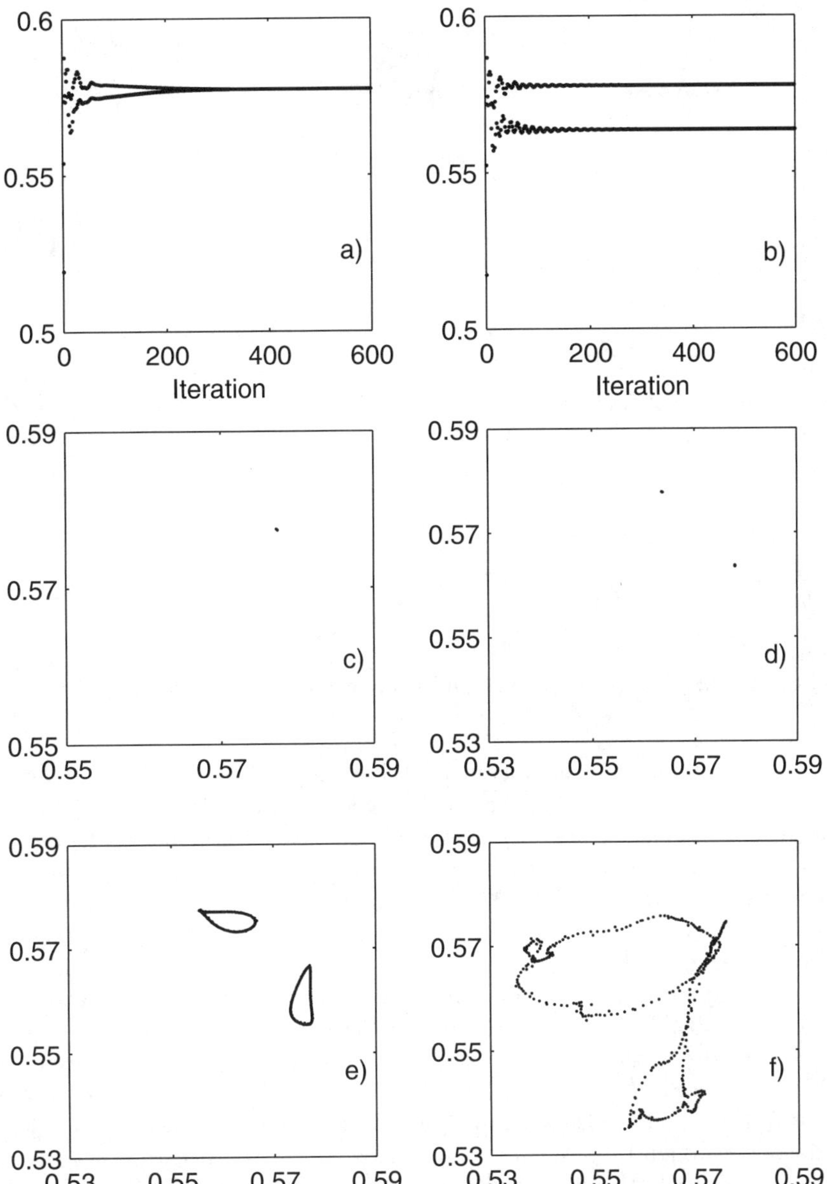

Fig.7. Transition to unequivocal fixed point via flip (or period doubling) bifurcation. a) and b): $E(l+1)$ vs l. c), d), e) and f): $E(l+1)$ vs $E(l)$. The values of SNR are: a) -6.0 dB; b) -5.96 dB; c) -6.0 dB; d) -5.96 dB; e) -5.94 dB; f) -5.84 dB. Figures a) and b), or c) and d), show a flip bifurcation. Figures d) and e) indicate a Neimark-Sacker bifurcation, while e) and f), torus breakdown route to chaos.

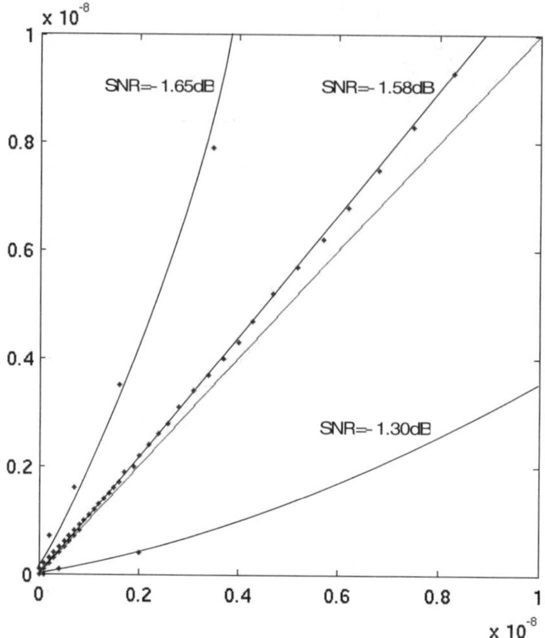

Fig.8. Stability analysis of the unequivocal fixed point at the origin.

analyze the stability of the unequivocal fixed point. Figure 8 shows $E(l+1)$ versus $E(l)$ for three different values of SNRs: -1.65 dB, -1.58 dB, and -1.30 dB. Note that we are very close (10^{-8}) to the fixed point at the origin. The two curves which are above the line $E(l+1) = E(l)$ have slopes at the origin greater than 1, indicating, for these values of SNRs: -1.65 dB and -1.58 dB, that the fixed point at origin is unstable. However, the third curve, which is below the line $E(l+1) = E(l)$, has slope smaller than 1. This is a numerical confirmation that the unequivocal fixed point becomes stable for SNR $\approx$ -1.5 dB. Although the unequivocal fixed point is stable in the region -1.5 dB to 0 dB, the algorithm does not seem to converge to it from the initial condition. This is due to the fact that the basin of attraction of this point in this region of SNR is too small. In the region 0 dB to 1 dB, which corresponds roughly to the waterfall region, the turbo decoding algorithm converges either to the chaotic invariant set or to the unequivocal fixed point, after a long transient behavior indicating the existence of a chaotic non-attracting invariant set in the vicinity of the unequivocal fixed point. In some cases, the algorithm spends a few thousand iterations before reaching the fixed point solution.

We have developed a simple adaptive control mechanism to reduce the long transient behavior in the decoding algorithm. For example, let us consider the trajectory shown in Fig. 3g), which converges to the unequivocal

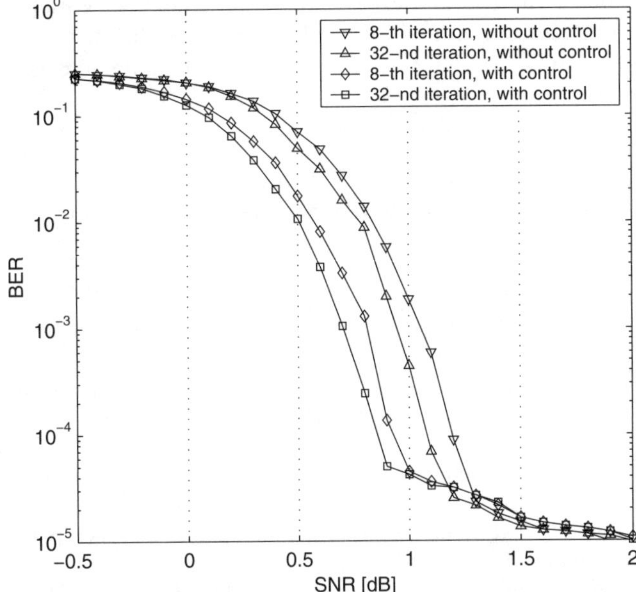

Fig.9. The performance of the classical turbo codes (interleaver length 1024) with and without control of the transient chaos.

fixed point after 577 iterations. When the control is applied, the algorithm approaches the unequivocal fixed point in 13 iterations only. Due to the space limitation, we will explain in detail our algorithm elsewhere. The results in terms of BER performance are summarized in Fig. 9. On average, the turbo decoding algorithm with control shows gain of 0.25 dB–0.3 dB comparing to the case of classical turbo decoding algorithm (without control). Note that the turbo decoding algorithm with control, which stops after 8th iteration, shows better performance than the classical turbo decoding algorithm (without control), which stops after 32nd iterations. Thus, in this case, the turbo decoding algorithm is 4 times faster and shows approximately 0.2 dB gain with respect to the classical one. We stress that our adaptive control algorithm is very simple, and can be easily implemented on both software and hardware, without increasing the complexity of the decoding algorithm.

6 Conclusions

The turbo decoding algorithm can be viewed as a high-dimensional dynamical system parameterized by a large number of parameters. In this work, we have shown that the turbo decoding algorithm exhibits a whole range of phenomena known to occur in nonlinear systems. These include the existence of multiple fixed points, oscillatory behavior, bifurcations, chaos and transient chaos. As an application of the theory developed, we have devised a simple technique to control transient chaos in the waterfall region of the turbo decoding algorithm. This results in a faster convergence and a significant gain in terms of BER performance.

Acknowledgments.
This work was supported in part by STMicroelectronics, Inc., UC DiMi program, and by the ARO.

References

1. Berrou, C., Glavieux, A., Thitimajshima, P. (1993) Near Shannon limit error-correcting coding and decoding: Turbo-Codes. In Proc. IEEE Int. Comm. Conf., 1064–1070
2. MacKay, D. J. C., Neal, R. M. (1996) Near Shannon limit performance of Low Density Parity Check Codes. Electr. Lett., **32**:1645–1646
3. MacKay, D. J. C. (1999) Good error correcting codes based on very sparse matrices. IEEE Trans. Infor. Theory, **45**:399–431
4. Richardson, T., Urbanke, R. (2001) The Capacity of Low-Density Parity Check Codes under message-passing decoding. IEEE Trans. Infor. Theory, **47**(2):599–618
5. Shannon, C. E. (1948) A mathematical theory of communication. Bell Syst. Tech. J., **27**:379–423; 623–56
6. Wiggins, S., John, F. (1990) Introduction to Applied Nonlinear Dynamical Systems and Chaos. 3rd ed., New York: Springer-Verlag
7. Gallager, R. G. (1962) Low Density Parity Check Codes. IRE Trans. Infor. Theory, **8**:21–28
8. Wiberg, N. (1996) Codes and decoding on general graphs. Dissertation no. 440, Dept. of Electrical Engineering, Linkoping University, Sweden
9. Benedetto, S., Montorsi, G. (1996) Unveiling turbo codes: some results on parallel concatenated coding schemes. IEEE Trans. Infor. Theory, **42**:409–428; Benedetto, S., Divsalar, D., Montorsi G., Pollara, F. (1998) Serial concatenation of interleaved codes: performance analysis, design and iterative decoding. IEEE Trans. Infor. Theory, **44**:909–926
10. Perez, L., Seghers, J., Costello, D. (1996) A Distance Spectrum Intrepretation of Turbo Codes. IEEE Trans. Infor. Theory, **42**:1690–1709
11. Richardson, T. (2000) The geometry of turbo-decoding dynamics. IEEE Trans. Infor. Theory, **46**:9–23
12. Agrawal, D., Vardy, A. (2001) The turbo decoding algorithm and its phase trajectories. IEEE Trans. Infor. Theory, **47**:699–722
13. Kuznetsov, Y. A. (1995) Elements of Applied Bifurcation Theory. SNew York: pringer-Verlag
14. Ott, E. (1993) Chaos in Dynamical Systems. New York: Cambridge Univ. Press
15. Bahl, L. R., Cocke, J., Jelinek, F., Raviv, J. (1974) Optimal decoding of linear codes for minimizing symbol error rate. IEEE Trans. Infor. Theory, **20**:284–287

Lecture Notes in Control and Information Sciences

Edited by M. Thoma and M. Morari
2000–2003 Published Titles:

Vol. 250: Corke, P.; Trevelyan, J. (Eds)
Experimental Robotics VI
552 pp. 2000 [1-85233-210-7]

Vol. 251: van der Schaft, A.; Schumacher, J.
An Introduction to Hybrid Dynamical Systems
192 pp. 2000 [1-85233-233-6]

Vol. 252: Salapaka, M.V.; Dahleh, M.
Multiple Objective Control Synthesis
192 pp. 2000 [1-85233-256-5]

Vol. 253: Elzer, P.F.; Kluwe, R.H.; Boussoffara, B.
Human Error and System Design and Management
240 pp. 2000 [1-85233-234-4]

Vol. 254: Hammer, B.
Learning with Recurrent Neural Networks
160 pp. 2000 [1-85233-343-X]

Vol. 255: Leonessa, A.; Haddad, W.H.; Chellaboina V.
Hierarchical Nonlinear Switching Control Design with Applications to Propulsion Systems
152 pp. 2000 [1-85233-335-9]

Vol. 256: Zerz, E.
Topics in Multidimensional Linear Systems Theory
176 pp. 2000 [1-85233-336-7]

Vol. 257: Moallem, M.; Patel, R.V.; Khorasani, K.
Flexible-link Robot Manipulators
176 pp. 2001 [1-85233-333-2]

Vol. 258: Isidori, A.; Lamnabhi-Lagarrigue, F.; Respondek, W. (Eds)
Nonlinear Control in the Year 2000 Volume 1
616 pp. 2001 [1-85233-363-4]

Vol. 259: Isidori, A.; Lamnabhi-Lagarrigue, F.; Respondek, W. (Eds)
Nonlinear Control in the Year 2000 Volume 2
640 pp. 2001 [1-85233-364-2]

Vol. 260: Kugi, A.
Non-linear Control Based on Physical Models
192 pp. 2001 [1-85233-329-4]

Vol. 261: Talebi, H.A.; Patel, R.V.; Khorasani, K.
Control of Flexible-link Manipulators Using Neural Networks
168 pp. 2001 [1-85233-409-6]

Vol. 262: Dixon, W.; Dawson, D.M.; Zergeroglu, E.; Behal, A.
Nonlinear Control of Wheeled Mobile Robots
216 pp. 2001 [1-85233-414-2]

Vol. 263: Galkowski, K.
State-space Realization of Linear 2-D Systems with Extensions to the General nD ($n>2$) Case
248 pp. 2001 [1-85233-410-X]

Vol. 264: Baños, A.; Lamnabhi-Lagarrigue, F.; Montoya, F.J
Advances in the Control of Nonlinear Systems
344 pp. 2001 [1-85233-378-2]

Vol. 265: Ichikawa, A.; Katayama, H.
Linear Time Varying Systems and Sampled-data Systems
376 pp. 2001 [1-85233-439-8]

Vol. 266: Stramigioli, S.
Modeling and IPC Control of Interactive Mechanical Systems – A Coordinate-free Approach
296 pp. 2001 [1-85233-395-2]

Vol. 267: Bacciotti, A.; Rosier, L.
Liapunov Functions and Stability in Control Theory
224 pp. 2001 [1-85233-419-3]

Vol. 268: Moheimani, S.O.R. (Ed)
Perspectives in Robust Control
390 pp. 2001 [1-85233-452-5]

Vol. 269: Niculescu, S.-I.
Delay Effects on Stability
400 pp. 2001 [1-85233-291-316]

Vol. 270: Nicosia, S. et al.
RAMSETE
294 pp. 2001 [3-540-42090-8]

Vol. 271: Rus, D.; Singh, S.
Experimental Robotics VII
585 pp. 2001 [3-540-42104-1]

Vol. 272: Yang, T.
Impulsive Control Theory
363 pp. 2001 [3-540-42296-X]

Vol. 273: Colonius, F.; Grüne, L. (Eds)
Dynamics, Bifurcations, and Control
312 pp. 2002 [3-540-42560-9]

Vol. 274: Yu, X.; Xu, J.-X. (Eds)
Variable Structure Systems: Towards the 21st Century
420 pp. 2002 [3-540-42965-4]

Vol. 275: Ishii, H.; Francis, B.A.
Limited Data Rate in Control Systems with Networks
171 pp. 2002 [3-540-43237-X]

Vol. 276: Bubnicki, Z.
Uncertain Logics, Variables and Systems
142 pp. 2002 [3-540-43235-3]

Vol. 277: Sasane, A.
Hankel Norm Approximation for Infinite-Dimensional Systems
150 pp. 2002 [3-540-43327-9]

Vol. 278: Chunling D. and Lihua X. (Eds)
H_∞ Control and Filtering of
Two-dimensional Systems
161 pp. 2002 [3-540-43329-5]

Vol. 279: Engell, S.; Frehse, G.; Schnieder, E. (Eds)
Modelling, Analysis, and Design of Hybrid Systems
516 pp. 2002 [3-540-43812-2]

Vol. 280: Pasik-Duncan, B. (Ed)
Stochastic Theory and Control
564 pp. 2002 [3-540-43777-0]

Vol. 281: Zinober A.; Owens D. (Eds)
Nonlinear and Adaptive Control
416 pp. 2002 [3-540-43240-X]

Vol. 282: Schröder, J.
Modelling, State Observation and
Diagnosis of Quantised Systems
368 pp. 2003 [3-540-44075-5]

Vol. 283: Fielding, Ch. et al. (Eds)
Advanced Techniques for Clearance of
Flight Control Laws
480 pp. 2003 [3-540-44054-2]

Vol. 284: Johansson, M.
Piecewise Linear Control Systems
216 pp. 2003 [3-540-44124-7]

Vol. 285: Wang, Q.-G.
Decoupling Control
373 pp. 2003 [3-540-44128-X]

Vol. 286: Rantzer, A. and Byrnes C.I. (Eds)
Directions in Mathematical Systems
Theory and Optimization
399 pp. 2003 [3-540-00065-8]

Vol. 287: Mahmoud, M.M.; Jiang, J. and Zhang, Y.
Active Fault Tolerant Control Systems
239 pp. 2003 [3-540-00318-5]

Vol. 288: Taware, A. and Tao, G.
Control of Sandwich Nonlinear Systems
230 pp. [3-540-44115-8]

Vol. 289: Giarré, L. and Bamieh, B.
Multidisciplinary Research in Control
237 pp. [3-540-00917-5]

Vol. 290: Borrelli, F.
Constrained Optimal Control
of Linear and Hybrid Systems
237 pp. [3-540-00257-X]

Vol. 291: Xu, J.-X. and Tan, Y.
Linear and Nonlinear Iterative Learning Control
189 pp. [3-540-40173-3]

Vol. 292: Chen, G. and Yu, X.
Chaos Control: Theory and Applications
380 pp. [3-540-40405-8]

Vol. 294: Benvenuti, L.; De Santis, A.; Farina, L. (Eds)
Positive Systems: Theory and Applications (POSTA 2003)
414 pp. [3-540-40342-6]

Vol. 295: Kang, W.; Xiao, M.; Borges, C. (Eds)
New Trends in Nonlinear Dynamics and Control,
and their Applications
365 pp. [3-540-10474-0]
(forthcoming)

Vol. 296: Matsuo, T.; Hasegawa, Y. (Eds)
Realization Theory of Discrete-Time Dynamical Systems
237 pp. [3-540-40675-1]
(forthcoming)

Printing and Binding: Strauss Offsetdruck GmbH